TEXTBOOK OF DYNAMICS
2nd Edition

ELLIS HORWOOD SERIES IN MATHEMATICS AND ITS APPLICATIONS

Series Editor: Professor G. M. BELL, Chelsea College, University of London
(and within the same series)

Statistics and Operational Research
Editor: B. W. CONOLLY, Chelsea College, University of London

Baldock, G. R. & Bridgeman, T. **MATHEMATICAL THEORY OF WAVE MOTION**
Beaumont, G. P. **INTRODUCTORY APPLIED PROBABILITY**
Burghes, D. N. & Wood, A. D. **MATHEMATICAL MODELS IN SOCIAL MANAGEMENT AND LIFE SCIENCES**
Burghes, D. N. **MODERN INTRODUCTION TO CLASSICAL MECHANICS AND CONTROL**
Burghes, D. N. & Graham, A. **CONTROL AND OPTIMAL CONTROL**
Burghes, D. N., Huntley, I. & McDonald, J. **APPLYING MATHEMATICS**
Butkovskiy, A. G. **GREEN'S FUNCTIONS AND TRANSFER FUNCTIONS HANDBOOK**
Butkovskiy, A. G. **STRUCTURE OF DISTRIBUTED SYSTEMS**
Chorlton, F. **TEXTBOOK OF DYNAMICS**
Chorlton, F. **VECTOR AND TENSOR METHODS**
Conolly, B. **TECHNIQUES IN OPERATIONAL RESEARCH**
Vol. 1: QUEUEING SYSTEMS
Vol. 2: MODELS, SEARCH, RANDOMIZATION
Dunning-Davies, J. **MATHEMATICAL METHODS FOR MATHEMATICS, PHYSICAL SCIENCE AND ENGINEERING**
Eason, G., Coles, C. W., Gettinby, G. **MATHEMATICS FOR THE BIOSCIENCES**
Exton, H. **HANDBOOK OF HYPERGEOMETRIC INTEGRALS**
Exton, H. **MULTIPLE HYPERGEOMETRIC FUNCTIONS**
Faux, I. D. & Pratt, M. J. **COMPUTATIONAL GEOMETRY FOR DESIGN AND MANUFACTURE**
Goult, R. J. **APPLIED LINEAR ALGEBRA**
Graham, A. **KRONECKER PRODUCTS AND MATRIX CALCULUS: WITH APPLICATIONS**
Graham, A. **MATRIX THEORY AND APPLICATIONS FOR ENGINEERS AND MATHEMATICIANS**
Griffel, D. H. **APPLIED FUNCTIONAL ANALYSIS**
Hoskins, R. F. **GENERALISED FUNCTIONS**
Hunter, S. C. **MECHANICS OF CONTINUOUS MEDIA, 2nd (Revised) Edition**
Huntley, I. & Johnson, R. M. **LINEAR AND NON-LINEAR DIFFERENTIAL EQUATIONS**
Jones, A. J. **GAME THEORY**
Kemp, K. W. **COMPUTATIONAL STATISTICS**
Kosinski, W. **FIELD SINGULARITIES AND WAVE ANALYSIS IN CONTINUUM MECHANICS**
Marichev, O. I. **INTEGRALS OF HIGHER TRANSCENDENTAL FUNCTIONS**
Meek, B. L. & Fairthorne, S. **USING COMPUTERS**
Muller-Pfeiffer, E. **SPECTRAL THEORY OF ORDINARY DIFFERENTIAL OPERATORS**
Nonweiler, T. R. F. **COMPUTATIONAL MATHEMATICS: An Introduction to Numerical Analysis**
Oliviera-Pinto, F. **SIMULATION CONCEPTS IN MATHEMATICAL MODELLING**
Oliviera-Pinto, F. & Conolly, B. W. **APPLICABLE MATHEMATICS OF NON-PHYSICAL PHENOMENA**
Scorer, R. S. **ENVIRONMENTAL AERODYNAMICS**
Smith, D. K. **NETWORK OPTIMISATION PRACTICE: A Computational Guide**
Stoodley, K. D. C., Lewis, T. & Stainton, C. L. S. **APPLIED STATISTICAL TECHNIQUES**
Sweet, M. V. **ALGEBRA, GEOMETRY AND TRIGONOMETRY FOR SCIENCE STUDENTS**
Temperley, H. N. V. & Trevena, D. H. **LIQUIDS AND THEIR PROPERTIES**
Temperley, H. N. V. **GRAPH THEORY AND APPLICATIONS**
Twizell, E. H. **COMPUTATIONAL METHODS FOR PARTIAL DIFFERENTIAL EQUATIONS IN BIOMEDICINE**
Whitehead, J. R. **THE DESIGN AND ANALYSIS OF SEQUENTIAL CLINICAL TRIALS**

TEXTBOOK OF DYNAMICS

2nd Edition

F. CHORLTON, B.Sc., M.Sc., D.I.C., F.I.M.A.
The Department of Mathematics
The University of Aston in Birmingham

ELLIS HORWOOD LIMITED
Publishers · Chichester

Halsted Press: a division of
JOHN WILEY & SONS
New York · Brisbane · Chichester · Toronto

First published in 1983 by

ELLIS HORWOOD LIMITED
Market Cross House, Cooper Street, Chichester, West Sussex, PO19 1EB, England

The publisher's colophon is reproduced from James Gillison's drawing of the ancient Market Cross, Chichester.

Distributors:

Australia, New Zealand, South-east Asia:
Jacaranda-Wiley Ltd., Jacaranda Press,
JOHN WILEY & SONS INC.,
G.P.O. Box 859, Brisbane, Queensland 40001, Australia

Canada:
JOHN WILEY & SONS CANADA LIMITED
22 Worcester Road, Rexdale, Ontario, Canada.

Europe, Africa:
JOHN WILEY & SONS LIMITED
Baffins Lane, Chichester, West Sussex, England.

North and South America and the rest of the world:
Halsted Press: a division of
JOHN WILEY & SONS
605 Third Avenue, New York, N.Y. 10016, U.S.A.

British Library Cataloguing in Publication Data
Chorlton, F.
Textbook of dynamics. – 2nd ed. –
(Ellis Horwood series in mathematics and its applications)
1. Dynamics
I. Title
531'.11   QA845

Library of Congress Card No. 82-25484

ISBN 0-85312-390-X (Ellis Horwood Ltd., Publishers –Library Edn.)
ISBN 0-85312-601-1 (Ellis Horwood Ltd., Publishers – Student Edn.)
ISBN 0-470-27407-7 (Halsted Press – Library Edn.)
ISBN 0-470-27408-5 (Halsted Press – Student Edn.)

Printed in Great Britain by Unwin Brothers of Woking.

Table of Contents

Key to Abbreviations 8

Preface 9

Chapter 1 VECTOR METHODS
1.1 Vectors and scalars 11
1.2 Addition and subtraction of vectors 12
1.3 Use of co-ordinates 14
1.4 Scalar product of two vectors 14
1.5 Vector product of two vectors 16
1.6 Triple products 18
1.7 Vector moment about a point and scalar moment about an axis 19
1.8 Vector and scalar couples 21
1.9 Centroids 21
1.10 Differentiation of vectors with respect to scalars 22
1.11 The vector calculus 23
Exercise 1 24

Chapter 2 KINEMATICS OF PARTICLES AND RIGID BODIES
2.1 Velocity and acceleration of a particle along a curve 26
2.2 Motion in a plane—radial and transverse components 28
2.3 Relative velocity and acceleration 30
2.4 Kinematics of a rigid body rotating about a fixed point 33
2.5 Vector angular velocity 34
2.6 General motion of a rigid body 35
2.7 General rigid body motion as a screw motion 36
2.8 Composition of angular velocities 38
2.9 Moving axes 40
2.10 Instantaneous axis of rotation and instantaneous centre of rotation 44
Exercise 2 48

Chapter 3 NEWTON'S LAWS OF MOTION
3.1 Mass, momentum, force, Newton's Laws of Motion 53
3.2 Work, energy and power 54

3.3 Conservative forces—potential energy 55
3.4 Impulsive forces 58
3.5 Rectilinear particle motion 58
(i) Uniformly accelerated motion 59
(ii) Resisted motion 59
(iii) Simple harmonic motion 62
(iv) Damped and forced vibrations 63
3.6 Elastic strings and springs 65
3.7 Changing mass problems 70
3.8 Detailed treatment of rocket motion 73
(i) Performance in flight of a single-stage rocket 73
(ii) Performance of a two-stage rocket 75
(iii) Performance of a multi-stage rocket 77
Exercise 3 80

Chapter 4 PARTICLE DYNAMICS
4.1 Problems in two and three dimensions 84
4.2 Projectile motion under gravity 84
4.3 Constrained particle motion 93
4.4 Angular momentum of a particle 98
4.5 The cycloid and its dynamical properties 103
Exercise 4 105

Chapter 5 ORBITAL MOTION
5.1 Motion of a particle under a central force 110
5.2 Use of reciprocal polar co-ordinates 111
5.3 Use of pedal co-ordinates and equations 113
5.4 Kepler's Laws of Planetary Motion and Newton's Law of Gravitation 117
5.5 Disturbed orbits 120
(i) Tangential blow. Motion in medium of small resistance 120
(ii) Normal blow 121
5.6 Elliptic harmonic motion 122
5.7 Satellite orbits 123
(i) Communication satellites 126
(ii) Evaluation of satellite orbit from initial conditions 127
Exercise 5 129

Chapter 6 MOTION OF A SYSTEM OF PARTICLES
6.1 Linear momentum of a system of particles 134
6.2 Angular momentum and rate of change of angular momentum of a system 136
6.3 Use of centroid 137
6.4 Moving origins 138
6.5 Impulsive forces 140
6.6 Elastic impact 144
Exercise 6 146

Chapter 7 INTRODUCTION TO RIGID BODY DYNAMICS
7.1 Moments and products of inertia 151
7.2 The theorems of parallel and perpendicular axes 152
7.3 Angular momentum of a rigid body about a fixed point and about fixed axes 152
7.4 Principal axes 153
7.5 Kinetic energy of a rigid body rotating about a fixed point 157
7.6 Momental ellipsoid—equimomental systems 157
7.7 Coplanar distributions 160
7.8 General motion of a rigid body 164
Exercise 7 166

Chapter 8 TWO-DIMENSIONAL RIGID BODY DYNAMICS
8.1 Introduction 168
8.2 Problems illustrating the laws of motion 168
8.3 Problems illustrating the law of conservation of angular momentum 172
8.4 Problems illustrating the law of conservation of energy 175
8.5 Problems illustrating impulsive motion 177
8.6 Further examples 180
Exercise 8 184

Chapter 9 THREE-DIMENSIONAL RIGID BODY DYNAMICS
9.1 Introduction 191
9.2 Euler's dynamical equations for the motion of a rigid body about a fixed point 191
9.3 Further properties of rigid motion under no forces 193
9.4 Some problems on general three-dimensional rigid body motion 194
9.5 The rotating Earth 202
Exercise 9 205

Chapter 10 GENERALIZED CO-ORDINATES
10.1 Note on dynamical methods 211
10.2 Preliminary notions 211
10.3 Generalized velocities 214
10.4 Virtual work and generalized forces 214
10.5 Derivation of Lagrange's equations for a holonomic system 215
10.6 Worked examples 216
10.7 Case of conservative forces 223
10.8 Generalized components of momentum and impulse 224
10.9 Lagrange's equations for impulsive forces 226
10.10 Kinetic energy as a quadratic function of velocities 229
10.11 Equilibrium configurations for conservative holonomic dynamical systems 231
10.12 Theory of small oscillations of conservative holonomic dynamical systems 235
10.13 Eulerian angles 241

10.14 Motion of a symmetrical top 243
10.15 Motion of a sphere on a sphere 247
10.16 The gyroscopic compass 248
Exercise 10 249

Chapter 11 VARIATIONAL METHODS IN MECHANICS
11.1 The calculus of variations 255
11.2 The brachistochrone problem 257
11.3 Extensions of the variational method 258
11.4 Hamilton's principle 259
11.5 The principle of least action 259
11.6 Distinctions between Hamilton's principle and the principle of least action 261
Exercise 11 262

APPENDIX I
Table showing centroid positions for standard distributions 264

APPENDIX II
Table showing square of the radius of gyration for standard distributions 265

Solutions 266

Index 269

KEY TO ABBREVIATIONS

A.M.	angular momentum	P.A.	principal axis
A.O.S.	axis of symmetry	P.E.	potential energy
C.A.	central axis	P.I.	product of inertia
D.Cs.	direction cosines	rad.	radian
D.Rs.	direction ratios	S.H.M.	simple harmonic motion
K.E.	kinetic energy	S.L.R.	semi latus rectum
L.R.	latus rectum	stat.	stationary
M.I.	moment of inertia		

Author's Preface

This book is a modern treatment of the entire field of classical dynamics based on vectorial methods and on the analytical developments of Lagrange. It is suitable for honours undergraduates in mathematics or mathematical physics and also for general science degree students offering mathematics as a principal subject. It should satisfy the needs of all scientists and engineers requiring a fuller understanding of the theoretical principles of higher mechanics.

Essentially the pattern of the First Edition is followed in which every effort was made to clarify the physical foundations of dynamics as well as formulating suitable mathematical models for solution. In this Second Edition, however, greater deliberation is placed upon the modelling process and to this end I must express my indebtedness to Professor David N. Burghes for his help and suggestions. So the reader will find added sections on the kinematics of plate tectonics, elementary shot putt analysis, a fairly detailed account of satellite orbits and staged rockets. In addition I have used the illuminating particle system analogy for obtaining the optimum location of an industrial warehouse—a model taught to me by Professor Burghes.

As before the work treats the dynamics not only of single particles but also of particle systems and rigid bodies. Some re-writing of sections has been undertaken to reinforce theoretical arguments where appropriate.

Examination questions drawn from various university sources have been used and I am grateful to the appropriate bodies for kind permission to use them.

In conclusion I have much pleasure in conveying my heartfelt thanks to the many colleagues and critics who expressed enthusiasm for the earlier work and I hope that the up-to-date reinforcements that I have mentioned will impart added value and interest to the Second Edition.

F. Chorlton
The University of Aston in Birmingham

Chapter 1

Vector Methods

1.1 VECTORS AND SCALARS

Physical measurements are mainly of two kinds: (i) *scalars* which have magnitude but no associated direction, and (ii) *vectors* which have both magnitude and an associated direction. Examples of scalars include mass, volume and energy. On the other hand, displacement, velocity and force—all of which require the stipulation of a direction for their complete specification—are vectorial quantities. Thus if we say that a train is travelling northeast at 60 ms^{-1}, the statement specifies completely the magnitude and direction of its velocity. If, on the other hand we merely say it is travelling at 60 ms^{-1} without specifying the direction of motion the statement only defines a physical magnitude which we term the speed.

Vectorial quantities may be represented in magnitude and direction by appropriately drawn line segments. Suppose the line segment $\overline{OA}$ represents a given vector in magnitude and direction (Fig. 1.1). If the vector does actually go through the point O, it is said to be *localized* at O: otherwise it is *non-localized.*

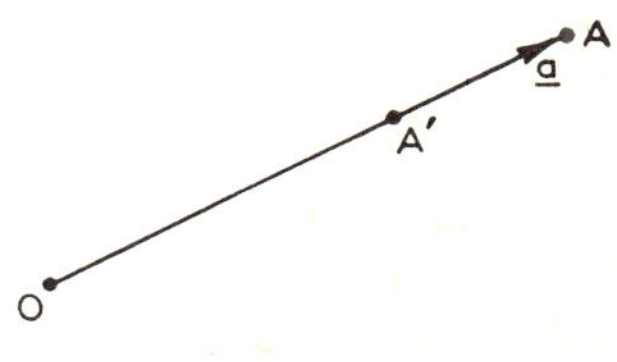

Fig. 1.1

It is convenient to have a single algebraic symbol to represent a vector such as $\overline{OA}$. Denote it by **a**. Then we write $\overline{OA} \equiv \mathbf{a}$. In full the statement means that the line segment directed from O to A represents a given vector **a** in magnitude and direction. This representation is incomplete since the position of the vector is unspecified. A further statement needs to be made if we are interested in this.

Now suppose we mark off a segment $\overline{OA'}$ in the direction $\overline{OA}$ and of unit length. Further, suppose $\overline{OA'} \equiv \hat{\mathbf{a}}$. Then $\hat{\mathbf{a}}$ is termed the *unit vector* in the direction of $\mathbf{a}$. The *magnitude* of $\mathbf{a}$ is denoted either by a or by $|\mathbf{a}|$, so that

$$\mathbf{a} = a\,\hat{\mathbf{a}} = |\mathbf{a}|\hat{\mathbf{a}} \ ,$$

since knowledge of a, $\hat{\mathbf{a}}$ define $\mathbf{a}$ in magnitude and direction.

The *negative* of the vector $\mathbf{a}$ is denoted by $-\mathbf{a}$ and is defined via the relationship $\overline{AO} \equiv -\mathbf{a}$.

A *zero vector* is one having zero magnitude, and denoted by $\mathbf{0}$.

1.2 ADDITION AND SUBTRACTION OF VECTORS

Suppose we wish to add together two given vectors $\mathbf{a}$, $\mathbf{b}$. Through any chosen origin O, draw $\overline{OA} \equiv \mathbf{a}$, $\overline{OB} \equiv \mathbf{b}$ (Fig. 1.2). It is immaterial whether or not the vectors actually pass through O. The construction merely means drawing line segments $\overline{OA}$, $\overline{OB}$ parallel to $\mathbf{a}$ and $\mathbf{b}$ and of appropriate lengths.

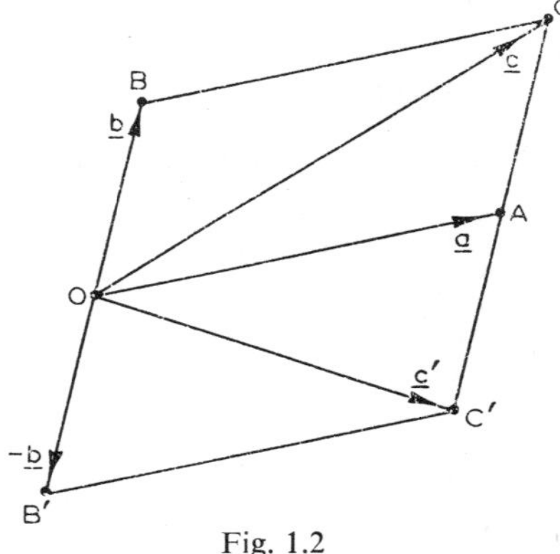

Fig. 1.2

Now complete the parallelogram $OACB$ and let $\overline{OC} \equiv \mathbf{c}$. Then we say that $\mathbf{c}$ is the sum of $\mathbf{a}$ and $\mathbf{b}$ and we write

$$\mathbf{c} = \mathbf{a} + \mathbf{b}$$

or

$$\overline{OC} = \overline{OA} + \overline{OB}.$$

This is the *Parallelogram Law of Vector Addition*. Since $\overline{AC} \equiv \mathbf{b}$, we could equally write

$$\overline{OC} = \overline{OA} + \overline{AC}$$

which form comprises the *Triangle Law of Addition*. This latter form is very useful in that the form $\overline{OA} + \overline{AC} = \overline{OC}$ shows that the same letter occurring in adjacent positions in summations is annihilated so that the operation $\overline{OA} + \overline{AC}$ may be performed without drawing a diagram.

Now suppose we draw $\overline{OB'} \equiv -\mathbf{b}$ (i.e. $\overline{OB'} = \overline{BO}$). Complete the parallelogram $OAC'B'$ and let $OC' \equiv \mathbf{c}'$. Then

$$\mathbf{c}' = \mathbf{a} + (-\mathbf{b}).$$

This last statement is more simply written as

$$\mathbf{c}' = \mathbf{a} - \mathbf{b} \ ,$$

thereby giving a method for finding the difference between two vectors.

The process of addition may be extended to several vectors. Thus to add the n vectors $\mathbf{a}_1, \mathbf{a}_2, \ldots, \mathbf{a}_n$, first choose an origin O and draw $\overline{OA_1} \equiv \mathbf{a}_1$. Then draw $\overline{A_1A_2} \equiv \mathbf{a}_2$, $\overline{A_2A_3} \equiv \mathbf{a}_3, \ldots, \overline{A_{n-1}A_n} \equiv \mathbf{a}_n$ (Fig. 1.3).

Then

$$\begin{aligned}
&\mathbf{a}_1 + \mathbf{a}_2 + \mathbf{a}_3 + \ldots + \mathbf{a}_n \\
&\equiv \overline{OA_1} + \overline{A_1A_2} + \overline{A_2A_3} + \ldots + \overline{A_{n-1}A_n} \\
&= \overline{OA_2} + \overline{A_2A_3} + \ldots + \overline{A_{n-1}A_n} \\
&= \overline{OA_3} + \overline{A_3A_4} + \ldots + \overline{A_{n-1}A_n} \\
&= \ldots\ldots\ldots\ldots\ldots \\
&= \overline{OA_{n-1}} + \overline{A_{n-1}A_n} \\
&= \overline{OA_n} \ ,
\end{aligned}$$

showing that $\overline{OA_n}$ represents the required sum. This is the *polygon of vectors*.

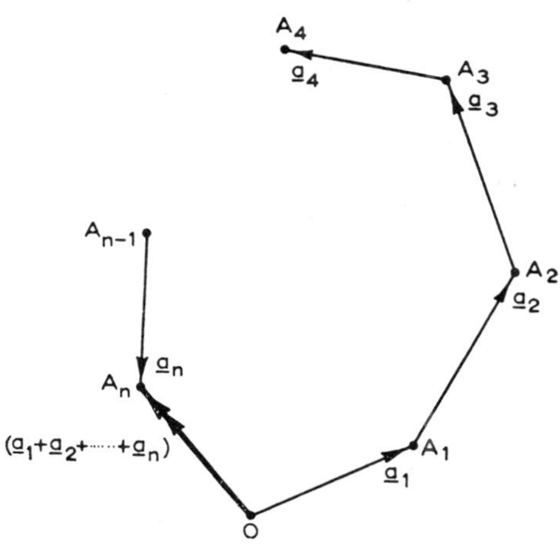

Fig. 1.3

1.3 USE OF CO-ORDINATES

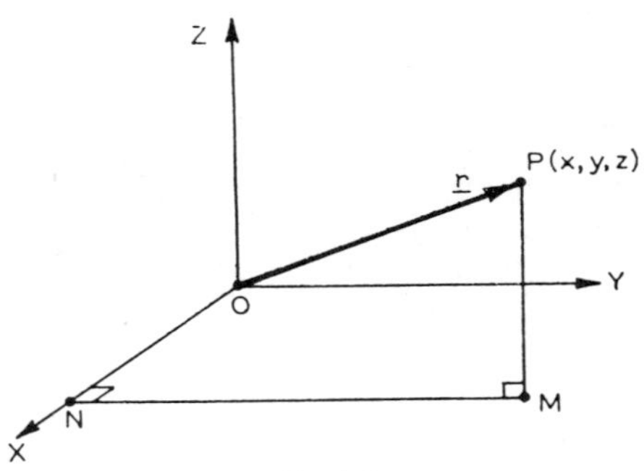

Fig. 1.4

Suppose $\overline{OP} \equiv \mathbf{r}$ (Fig. 1.4). Through O draw three mutually perpendicular lines OX, OY, OZ to form a right-handed co-ordinate frame in the sense that the rotation of a right-handed screw from OY to OZ advances it along $\overline{OX}$; a rotation from OZ to OX advances it along $\overline{OY}$; and a rotation from OX to OY advances it along $\overline{OZ}$. Let PM be the perpendicular from P on to the plane XOY and MN the perpendicular from M on to the axis OX. Let $ON = x$, $NM = y$, $MP = z$. Then (x, y, z) are called the *Cartesian co-ordinates* of P with respect to the frame. Clearly x, y, z are the respective distances of P from the co-ordinate planes YOZ, ZOX, XOY.

Then

$$\overline{ON} \equiv x\mathbf{i}\ , \quad \overline{NM} \equiv y\mathbf{j}\ , \quad \overline{MP} \equiv z\mathbf{k}\ .$$

Since

$$\overline{OP} = \overline{ON} + \overline{NM} + \overline{MP}\ ,$$

$$\overline{OP} \equiv \mathbf{r} = x\mathbf{i} + y\mathbf{j} + z\mathbf{k}\ . \tag{1}$$

Further,

$$OP^2 = OM^2 + MP^2 = ON^2 + NM^2 + MP^2\ ,$$

$$\therefore\ r^2 = x^2 + y^2 + z^2\ . \tag{2}$$

The statement (1) will sometimes be more briefly written $\mathbf{r} = [x, y, z]$, meaning that x, y, z are the components of $\mathbf{r}$. The unit vector in the direction of $\mathbf{r}$ is $\hat{\mathbf{r}}$ where

$$\hat{\mathbf{r}} = \mathbf{r}/r = [x/r, y/r, z/r]\ . \tag{3}$$

1.4 SCALAR PRODUCT OF TWO VECTORS

In Fig. 1.5, $\overline{OA} \equiv \mathbf{a}$, $\overline{OB} \equiv \mathbf{b}$, $\angle AOB = \theta$. Then the *scalar product* of $\mathbf{a}$ by $\mathbf{b}$ is defined to be $ab \cos \theta$ and we write it as $\mathbf{a}.\mathbf{b}$ so that

$$\mathbf{a}.\mathbf{b} = ab \cos \theta\ .$$

From this definition,

$$\mathbf{b}.\mathbf{a} = ba \cos (2\pi - \theta) = ba \cos \theta = \mathbf{a}.\mathbf{b} \ .$$

Thus scalar multiplication is *commutative*.

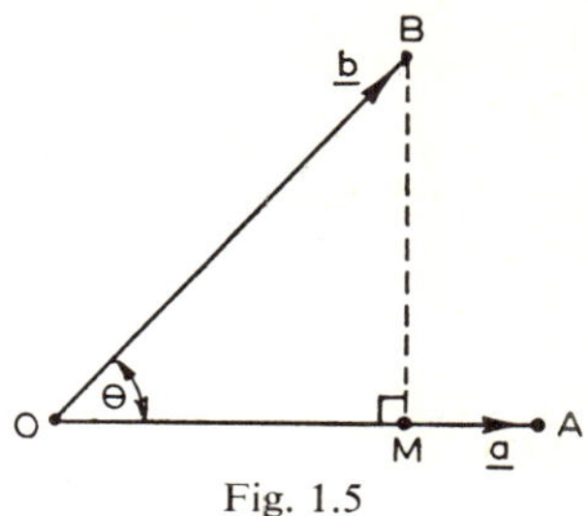

Fig. 1.5

It follows that

$$\hat{\mathbf{a}}.\mathbf{b} = \mathbf{b}.\hat{\mathbf{a}} = b \cos \theta = OM \ ,$$

where OM is the projection of OB on OA.

For any vector **a** we have, writing $\mathbf{a}^2$ for **a**.**a**,

$$\mathbf{a}^2 = \mathbf{a}.\mathbf{a} = (a)(a)(\cos 0) = a^2 \ .$$

The reader will readily see that for unit vectors **i**, **j**, **k** in the co-ordinate axes,

$$\mathbf{i}^2 = 1, \quad \mathbf{j}^2 = 1 \ , \quad \mathbf{k}^2 = 1 \ ,$$
$$\mathbf{i}.\mathbf{j} = 0, \quad \mathbf{j}.\mathbf{k} = 0 \ , \quad \mathbf{k}.\mathbf{i} = 0 \ .$$

We now prove that for given vectors **a**, **b**, **c**, the distributive law for scalar products holds, viz.

$$\mathbf{a}.(\mathbf{b} + \mathbf{c}) = \mathbf{a}.\mathbf{b} + \mathbf{a}.\mathbf{c} \ .$$

(This result makes the vector method useful in analytical work.)

Proof

In Fig. 1.6, let $\overline{OA} \equiv \mathbf{a}$, $\overline{OB} \equiv \mathbf{b}$, $\overline{OC} \equiv \mathbf{c}$.

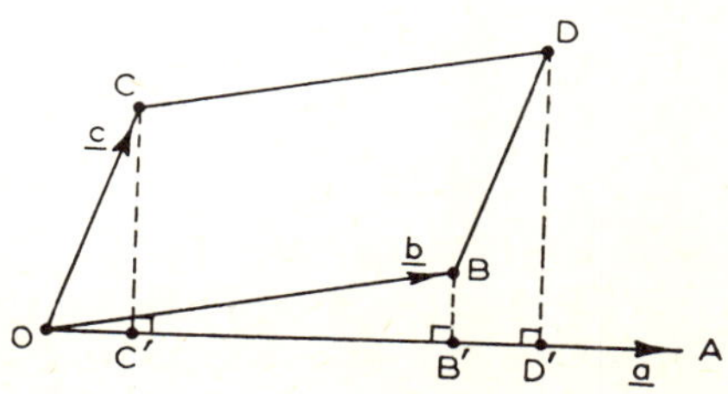

Fig. 1.6

Complete the parallelogram $OBDC$ so that $\overline{OD} \equiv \mathbf{b}+\mathbf{c}$. Let B', C', D' be the orthogonal projections of B, C, D on OA. Then, from the geometry of the figure,

$$OD' = OB' + B'D' = OB' + OC'$$
$$\therefore\ \hat{\mathbf{a}}.(\mathbf{b}+\mathbf{c}) = \hat{\mathbf{a}}.\mathbf{b} + \hat{\mathbf{a}}.\mathbf{c}\ ,$$

since OB' is the projection of $\mathbf{b}$ on OA, etc. Multiplying this last result through by the magnitude a now produces the stated result.

The usefulness of the distributive law is realized when vectors are expressed in Cartesian components. Thus for given vectors $\mathbf{a}=[a_1,a_2,a_3]$, $\mathbf{b}=[b_1,b_2,b_3]$,

$$\begin{aligned}\mathbf{a}.\mathbf{b} &= (a_1\mathbf{i}+a_2\mathbf{j}+a_3\mathbf{k}).(b_1\mathbf{i}+b_2\mathbf{j}+b_3\mathbf{k})\\ &= a_1\mathbf{i}.(b_1\mathbf{i}+b_2\mathbf{j}+b_3\mathbf{k})+a_2\mathbf{j}.(b_1\mathbf{i}+b_2\mathbf{j}+b_3\mathbf{k})+a_3\mathbf{k}.(b_1\mathbf{i}+b_2\mathbf{j}+b_3\mathbf{k})\\ &= \underline{a_1b_1+a_2b_2+a_3b_3}\ .\end{aligned}$$

If $[l_1, m_1, n_1]$, $[l_2, m_2, n_2]$ be the direction cosines of two unit vectors $OA_1 \equiv \hat{\mathbf{a}}_1$, $OA_2 \equiv \hat{\mathbf{a}}_2$ where $\angle A_1OA_2 = \theta$, then $\hat{\mathbf{a}}_1.\hat{\mathbf{a}}_2 = \cos\theta$. But since

$$\hat{\mathbf{a}}_1 = [l_1, m_1, n_1]\ ,\quad \hat{\mathbf{a}}_2 = [l_2, m_2, n_2]\ ,$$
$$\hat{\mathbf{a}}_1.\hat{\mathbf{a}}_2 = l_1l_2 + m_1m_2 + n_1n_2\ ,$$
$$\therefore\ \underline{\cos\theta = l_1l_2 + m_1m_2 + n_1n_2}\ .$$

We finish this section by noting that for any vector $\mathbf{F}$,

$$\mathbf{F} = (\mathbf{F}.\mathbf{i})\mathbf{i} + (\mathbf{F}.\mathbf{j})\mathbf{j} + (\mathbf{F}.\mathbf{k})\mathbf{k} = [\mathbf{F}.\mathbf{i},\ \mathbf{F}.\mathbf{j},\ \mathbf{F}.\mathbf{k}].$$

[Take $\mathbf{F} = \mathbf{F}_1\mathbf{i} + \mathbf{F}_2\mathbf{j} + \mathbf{F}_3\mathbf{k}$ and show $\mathbf{F}_1 = \mathbf{F}.\mathbf{i}$, etc.]

1.5 VECTOR PRODUCT OF TWO VECTORS

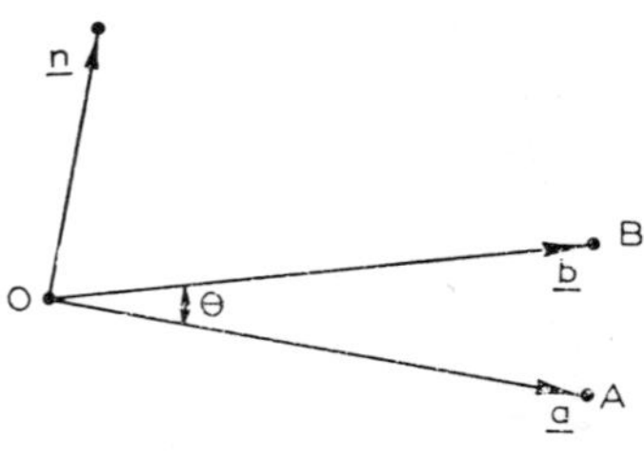

Fig. 1.7

In Fig. 1.7, $\overline{OA} \equiv \mathbf{a}$, $\overline{OB} \equiv \mathbf{b}$, $\angle AOB = \theta$ and $\mathbf{n}$ is the unit vector through O normal to the plane AOB in the sense of a right-handed screw turn from $\mathbf{a}$ to $\mathbf{b}$. Then the *vector product* of $\mathbf{a}$ by $\mathbf{b}$ is defined to be the vector whose magnitude is $ab \sin\theta$ and whose direction is $\mathbf{n}$. We denote this vector product by $\mathbf{a}_\wedge\mathbf{b}$ so that

$$\mathbf{a}_\wedge\mathbf{b} = (ab \sin\theta)\mathbf{n} \ .$$

$$\therefore \ \mathbf{b}_\wedge\mathbf{a} = (ba \sin\theta)(-\mathbf{n}) = -(\mathbf{a}_\wedge\mathbf{b})$$

since $-\mathbf{n}$ is the unit vector in the sense of a right-handed turn from $\mathbf{b}$ to $\mathbf{a}$ through θ. Thus vector multiplication is *non-commutative*.

The reader will readily show that for any vector $\mathbf{a}$, $\mathbf{a}_\wedge\mathbf{a} = \mathbf{0}$. Further, for the unit vectors $\mathbf{i}$, $\mathbf{j}$, $\mathbf{k}$,

$$\mathbf{i}_\wedge\mathbf{i} = \mathbf{j}_\wedge\mathbf{j} = \mathbf{k}_\wedge\mathbf{k} = \mathbf{0} \ ;$$
$$\mathbf{i}_\wedge\mathbf{j} = \mathbf{k},\ \mathbf{j}_\wedge\mathbf{k} = \mathbf{i},\ \mathbf{k}_\wedge\mathbf{i} = \mathbf{j} \ ;$$
$$\mathbf{j}_\wedge\mathbf{i} = -\mathbf{k},\ \mathbf{k}_\wedge\mathbf{j} = -\mathbf{i},\ \mathbf{i}_\wedge\mathbf{k} = -\mathbf{j} \ .$$

We now establish the distributive law for vector products in the form

$$\mathbf{a}_\wedge(\mathbf{b}+\mathbf{c}) = \mathbf{a}_\wedge\mathbf{b} + \mathbf{a}_\wedge\mathbf{c} \ .$$

Proof

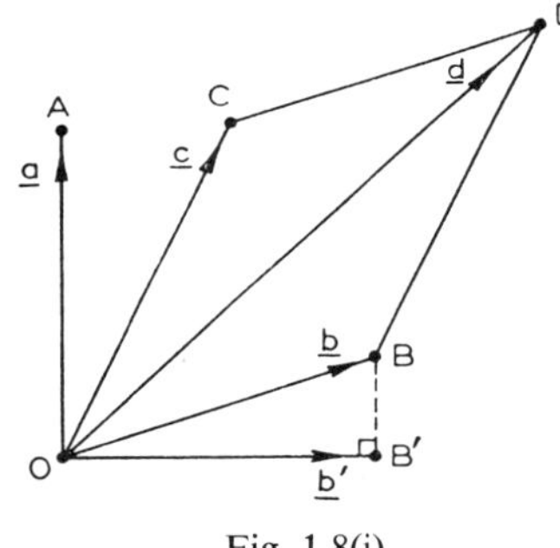

Fig. 1.8(i)

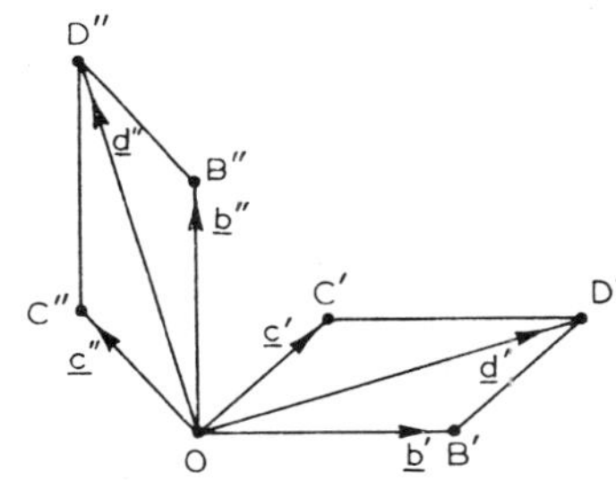

Fig. 1.8(ii)

Let $\overline{OA} \equiv \mathbf{a}$, $\overline{OB} \equiv \mathbf{b}$, $\overline{OC} \equiv \mathbf{c}$ (Fig. 1.8(i)). Complete the parallelogram $OBDC$ so that $\overline{OD} \equiv \mathbf{d} = \mathbf{b}+\mathbf{c}$. Let Π be the plane through O perpendicular to $\mathbf{a}$ and let B', C', D' be the orthogonal projections of B, C, D on Π and write

$$\overline{OB'} \equiv \mathbf{b}',\ \overline{OC'} \equiv \mathbf{c}',\ \overline{OD'} \equiv \mathbf{d}' \ .$$

In Fig. 1.8(i), only $\overline{OB'}$ is shown for clarity, but Fig. 1.8(ii) is a plan of the plane Π. Clearly the parallelogram $OBDC$ projects into a parallelogram $OB'D'C'$ so that

$$\mathbf{d}' = \mathbf{b}' + \mathbf{c}' \ . \tag{1}$$

From the definition of vector product it is apparent that

$$\mathbf{a}_\wedge\mathbf{b} = \mathbf{a}_\wedge\mathbf{b}',\ \mathbf{a}_\wedge\mathbf{c} = \mathbf{a}_\wedge\mathbf{c}',\ \mathbf{a}_\wedge\mathbf{d} = \mathbf{a}_\wedge\mathbf{d}' \ . \tag{2}$$

Let $\mathbf{b}'' = \mathbf{a}_\wedge\mathbf{b}'$, $\mathbf{c}'' = \mathbf{a}_\wedge\mathbf{c}'$, $\mathbf{d}'' = \mathbf{a}_\wedge\mathbf{d}'$ and $\overline{OB''} \equiv \mathbf{b}''$, $\overline{OC''} \equiv \mathbf{c}''$, $\overline{OD''} \equiv \mathbf{d}''$. The effect of pre-vector multiplying each of the coplanar vectors $\mathbf{b}'$, $\mathbf{c}'$, $\mathbf{d}'$ by $\mathbf{a}$ is to rotate each of them through a right angle in the same sense in the plane Π

and to multiply their magnitudes by a. Thus $OB''D''C''$ is a parallelogram and so

$$\mathbf{d}'' = \mathbf{b}'' + \mathbf{c}'' \ . \tag{3}$$

From (3),

$$\mathbf{a}_\wedge\mathbf{d}' = \mathbf{a}_\wedge\mathbf{b}' + \mathbf{a}_\wedge\mathbf{c}' \ ,$$

i.e.

$$\mathbf{a}_\wedge\mathbf{d} = \mathbf{a}_\wedge\mathbf{b} + \mathbf{a}_\wedge\mathbf{c}$$

or

$$\underline{\mathbf{a}_\wedge(\mathbf{b}+\mathbf{c}) = \mathbf{a}_\wedge\mathbf{b} + \mathbf{a}_\wedge\mathbf{c}} \ .$$

This result enables vector products to be evaluated in terms of components. Thus if $\mathbf{a} = [a_1, a_2, a_3]$, $\mathbf{b} = [b_1, b_2, b_3]$,

$$\begin{aligned}\mathbf{a}_\wedge\mathbf{b} &= (a_1\mathbf{i} + a_2\mathbf{j} + a_3\mathbf{k})_\wedge(b_1\mathbf{i} + b_2\mathbf{j} + b_3\mathbf{k}) \\ &= a_1\mathbf{i}_\wedge(b_1\mathbf{i} + b_2\mathbf{j} + b_3\mathbf{k}) + a_2\mathbf{j}_\wedge(b_1\mathbf{i} + b_2\mathbf{j} + b_3\mathbf{k}) + a_3\mathbf{k}_\wedge(b_1\mathbf{i} + b_2\mathbf{j} + b_3\mathbf{k}) \\ &= a_1b_2\mathbf{k} - a_1b_3\mathbf{j} - a_2b_1\mathbf{k} + a_2b_3\mathbf{i} + a_3b_1\mathbf{j} - a_3b_2\mathbf{i} \\ &= (a_2b_3 - a_3b_2)\mathbf{i} + (a_3b_1 - a_1b_3)\mathbf{j} + (a_1b_2 - a_2b_1)\mathbf{k} \ .\end{aligned}$$

This result may be more conveniently expressed in the determinantal form

$$\mathbf{a}_\wedge\mathbf{b} = \begin{vmatrix} \mathbf{i} & \mathbf{j} & \mathbf{k} \\ a_1 & a_2 & a_3 \\ b_1 & b_2 & b_3 \end{vmatrix}$$

Thus for two unit vectors $\overline{OA_1} \equiv \hat{\mathbf{a}}_1$, $\overline{OA_2} \equiv \hat{\mathbf{a}}_2$, if $\angle A_1OA_2 = \theta$, and $\hat{\mathbf{a}}_1 = [l_1, m_1, n_1]$, $\hat{\mathbf{a}}_2 = [l_2, m_2, n_2]$, then

$$\begin{aligned}\hat{\mathbf{a}}_{1\wedge}\hat{\mathbf{a}}_2 &= \begin{vmatrix} \mathbf{i} & \mathbf{j} & \mathbf{k} \\ l_1 & m_1 & n_1 \\ l_2 & m_2 & n_2 \end{vmatrix} \\ &= [(m_1n_2 - m_2n_1), (n_1l_2 - n_2l_1), (l_1m_2 - l_2m_1)] \ ;\end{aligned}$$

$$\therefore \ \sin\theta = |\hat{\mathbf{a}}_{1\wedge}\hat{\mathbf{a}}_2| = \sqrt{\{(m_1n_2 - m_2n_1)^2 + (n_1l_2 - n_2l_1)^2 + (l_1m_2 - l_2m_1)^2\}} \ .$$

1.6 TRIPLE PRODUCTS

The quantity $\mathbf{a}.(\mathbf{b}_\wedge\mathbf{c})$ is the scalar product of the vectors $\mathbf{a}$ and $\mathbf{b}_\wedge\mathbf{c}$: it is called a *scalar triple product* and is denoted by $[\mathbf{a}, \mathbf{b}, \mathbf{c}]$. If we denote the components of $\mathbf{a}$ by $[a_1, a_2, a_3]$, etc., the reader will easily show that

$$[\mathbf{a}, \mathbf{b}, \mathbf{c}] = \begin{vmatrix} a_1 & a_2 & a_3 \\ b_1 & b_2 & b_3 \\ c_1 & c_2 & c_3 \end{vmatrix}.$$

From elementary determinantal properties it readily follows that

$$[\mathbf{a}, \mathbf{b}, \mathbf{c}] = [\mathbf{b}, \mathbf{c}, \mathbf{a}] = [\mathbf{c}, \mathbf{a}, \mathbf{b}] \ ,$$

but $[\mathbf{c}, \mathbf{b}, \mathbf{a}] = -[\mathbf{a}, \mathbf{b}, \mathbf{c}]$, etc. Thus cyclic permuting of the letters leaves the scalar triple product unchanged, but anti-cyclic permuting changes its sign.

The quantity $\mathbf{a}_\wedge(\mathbf{b}_\wedge\mathbf{c})$, being the vector product of $\mathbf{a}$ with $(\mathbf{b}_\wedge\mathbf{c})$, is vectorial and is called a *vector triple product*. We now prove the result

$$\mathbf{a}_\wedge(\mathbf{b}_\wedge\mathbf{c}) = (\mathbf{a}.\mathbf{c})\mathbf{b} - (\mathbf{a}.\mathbf{b})\mathbf{c} \ .$$

Proof

With the previous notation,

$$\mathbf{b}_\wedge\mathbf{c} = \begin{vmatrix} \mathbf{i} & \mathbf{j} & \mathbf{k} \\ b_1 & b_2 & b_3 \\ c_1 & c_2 & c_3 \end{vmatrix} = \Sigma(b_2c_3 - b_3c_2)\mathbf{i} \ ,$$

the summation being taken cyclically over **i**, **j**, **k**; 1, 2, 3.

$$\therefore \ \mathbf{a}_\wedge(\mathbf{b}_\wedge\mathbf{c}) = \begin{vmatrix} \mathbf{i} & \mathbf{j} & \mathbf{k} \\ a_1 & a_2 & a_3 \\ (b_2c_3 - b_3c_2) & (b_3c_1 - b_1c_3) & (b_1c_2 - b_2c_1) \end{vmatrix}$$

$$= \Sigma\{a_2(b_1c_2 - b_2c_1) - a_3(b_3c_1 - b_1c_3)\}\mathbf{i}$$
$$= \Sigma(a_2c_2 + a_3c_3)b_1\mathbf{i} - \Sigma(a_2b_2 + a_3b_3)c_1\mathbf{i}$$
$$= \Sigma(a_1c_1 + a_2c_2 + a_3c_3)b_1\mathbf{i} - \Sigma(a_1b_1 + a_2b_2 + a_3b_3)c_1\mathbf{i}$$
$$= (a_1c_1 + a_2c_2 + a_3c_3)\Sigma b_1\mathbf{i} - (a_1b_1 + a_2b_2 + a_3b_3)\Sigma c_1\mathbf{i}$$
$$= \underline{(\mathbf{a}.\mathbf{c})\mathbf{b} - (\mathbf{a}.\mathbf{b})\mathbf{c}} \ .$$

1.7 VECTOR MOMENT ABOUT A POINT AND SCALAR MOMENT ABOUT AN AXIS

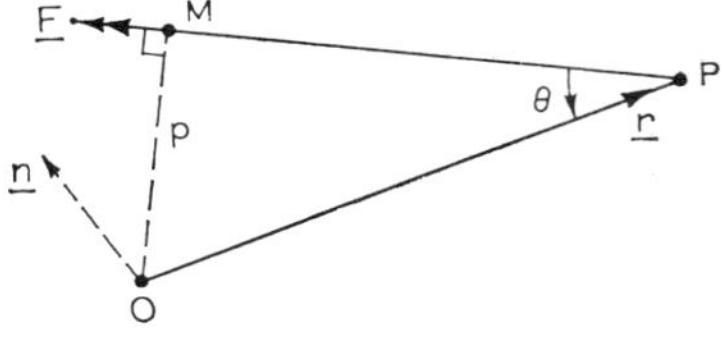

Fig. 1.9

Let **F** be a vector quantity acting at a point P whose position vector with respect to a chosen origin O is given by $\overline{OP} \equiv \mathbf{r}$. Then the *vector moment* **G** of **F** about O is defined to be

$$\mathbf{G} = \mathbf{r}_\wedge\mathbf{F}.$$

If $p = OM$ be the perpendicular from O to **F**, then denoting $\theta = \angle OPM$, from the definition of vector product,

$$\mathbf{G} = (rF \sin\theta)\mathbf{n} = (pF)\mathbf{n} \ ,$$

where **n** is the unit normal to the plane of OPM in the sense of a right-handed rotation from **r** to **F**.

If we identify **F** with a force, the magnitude of this moment $G = pF$ is clearly the moment of the force about the axis **n** through O. Thinking of OP as the arm of a vice fixed to O by means of a right-handed screw thread, it will readily be appreciated that the effect of this turning moment is to cause O to advance in the direction **n**. Thus we see that associated with the turning effect of the moment of this force about O is a tendency to advance O in the specific direction **n**. In other words, the moment of a force is strictly vectorial in kind and we shall speak of its (vector) moment about O as being $\mathbf{r}_\wedge\mathbf{F}$. The magnitude pF will be referred to as the *physical moment* about the axis **n** through O.

Since moments are vectorial, they may be combined as vectors. Thus if we have n forces $\mathbf{F}_k$ acting at n points P_k with position vectors $\overline{OP_k} \equiv \mathbf{r}_k$ $(k = 1, 2, \ldots, n)$, then the total vector moment about O will be

$$\mathbf{G} = \sum_{k=1}^{n} \mathbf{r}_{k\wedge}\mathbf{F}_k \ .$$

We now investigate the physical moment of a force about any axis. Let **F** be a force acting at a point P distant p from an axis specified by a unit vector $\hat{\mathbf{a}}$ (Fig. 1.10).

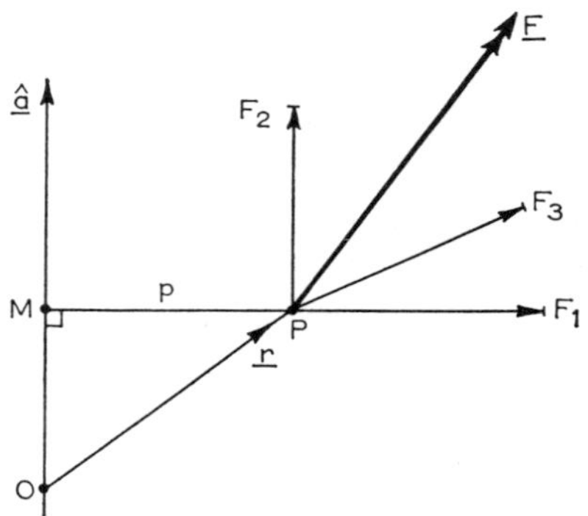

Fig. 1.10

Let P have position vector $\overline{OP} \equiv \mathbf{r}$ w.r.t. a point O shown on $\hat{\mathbf{a}}$. Let **F** have components $[F_1, F_2, F_3]$ in directions along $\overline{MP}$, parallel to $\hat{\mathbf{a}}$ and perpendicular to both these directions in the sense $\mathbf{r}_\wedge\mathbf{F}$, respectively. Since F_1 and F_2 have no turning effect about $\hat{\mathbf{a}}$, the physical moment of the force about $\hat{\mathbf{a}}$ is simply pF_3. Since $\hat{\mathbf{a}}_\wedge\mathbf{r}$ is a vector having magnitude $r \sin\theta$ or p, and direction that of the component F_3 of **F**, we see that

$$pF_3 = (r \sin\theta)F_3 = (\hat{\mathbf{a}}_\wedge\mathbf{r}).\mathbf{F} = [\hat{\mathbf{a}}, \mathbf{r}, \mathbf{F}].$$

Thus the physical moment of **F** about the axis $\hat{\mathbf{a}}$ is given by the scalar triple product $[\hat{\mathbf{a}}, \mathbf{r}, \mathbf{F}]$, where **r** is the position vector of **F** w.r.t. any chosen point O on $\hat{\mathbf{a}}$. We shall generalize this result to cover cases in which **F** is any vector localized at P, force or otherwise.

1.8 VECTOR AND SCALAR COUPLES

Two equal and opposite vectors $\pm\mathbf{F}$ not acting in the same straight line are said to comprise a *couple*. Let P_1, P_2 be any points chosen on F and $-\mathbf{F}$ and let their position vectors with respect to any chosen origin O be given by $\overline{OP_1}\equiv\mathbf{r}_1$, $\overline{OP_2}\equiv\mathbf{r}_2$, and let $\overline{P_2P_1}\equiv$ **s** (Fig. 1.11).

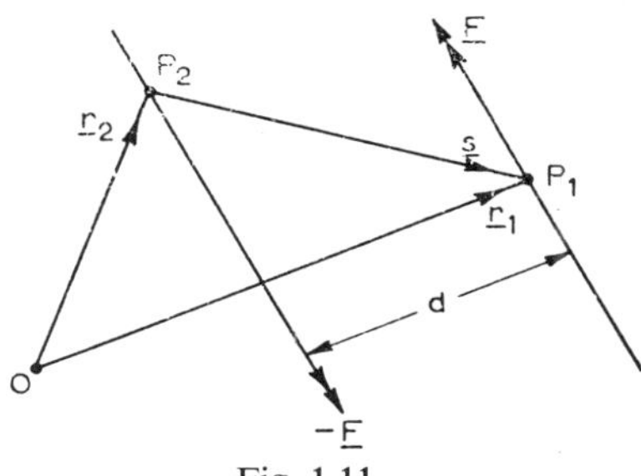

Fig. 1.11

The total vector moment about O is

$$\begin{aligned}\mathbf{G} &= \mathbf{r}_1\wedge\mathbf{F}+\mathbf{r}_2\wedge(-\mathbf{F})\\ &= (\mathbf{r}_1-\mathbf{r}_2)\wedge\mathbf{F}=\mathbf{s}\wedge\mathbf{F}\ .\end{aligned}$$

This latter is seen to be invariant for any choice of O. Thus the vector moment of a couple is independent of choice of origin. Further, if we denote by d the distance between the vectors $\pm\mathbf{F}$, we see the physical moment is

$$G=|\mathbf{s}\wedge\mathbf{F}|=dF\ ,$$

in accordance with elementary ideas of couples when the vectors $\pm\mathbf{F}$ are specifically thought of as forces. Further, the direction of **G** is the normal to the plane of $\pm\mathbf{F}$ in the sense of a right-handed rotation from **s** to $+\mathbf{F}$.

1.9 CENTROIDS

Suppose we have scalar magnitudes m_k associated with n points P_k of space having position vectors $\overline{OP_k}\equiv\mathbf{r}_k$ with respect to any chosen origin O ($k=1, \ldots, n$). Often in mechanics the quantities m_k will be point masses of the P_k. Then, provided $m_1+m_2+\ldots+m_n\neq 0$, the *centroid* or *mean centre* of the scalar magnitudes is defined to be the point G having position vector $\bar{\mathbf{r}}$ where $\overline{OG}\equiv\bar{\mathbf{r}}=(\sum_{k=1}^{n} m_k\mathbf{r}_k)/(\sum_{k=1}^{n} m_k)$.

Writing $\mathbf{r}_k=[x_k, y_k, z_k](k=1, \ldots, n)$, $\mathbf{r}=[\bar{x}, \bar{y}, \bar{z}]$, we see that the Cartesian co-ordinates of G are given by

$$\bar{x}=(\sum_{k=1}^{n} m_kx_k)/(\sum_{k=1}^{n} m_k),\text{ etc.,} \tag{1}$$

and that, for continuous systems, these summations in limiting form lead to the definite integral expressions

$$\bar{x} = (\int x\, dm)/\int dm, \text{ etc.,} \tag{2}$$

where the integrations extend throughout the entire system.

(2) is derived as follows. If we have point masses δm at co-ordinate distances x from the plane $x=0$, then from (1),

$$\bar{x} = (\Sigma x \delta m)/(\Sigma \delta m) \ ,$$

the summations being throughout the entire distribution. On taking the limit as $\delta m \rightarrow 0$, the result (2) follows.

We shall assume the reader is conversant with the methods for determining centroid positions of standard distributions. (See Appendix I.)

Suppose now we take a second origin O' where $\overline{O'O} \equiv \mathbf{s}$ and $\overline{O'P_k} \equiv \mathbf{r}_k{}'$ $(k=1, \ldots, n)$. Then $\mathbf{s} + \mathbf{r}_k = \mathbf{r}_k{}'$ $(k=1, \ldots, n)$. If we denote the centroid of the same system for O' by G', then

$$\begin{aligned} \overline{O'G'} &\equiv \Big(\sum_{k=1}^{n} m_k \mathbf{r}_k{}'\Big)\Big/\Big(\sum_{k=1}^{n} m_k\Big) \\ &= \mathbf{s} + \Big(\sum_{k=1}^{n} m_k \mathbf{r}_k\Big)\Big/\Big(\sum_{k=1}^{n} m_k\Big) \ . \end{aligned}$$

$$\therefore \ \overline{O'G'} = \overline{O'O} + \overline{OG} = \overline{O'G} \ .$$

This shows G', G coincide, so that the centroid of the system is a *unique point*.

1.10 DIFFERENTIATION OF VECTORS WITH RESPECT TO SCALARS

Just as in scalar calculus a dependent variable may be a function of an independent variable, having one or more differential coefficients with respect to the independent variable, so a vector quantity $\mathbf{r}$, say, may be a function of a scalar t. Such a functional dependence may be denoted by $\mathbf{r}(t)$. If, corresponding to the value $t+\delta t$ of the independent variable the vector function assumes the value $\mathbf{r}(t+\delta t) = \mathbf{r} + \delta\mathbf{r}$, then the differential coefficient, if it exists, is denoted by $d\mathbf{r}/dt$ and is defined to be

$$d\mathbf{r}/dt = \lim_{\delta t \to 0} (\delta \mathbf{r}/\delta t) \ .$$

If $\mathbf{r}$, $\mathbf{s}$ be functions of the scalar t, then

$$\begin{aligned} d(\mathbf{r}.\mathbf{s})/dt &= \lim_{\delta t \to 0} \{[(\mathbf{r}+\delta\mathbf{r}).(\mathbf{s}+\delta\mathbf{s}) - \mathbf{r}.\mathbf{s}]/\delta t\} \\ &= \lim_{\delta t \to 0} \{(\mathbf{r}.\delta\mathbf{s} + \mathbf{s}.\delta\mathbf{r} + \delta\mathbf{r}.\delta\mathbf{s})/\delta t\} \\ &= \underline{\mathbf{r}.(d\mathbf{s}/dt) + \mathbf{s}.(d\mathbf{r}/dt)} \ . \end{aligned}$$

It can similarly be shown that

$$d(\mathbf{r}\wedge\mathbf{s})/dt = \mathbf{r}\wedge(d\mathbf{s}/dt) + (d\mathbf{r}/dt)\wedge\mathbf{s} \ ,$$

where the stated order of vector multiplication must be strictly preserved.

1.11 THE VECTOR CALCULUS

For completion we append a brief résumé of the more common results of the vector calculus. For a more detailed treatment the reader is advised to consult *Vector and Tensor Methods*, by F. Chorlton (Ellis Horwood, 1976).

If at each point $P(x, y, z)$ of space, or of a region of space, a scalar function $\phi(x, y, z)$ is uniquely defined then the totality of its values throughout the region constitutes a *scalar field*. (An example of such a scalar field is afforded by the temperature at each point of a room.) Throughout the specified region, we may construct surfaces on which ϕ is constant, there being a unique iso-ϕ surface through each point. The reader conversant with the elements of analytical solid geometry will recognize that the vector having components $[\partial\phi/\partial x, \partial\phi/\partial y, \partial\phi/\partial z]$ at each point (x, y, z) of the surface $\phi=$ constant is normal to the surface. This vector is termed the *vector gradient* of the scalar quantity ϕ and is denoted either by grad ϕ or $\nabla\phi$, where $\nabla \equiv \mathbf{i}\partial/\partial x + \mathbf{j}\,\partial/\partial + \mathbf{k}\,\partial/\partial z$,

i.e. $$\text{grad}\,\phi = \nabla\phi = (\partial\phi/\partial x)\,\mathbf{i} + (\partial\phi/\partial y)\,\mathbf{j} + (\partial\phi/\partial z)\,\mathbf{k} \ . \qquad (1)$$

Now suppose that at each point $P(x, y, z)$ of space, or of a given region of space, a unique vector function $\mathbf{F}(x, y, z)$ may be defined where

$$\mathbf{F}(x, y, z) = F_1\mathbf{i} + F_2\mathbf{j} + F_3\mathbf{k} \qquad (2)$$

and $F_1 = F_1(x, y, z)$, etc. (Such a case is afforded by the earth's gravitational force $\mathbf{g}$ per unit mass taken throughout all space.) The totality of all such functions $\mathbf{F}$ constitutes a *vector field*.

The *divergence* of such a vector field $\mathbf{F}$ is a scalar quantity denoted either by div $\mathbf{F}$ or by $\nabla.\mathbf{F}$ and is defined to be

$$\text{div}\,\mathbf{F} = \nabla.\mathbf{F} = \partial F_1/\partial x + \partial F_2/\partial y + \partial F_3/\partial z \ , \qquad (3)$$

provided the first order derivatives on the R.H.S. of (3) exist uniquely.

The *curl* of the vector function $\mathbf{F}$ is vectorial in kind and is denoted either by curl $\mathbf{F}$ or by $\nabla\wedge\mathbf{F}$. It is defined to be

$$\text{curl}\,\mathbf{F} = \nabla\wedge\mathbf{F} = \begin{vmatrix} \mathbf{i} & \mathbf{j} & \mathbf{k} \\ \partial/\partial x & \partial/\partial y & \partial/\partial z \\ F_1 & F_2 & F_3 \end{vmatrix}$$

$$= (\partial F_3/\partial y - \partial F_2/\partial z)\mathbf{i} + (\partial F_1/\partial z - \partial F_3/\partial x)\mathbf{j} + (\partial F_2/\partial x - \partial F_1/\partial y)\mathbf{k} \ , \qquad (4)$$

provided all derivatives on the R.H.S. of (4) exist uniquely.

The forms (3) and (4) show that the divergence is a kind of scalar product and the curl is a kind of vector product. These quantities can only be formed from vector functions, just as the gradient can only be formed from scalar functions.

We observe that $\nabla\phi$, as defined, is vectorial: hence in general it has a divergence. This is written either as $\nabla.\nabla\phi$ or as div grad ϕ. In Cartesian form we have

$$\nabla.\nabla\phi = \partial^2\phi/\partial x^2 + \partial^2\phi/\partial y^2 + \partial^2\phi/\partial z^2 = \nabla^2\phi \tag{5}$$

where $\nabla^2 \equiv \partial^2/\partial x^2 + \partial^2/\partial y^2 + \partial^2/\partial z^2$. ∇^2 is called the *Laplacian*.

From the above definitions, the reader will be able to confirm the important identities,

$$\text{curl grad } \phi \equiv \mathbf{0}\ , \qquad \text{div curl } \mathbf{F} \equiv 0 \tag{6}$$

for all (x, y, z), whenever all scalar and vector functions and their first and second order derivatives exist uniquely.

EXERCISE 1

1. *Coplanar Vectors.* If three vectors **a**, **b**, **c** all lie in the same plane prove that in general one can find scalar magnitudes λ, μ such that

$$\mathbf{c} = \lambda\mathbf{a} + \mu\mathbf{b}\ .$$

2. If a linear relationship of the above type holds between any three vectors **a**, **b**, **c**, show that the vectors are all parallel to a plane. Show that the vectors $(2\mathbf{a} - 3\mathbf{b} + \mathbf{c})$, $(\mathbf{a} - 2\mathbf{b} + 4\mathbf{c})$, $(7\mathbf{a} - 12\mathbf{b} + 14\mathbf{c})$ are all parallel to the same plane.

3. If P_1, P_2 have co-ordinates $(-7, 2, 3)$, $(-8, 4, 5)$, prove that $\overline{P_1P_2} \equiv -\mathbf{i} + 2\mathbf{j} + 2\mathbf{k}$ and that the unit vector in the direction $\overline{P_1P_2}$ has components $[-\frac{1}{3}, \frac{2}{3}, \frac{2}{3}]$.

4. *Internal Division of a Line Segment.* Let $\overline{OP_1} \equiv \mathbf{r}_1 = [x_1, y_1, z_1]$, $\overline{OP_2} \equiv \mathbf{r}_2 = [x_2, y_2, z_2]$ and suppose P divides P_1P_2 internally in the ratio $P_1P:PP_2 = \lambda:\mu$. Show that the position vector **r** of P is given by

$$\overline{OP} \equiv \mathbf{r} = (\lambda\mathbf{r}_2 + \mu\mathbf{r}_1)/(\lambda + \mu)$$

and deduce that the co-ordinates of P are

$$\left(\frac{\lambda x_2 + \mu x_1}{\lambda + \mu}, \frac{\lambda y_2 + \mu y_1}{\lambda + \mu}, \frac{\lambda z_2 + \mu z_1}{\lambda + \mu}\right).$$

5. *External Division of a Line Segment.* If, in the previous case, the division of P_1P_2 be external and in the ratio $P_1P:PP_2 = \lambda:\mu$, show that

$$\overline{OP} \equiv \mathbf{r} = (\lambda\mathbf{r}_2 - \mu\mathbf{r}_1)/(\lambda - \mu)$$

and find the co-ordinates of P.

6. Prove that $(\mathbf{a}\wedge\mathbf{b})\wedge\mathbf{c} = (\mathbf{a}.\mathbf{c})\mathbf{b} - (\mathbf{b}.\mathbf{c})\mathbf{a}$.

7. Prove that

$$(\mathbf{P}\wedge\mathbf{Q}).(\mathbf{R}\wedge\mathbf{S}) = (\mathbf{P}.\mathbf{R})(\mathbf{Q}.\mathbf{S}) - (\mathbf{P}.\mathbf{S})(\mathbf{Q}.\mathbf{R})\ .$$

8. Prove that $\mathbf{P}_\wedge(\mathbf{Q}_\wedge\mathbf{R})+\mathbf{Q}_\wedge(\mathbf{R}_\wedge\mathbf{P})+\mathbf{R}_\wedge(\mathbf{P}_\wedge\mathbf{Q})=\mathbf{0}$.

9. *Test of Perpendicularity of Two Vectors.* If $\mathbf{a}.\mathbf{b}=0$ and $\mathbf{a}$ and $\mathbf{b}$ are non zero vectors, prove that $\mathbf{a}$ and $\mathbf{b}$ are perpendicular.

10. *Concurrence of Perpendicular Bisectors of the Sides of a Triangle.* Let the perpendicular bisectors of the sides BC, CA of a triangle ABC meet in O and let $\overline{OA}\equiv\mathbf{a}$, $\overline{OB}\equiv\mathbf{b}$, $\overline{OC}\equiv\mathbf{c}$. Let D, E, F be the mid-points of BC, CA, AB. Show that $\overline{OD}\equiv\frac{1}{2}(\mathbf{b}+\mathbf{c})$, $\overline{OE}\equiv\frac{1}{2}(\mathbf{c}+\mathbf{a})$, $\overline{OF}\equiv\frac{1}{2}(\mathbf{a}+\mathbf{b})$. Show further that $\mathbf{b}^2-\mathbf{c}^2=0=\mathbf{c}^2-\mathbf{a}^2$ and hence infer that OF is perpendicular to AB.

11. Prove that

$$\begin{aligned}(\mathbf{P}_\wedge\mathbf{Q})_\wedge(\mathbf{R}_\wedge\mathbf{S}) &= [\mathbf{P},\,\mathbf{Q},\,\mathbf{S}]\mathbf{R}-[\mathbf{P},\,\mathbf{Q},\,\mathbf{R}]\mathbf{S}\\ &= [\mathbf{P},\,\mathbf{R},\,\mathbf{S}]\mathbf{Q}-[\mathbf{Q},\,\mathbf{R},\,\mathbf{S}]\mathbf{P}\ .\end{aligned}$$

12. A force having components $[-1,\ 2,\ 1]$ acts through the point $(1,\ 1,\ -2)$. Find the vector moment about $(0,\ 1,\ -1)$ and the physical moment about the axis through this point having direction cosines $[+\frac{1}{3},\ -\frac{2}{3},\ +\frac{2}{3}]$.

13. *Reduction of a System of Forces.* Suppose n forces $\mathbf{F}_k$ act through points P_k having position vectors $\overline{OP_k}\equiv\mathbf{r}_k$ $(k=1,\ldots,n)$. At O, introduce equal and opposite forces $\pm\mathbf{F}_k$. Then $+\mathbf{F}_k$ at P_k and $-\mathbf{F}_k$ at O constitute a couple of vector moment $\mathbf{r}_{k\wedge}\mathbf{F}_k$. By treating each force in the same way, show that the original system is reduced to a single force $\mathbf{R}$ at O together with a couple $\mathbf{G}$ where

$$\mathbf{R}=\sum_{k=1}^{n}\mathbf{F}_k\ ,\quad \mathbf{G}=\sum_{k=1}^{n}(\mathbf{r}_{k\wedge}\mathbf{F}_k)\ .$$

14. Forces $k\,\overline{BC}$, $k\,\overline{CA}$, $k\,\overline{AB}$ act along the sides of a triangle ABC. Show that in general they reduce to a couple.

15. Prove that the centroid of a uniform triangular lamina coincides with that of three particles of equal mass placed at its vertices. Hence or otherwise show that the centroid of a uniform quadrilateral lamina is the same as that of equal particles each of mass m at its vertices, together with a particle of mass $-m$ at the intersection of its diagonals.

16. If $\hat{\mathbf{a}}(t)$ be a unit vector which is a function of a scalar t, prove that in general $\hat{\mathbf{a}}$ and $d\hat{\mathbf{a}}/dt$ are at right angles.

17. If $\mathbf{r}=[x,\,y,\,z]$ prove that

$$\text{(i) div}\,\mathbf{r}=3\ ;\quad \text{(ii) curl}\,\mathbf{r}=\mathbf{0}\ .$$

Find a function $\phi(x,\,y,\,z)$ such that $\mathbf{r}=\text{grad}\,\phi$.

18. Prove the results

$$\text{(i)}\ \nabla(\phi_1\phi_2)=\phi_1\nabla\phi_2+\phi_2\nabla\phi_1\ ;$$
$$\text{(ii)}\ \nabla.(\phi\mathbf{F})=\phi\,\nabla.\mathbf{F}+\mathbf{F}.(\nabla\phi)\ .$$

Chapter 2

Kinematics of Particles and Rigid Bodies

2.1 VELOCITY AND ACCELERATION OF A PARTICLE ALONG A CURVE

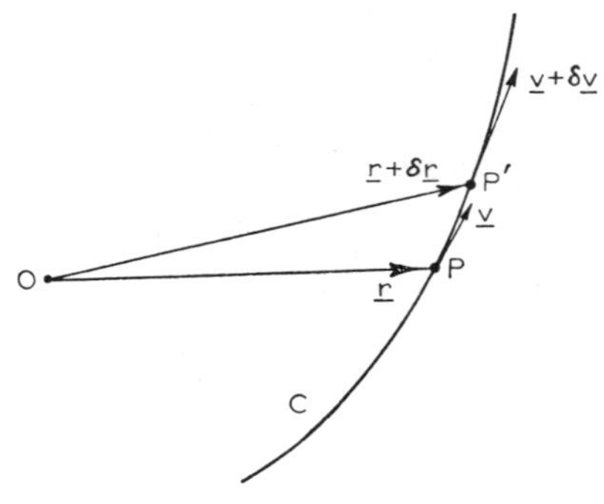

Fig. 2.1

Consider a particle moving along a curve C, not necessarily plane (Fig. 2.1). Let P, P' be its positions at times t, $t+\delta t$, where P and P' have position vectors $\overline{OP} \equiv \mathbf{r}$, $\overline{OP'} \equiv \mathbf{r}+\delta\mathbf{r}$, referred to a fixed origin O. Suppose $\mathbf{v}$, $\mathbf{v}+\delta\mathbf{v}$ are the velocities of the particle at P and P': these are necessarily tangential to C at P, P'. Since $\overline{PP'} \equiv \delta\mathbf{r}$ is the movement of the particle in time δt, the velocity $\mathbf{v}$ at P is $\mathbf{v} = \lim_{\delta t\to 0} (\delta\mathbf{r}/\delta t)$, provided the limit exists, and is denoted by $d\mathbf{r}/dt$, in accordance with the calculus convention. If the arc length of PP' be denoted by δs, then we have

$$\delta\mathbf{r}/\delta t = (\delta\mathbf{r}/\delta s) \times (\delta s/\delta t) \ ,$$

or letting $\delta t \to 0$,

$$d\mathbf{r}/dt = (d\mathbf{r}/ds) \times (ds/dt) \ ,$$

provided the limits exist.

To discern the meaning of $d\mathbf{r}/ds$, we note that $\delta\mathbf{r}/\delta s$ is a vector in the direction $\overline{PP'}$ which tends ultimately to that of the tangent at P to the curve C in the direction of motion of the particle. Let $\mathbf{t}$ denote the unit vector in this tangent. We further note that the magnitude of $\delta\mathbf{r}/\delta s$ is

$$|\delta\mathbf{r}|/\delta s = \text{chord } PP'/\text{arc } PP'.$$

Since the chord and arc elements become more nearly equal in length as $P' \to P$, we may assume that the limiting value of this last ratio is unity. Then $d\mathbf{r}/ds$ has unit magnitude and direction $\mathbf{t}$, so that $d\mathbf{r}/ds = \mathbf{t}$. Hence

$$\mathbf{v} = d\mathbf{r}/dt = (ds/dt)\mathbf{t} = v\mathbf{t} \;,$$

where $v = ds/dt$. Thus the velocity $\mathbf{v}$ has magnitude $v = ds/dt = \dot{s}$ and direction $\mathbf{t}$. It is sometimes convenient to denote $(ds/dt)\mathbf{t}$ by $d\mathbf{s}/dt$ or by $\dot{\mathbf{s}}$.

Just as the velocity $\mathbf{v}$ of the particle at P is the time rate of change of the position vector $\mathbf{r}$, so the *acceleration* $\mathbf{f}$ is defined to be the time rate of change of the velocity $\mathbf{v}$. Hence

$$\mathbf{f} = d\mathbf{v}/dt = d(v\mathbf{t})/dt = \dot{v}\mathbf{t} + v\, d\mathbf{t}/dt \;.$$

Now $d\mathbf{t}/dt = (d\mathbf{t}/ds)(ds/dt) = v d\mathbf{t}/ds$. To evaluate $d\mathbf{t}/ds$ consider $\lim\limits_{P' \to P} (\delta\mathbf{t}/\delta s)$.

Since $\mathbf{t}$ denotes the unit vector in the tangent at P to C in the direction of $\mathbf{v}$, we interpret $\mathbf{t} + \delta\mathbf{t}$ to mean the corresponding unit tangential vector at P'. Choosing any fixed point O', draw $\overline{O'T} \equiv \mathbf{t}$, $\overline{O'T'} \equiv \mathbf{t} + \delta\mathbf{t}$ (Fig. 2.2).

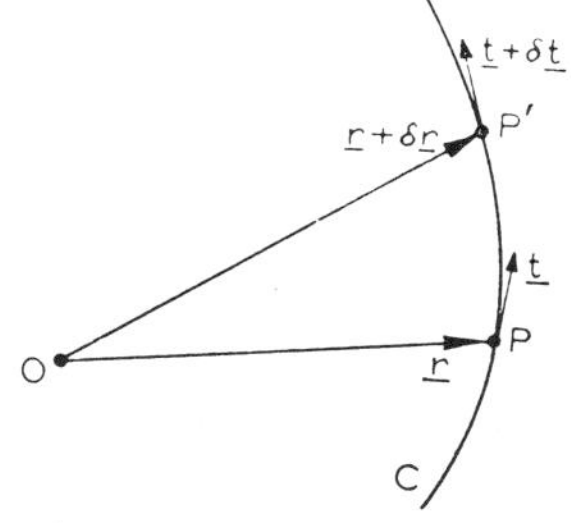

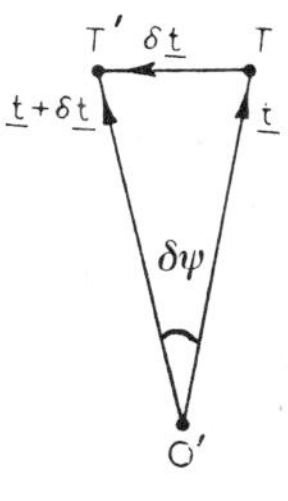

Fig. 2.2

The triangle $O'TT'$ is isosceles, $O'T$, $O'T'$ each being of unit length. Let $\delta\psi = \angle T'O'T$. As $\delta t \to 0$, $P' \to P$, so that $T' \to T$ and the direction $\delta\mathbf{t}$ $(\equiv \overline{TT'})$ tends to become perpendicular to $\mathbf{t}$. Suppose $\mathbf{n}$ denotes the unit vector in this limiting direction. Such a direction specifies that of the *principal normal* to C at P. Now $|\delta\mathbf{t}| = TT' = 2\sin(\delta\psi/2)$, and so $|d\mathbf{t}/ds| = \lim |\delta\mathbf{t}/\delta s| = \lim [\sin(\delta\psi/2)/(\delta s/2)] = \lim [\delta\psi/\delta s] = d\psi/ds$. The value of $d\psi/ds$ is termed the *curvature* of C at P and is denoted by κ. We choose κ to be essentially *positive*. Hence we have shown

$$d\mathbf{t}/ds = \kappa\mathbf{n}$$

and further

$$\mathbf{f} = \dot{v}\mathbf{t} + \kappa v^2\mathbf{n} \;.$$

The above results apply for both twisted and plane curves. In the latter case, **n** is measured inwards towards the centre of curvature and, denoting by ρ the radius of curvature, **f** *has a tangential component of $\dot{v}$ in the direction of s increasing and a normal component v^2/ρ towards the centre of curvature*.

Example

A particle is constrained to move along a circular helix so that its co-ordinates at time t are $(a \cos \theta,\ a \sin \theta,\ a\,\theta \tan \alpha)$, where a, α are constants and $0 < \alpha < \frac{\pi}{2}$. The speed increases linearly with t from zero at $t=0$ to V at $t=T$. Find the acceleration at any time $t<T$, the motion taking place in the sense of θ increasing, and starting from the point $(a, 0, 0)$.

At time t $(0<t<T)$, the position vector is

$$\mathbf{r}=a(\cos\theta\,\mathbf{i}+\sin\theta\,\mathbf{j}+\theta\tan\alpha\,\mathbf{k}) \;;$$

$$\therefore\ d\mathbf{r}/dt=(d\mathbf{r}/d\theta)(d\theta/dt)=a\dot{\theta}(-\sin\theta\mathbf{i}+\cos\theta\mathbf{j}+\tan\alpha\mathbf{k}).$$

Since the magnitude of $(-\sin\theta\mathbf{i}+\cos\theta\mathbf{j}+\tan\alpha\mathbf{k})$ is $\sec\alpha$,

$$d\mathbf{r}/dt=(a\dot{\theta}\sec\alpha)(-\sin\theta\cos\alpha\mathbf{i}+\cos\theta\cos\alpha\mathbf{j}+\sin\alpha\mathbf{k}) \;,$$

i.e. $(Vt/T)\mathbf{t}=a\dot{\theta}\sec\alpha(-\sin\theta\cos\alpha\mathbf{i}+\cos\theta\cos\alpha\mathbf{j}+\sin\alpha\mathbf{k})$, for $0<t<T$. Hence

$$\begin{cases} Vt/T=a\,\dot{\theta}\sec\alpha \;, \\ \mathbf{t}=-\sin\theta\cos\alpha\mathbf{i}+\cos\theta\cos\alpha\mathbf{j}+\sin\alpha\mathbf{k} \;. \end{cases}$$

From the first relation, $\theta=Vt^2\cos\alpha/2aT$. From the second,

$$\begin{aligned} d\mathbf{t}/ds &= (d\mathbf{t}/d\theta)(d\theta/ds) \\ &= (-\cos\theta\cos\alpha\mathbf{i}-\sin\theta\cos\alpha\mathbf{j})d\theta/ds \;. \end{aligned}$$

Now $d\theta/ds=(d\theta/dt)/(ds/dt)=(Vt\cos\alpha/aT)/(Vt/T)=(\cos\alpha/a)$.

$$\therefore\ d\mathbf{t}/ds=(-\cos\theta\mathbf{i}-\sin\theta\mathbf{j})(\cos^2\alpha/a) \;,$$

$$\therefore\ \kappa=\cos^2\alpha/a \;,$$

$$\mathbf{n}=-\cos\theta\mathbf{i}-\sin\theta\mathbf{j} \;.$$

Thus at any time $t(0<t<T)$, the particle moves with acceleration having components V/T along the tangent and $\kappa v^2=V^2t^2\cos^2\alpha/aT^2$ along the principal normal.

2.2 MOTION IN A PLANE—RADIAL AND TRANSVERSE COMPONENTS

Let us now consider the motion of a particle in a plane and specified at time t by polar co-ordinates (r, θ). If $\mathbf{r}$ denote the position vector $\overline{OP}$ at time t (Fig. 2.3), then $\mathbf{r}=r\hat{\mathbf{r}}$, where $\hat{\mathbf{r}}$ denotes the unit vector in $\overline{OP}$ in the sense of r increasing.

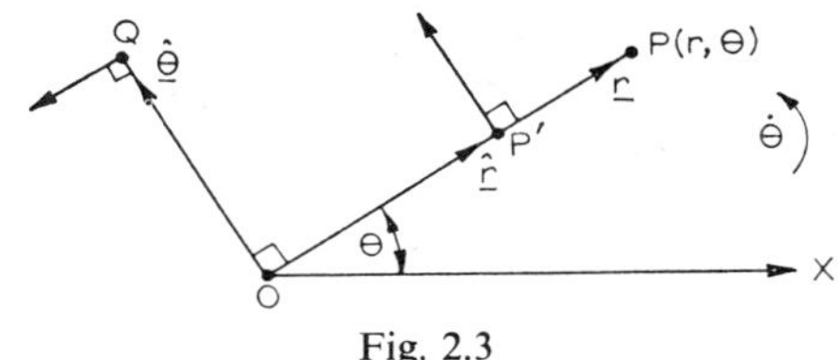

Fig. 2.3

The velocity of P is $\mathbf{v}$ where

$$\mathbf{v}=d\mathbf{r}/dt=d(r\hat{\mathbf{r}})/dt=\dot{r}\hat{\mathbf{r}}+rd\hat{\mathbf{r}}/dt \ .$$

To find $d\hat{\mathbf{r}}/dt$, we observe that it signifies the velocity of the point P' in OP where position vector $\overline{OP'}\equiv\hat{\mathbf{r}}$. Now $\overline{OP'}$ is of fixed length, but its direction varies with θ. Let $\hat{\boldsymbol{\theta}}$ denote the unit vector in the *transverse* direction, i.e. in the plane of POX, perpendicular to OP and in the sense of θ increasing. In the diagram, $\overline{OQ}\equiv\hat{\boldsymbol{\theta}}$. The motion of P' is along $\hat{\boldsymbol{\theta}}$. Since the frame POQ rotates with angular velocity $\dot{\theta}$ about the normal to POQ and in the sense of θ increasing, we see that the velocity of P' is

$$d\hat{\mathbf{r}}/dt=\dot{\theta}\,\hat{\boldsymbol{\theta}} \ .$$

$$\therefore\ \mathbf{v}=\dot{r}\hat{\mathbf{r}}+r\dot{\theta}\,\hat{\boldsymbol{\theta}} \ . \qquad (1)$$

Thus P moves with radial and transverse velocity components $[\dot{r}, r\dot{\theta}]$.

Further, the acceleration $\mathbf{f}$ of (1) is given by

$$\mathbf{f}=d\mathbf{v}/dt=\ddot{r}\hat{\mathbf{r}}+\dot{r}\ d\hat{\mathbf{r}}/dt+\hat{\boldsymbol{\theta}}d(r\dot{\theta})/dt+r\dot{\theta}d\hat{\boldsymbol{\theta}}/dt \ .$$

From Fig. 2.3, we see $d\hat{\boldsymbol{\theta}}/dt=-\dot{\theta}\hat{\mathbf{r}}$, being the velocity of Q.

$$\therefore\ \mathbf{f}=(\ddot{r}-r\dot{\theta}^2)\mathbf{r}+(r\ddot{\theta}+2\dot{r}\dot{\theta})\hat{\boldsymbol{\theta}} \ . \qquad (2)$$

Thus P moves with radial and transverse acceleration components

$$f_r=\ddot{r}-r\dot{\theta}^2 \ ,$$
$$f_\theta=r\ddot{\theta}+2\dot{r}\dot{\theta}=(1/r)d(r^2\dot{\theta})/dt \ .$$

A method for obtaining these results using complex numbers is given in Exercise 2, No. 1.

Example

A particle is constrained to move along the equiangular spiral $r=ae^{b\theta}$ so that the radius vector moves with constant angular velocity ω. Determine the velocity and acceleration components.

$$r=a\,e^{b\omega t}, \text{ taking } \theta=\omega t \ .$$
$$\therefore\ \dot{r}=\omega\ abe^{b\omega t}, \qquad \dot{\theta}=\omega \ ;$$
$$\ddot{r}=\omega^2ab^2e^{b\omega t}, \qquad \ddot{\theta}=0 \ .$$
$$\therefore\ f_r=\ddot{r}-r\dot{\theta}^2=(b^2-1)\omega^2ae^{b\omega t} \ ,$$
$$f_\theta=r\ddot{\theta}+2\dot{r}\dot{\theta}=2ab\omega^2e^{b\omega t} \ .$$

Thus the radial and transverse velocity components are $[\omega abe^{b\omega t}, a\omega e^{b\omega t}]$ and those of acceleration are $[(b^2-1)a\omega^2 e^{b\omega t}, 2ab\omega^2 e^{b\omega t}]$.

2.3 RELATIVE VELOCITY AND ACCELERATION

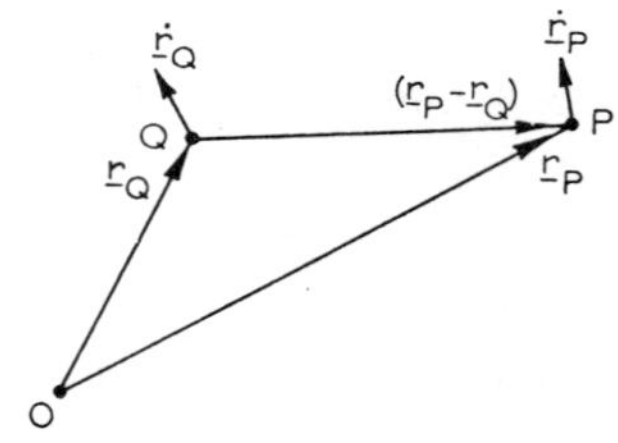

Fig. 2.4

Let P, Q be points having position vectors $\mathbf{r}_P$, $\mathbf{r}_Q$ with respect to a fixed origin O at time t. Then their velocities are $\dot{\mathbf{r}}_P$, $\dot{\mathbf{r}}_Q$ and their accelerations $\ddot{\mathbf{r}}_P$, $\ddot{r}_Q$. If (x_P, y_P, z_P), (x_Q, y_Q, z_Q) denote the Cartesian co-ordinates of P and Q at time t, referred to a set of *fixed* axes through O, then

$$\mathbf{r}_P = x_P\mathbf{i} + y_P\mathbf{j} + z_P\mathbf{k} \ ,$$

$$\dot{\mathbf{r}}_P = \dot{x}_P\mathbf{i} + \dot{y}_P\mathbf{j} + \dot{z}_P\mathbf{k} \ ,$$

since $d\mathbf{i}/dt = \mathbf{0}$, etc.,

$$\ddot{\mathbf{r}}_P = \ddot{x}_P\mathbf{i} + \ddot{y}_P\mathbf{j} + \ddot{z}_P\mathbf{k} \ .$$

Similar expressions hold for $\dot{\mathbf{r}}_Q$, $\ddot{\mathbf{r}}_Q$.

Now the position of P relative to Q at time t is $\mathbf{r}_P - \mathbf{r}_Q$, so that the velocity of P relative to Q is

$$\dot{\mathbf{r}}_P - \dot{\mathbf{r}}_Q = (\dot{x}_P - \dot{x}_Q)\mathbf{i} + (\dot{y}_P - \dot{y}_Q)\mathbf{j} + (\dot{z}_P - \dot{z}_Q)\mathbf{k} \ ,$$

and the acceleration of P relative to Q is

$$\ddot{\mathbf{r}}_P - \ddot{\mathbf{r}}_Q = (\ddot{x}_P - \ddot{x}_Q)\mathbf{i} + (\ddot{y}_P - \ddot{y}_Q)\mathbf{j} + (\ddot{z}_P - \ddot{z}_Q)\mathbf{k} \ .$$

Example 1—Pursuit curves.

An aircraft pursues a straight course with constant velocity V and is being chased by a guided missile moving with constant speed $2V$ and fitted with a homing device to ensure that its motion is always directed at the target. Initially the missile is at right angles to the course of the aircraft and distant R from it. Find the polar equation of the missile's 'pursuit curve' relative to the target, taking the course of the target as the initial line $\theta = 0$, and find the time taken by it to strike the target.

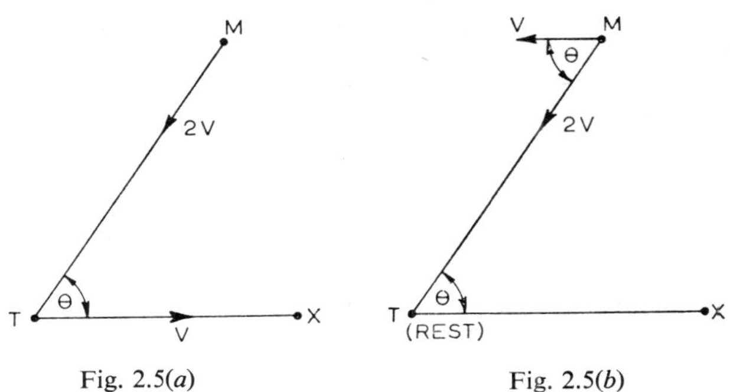

Fig. 2.5(*a*) Fig. 2.5(*b*)

Fig. 2.5(*a*) illustrates the situation at time t: Fig. 2.5(*b*) shows the motion of the missile M *relative* to the target T. Take $TM=r$, $\angle MTX=\theta$. Resolving in the radial and transverse directions in Fig. 2.5(*b*)

$$\begin{cases} \dot{r} = -V(2+\cos\theta) \ , \\ r\dot{\theta} = V\sin\theta \ ; \end{cases}$$

$$\therefore \ dr/r = -(2/\sin\theta + \cot\theta)d\theta \ ,$$

and so $$A/r = \tan^2(\theta/2)\sin\theta \ .$$

When $\theta=\pi/2$, $r=R$, so that

$$\underline{R/r = \sin\theta\tan^2(\theta/2) = 2\sin^3(\theta/2)/\cos(\theta/2) \ .}$$

Thus, $r=0$ when $\theta=\pi$.

$$\therefore \ \dot{\theta} = (V\sin\theta)/r = (V/R)\sin^2\theta\tan^2(\theta/2) = (4V/R)\sin^4(\theta/2) \ .$$

$$\therefore \ dt = (R/4V)\operatorname{cosec}^4(\theta/2)d\theta \ ,$$

$$\therefore \ \text{Required time} = (R/4V)\int_{\pi/2}^{\pi}\operatorname{cosec}^4(\theta/2)d\theta$$

$$= (R/4V)\int_{\pi/2}^{\pi}[1+\cot^2(\theta/2]\operatorname{cosec}^2(\theta/2)d\theta$$

$$\underline{=\tfrac{2}{3}R/V \ .}$$

Example 2—Hodograph.

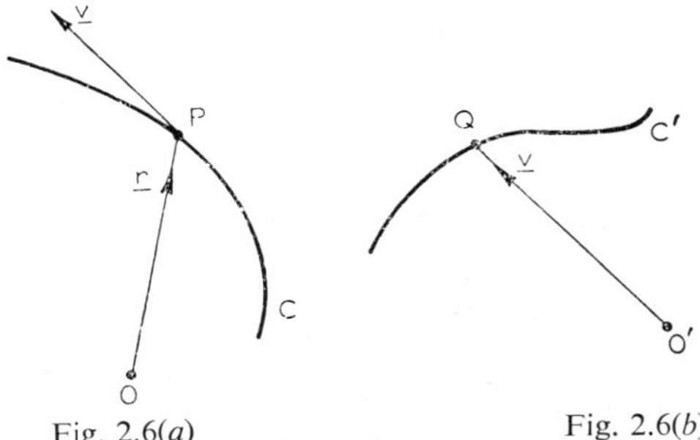

Fig. 2.6(*a*) Fig. 2.6(*b*)

Suppose $\overline{OP} \equiv \mathbf{r} = \mathbf{r}(t)$ at some time t, O being fixed. Then P moves along a curve C (Fig. 2.6(*a*)). Let $\mathbf{v} = d\mathbf{r}/dt$ be the velocity of P. In Fig. 2.6(*b*) through a fixed point O', we draw $\overline{O'Q} \equiv \mathbf{v}$. Then the locus C' of Q is called the *hodograph* of the velocity of P.

Thus if P describes the ellipse

$$x = a \cos nt, \; y = b \sin nt \; ,$$

where t signifies time,

$$\dot{x} = -na \sin nt, \; \dot{y} = bn \cos nt \; ,$$

and so the locus of $(\dot{x}, \dot{y})$ in the hodograph plane is

$$(\dot{x}/na)^2 + (\dot{y}/nb)^2 = 1 \; ,$$

an ellipse which is similar and similarly situated to the original ellipse.

Example 2(a)—Use of hodograph.

A particle moves in a plane in such a way that when its velocity at any instant is $\mathbf{V}$ its acceleration is $\mathbf{f} = k\mathbf{V} + \mathbf{a}$, where k is a constant scalar and $\mathbf{a}$ is a constant vector. Prove that the direction of its acceleration is constant.

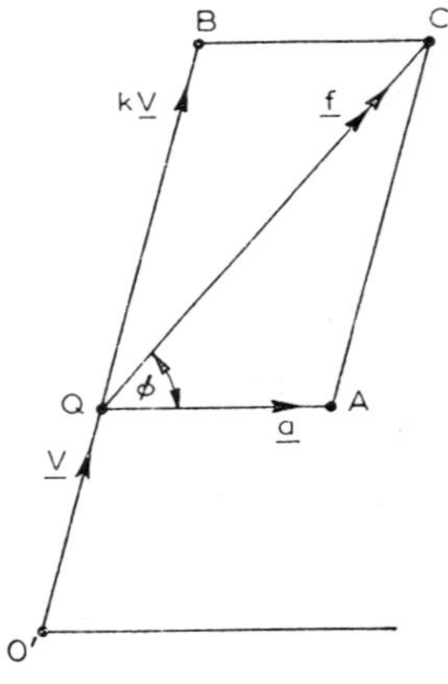

Fig. 2.7

In Fig. 2.7, $\overline{O'Q} \equiv \mathbf{V}$, $\overline{QA} \equiv \mathbf{a}$, $\overline{QB} \equiv k\mathbf{V}$. Complete the parallelogram $QACB$ so that $\overline{QC} \equiv \mathbf{f}$. Let $C\hat{Q}A = \phi$. Let $\mathbf{i}$, $\mathbf{j}$ be the unit vectors in the constant directions $\overline{QA}$ and the line in the plane $QACB$ perpendicular to QA. Suppose $\mathbf{V} = u\mathbf{i} + v\mathbf{j}$ and $\mathbf{a} = a\mathbf{i}$. Then

$$\mathbf{f} = (ku + a)\mathbf{i} + kv\mathbf{j}$$

and so $\tan \phi = kv/(ku + a)$. Since $\mathbf{f} = \dot{u}\mathbf{i} + \dot{v}\mathbf{j}$, $\tan \phi = \dot{v}/\dot{u}$, and so

$$\dot{v}/\dot{u} = kv/(ku + a) \; .$$

Integration leads at once to $kv = A(ku + a)$, where A is a constant. Hence $\tan\phi = A$, so that ϕ is constant.

2.4 KINEMATICS OF A RIGID BODY ROTATING ABOUT A FIXED POINT

A *rigid body* is defined to be a collection of particles, the distance between any two of which remains invariant.

First suppose such a body is moving quite generally in space. Let P_1, P_2, P_3 be any three non-collinear points fixed in the body and suppose that at any instant their co-ordinates referred to Cartesian axes fixed in space are (x_1, y_1, z_1), (x_2, y_2, z_2), (x_3, y_3, z_3). If we can fix all these points, then the rigid body is fixed in space. This apparently requires knowing nine co-ordinates. But the distances P_1P_2, P_2P_3, P_3P_1 are invariant. Thus we obtain three equations of the form

$$(x_2 - x_1)^2 + (y_2 - y_1)^2 + (z_2 - z_1)^2 = \text{const., etc.,}$$

and so there are actually only six independent co-ordinates. We say that such a rigid body motion has *six degrees of freedom*.

Now suppose we constrain the rigid body to rotate about a fixed point O. To fix O means prescribing three co-ordinates and so *a rigid body rotating about a fixed point O has but three degrees of freedom*. Two direction cosines are required to fix the direction of the axis of rotation through O and an angular rotation about this axis completely specifies the motion of such a body. Since the size and shape of the rigid body are not involved in specifying its three degrees of freedom, we may describe its motion by considering that of a sphere S fixed in the body and having its centre at O. If P is any point of the body and if OP meets S in P', then the velocity of P is clearly OP/OP' times that of P'.

Let A and B be two points on the surface of the sphere S. As the body rotates about O suppose that at some later instant A and B have moved to positions A', B' on the surface S. Draw the great circular arcs on S through A, B; A', B' and also those through A, A'; B, B' (Fig. 2.8). Let A'', B'' be the midpoints of the great circular arcs AA', BB'. Through A'', B'' draw the great circles on S at right angles to the arcs AA', BB', respectively and let these meet in C. Draw the great circular arcs CA, CA', CB, CB' on S. Since the arc CA'' is the perpendicular bisector of the arc AA', the arcs CA, CA' are equal. Similarly the arcs CB, CB' are equal. But from the definition of a rigid body, the arcs AB, $A'B'$ are equal. Hence the spherical triangles ABC, $A'B'C$ are congruent and so the portion of the sphere originally within the area of the spherical triangle ABC is now within the spherical triangle $A'B'C$. During the rotation about O the point C has remained fixed. This means that *a rotation about* OC *has taken place*.

Suppose now that the rotation about OC is $\delta\theta$ and that it has taken place in time δt. Then $\delta\theta$ is the angle between the planes AOC, $A'OC$ and between

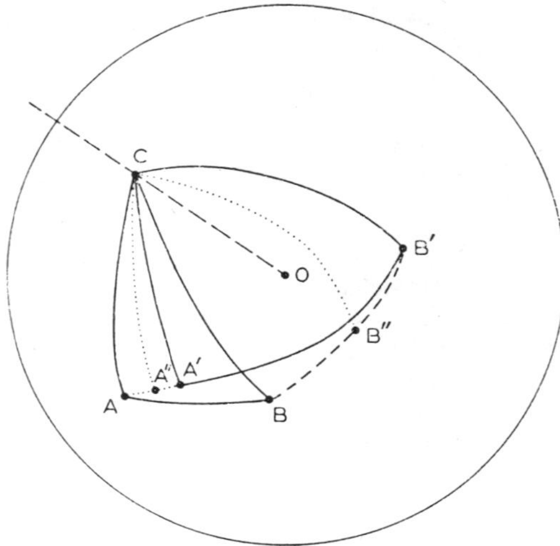

Fig. 2.8

BOC, *B′OC*. The mean angular velocity in the time δt is, then, $\delta\theta/\delta t$. If the observations are carried out at times t, $t+\delta t$, then the *instantaneous angular velocity* at time t about OC is $\lim_{\delta t\to 0} \delta\theta/\delta t = d\theta/dt = \omega$, say.

We have thus shown that *any rotation of a rigid body about a fixed point O is equivalent to a rotation about a definite axis through O.* Further, we have shown how the instantaneous angular velocity about this axis can be measured. This axis about which the body instantaneously rotates is called the *instantaneous axis of rotation.*

2.5 VECTOR ANGULAR VELOCITY

Let a rigid body rotate about a fixed point O and let ω be its angular velocity about the instantaneous axis OC at time t. Let P be a *point fixed in the body* such that $\overline{OP} \equiv \mathbf{r}$ and PM is the perpendicular from P on OC (Fig. 2.9). Since $PM = r \sin\theta$, the speed of P is $\omega r \sin\theta$ and the diagram shows the velocity $\mathbf{v}$ of P has the direction of $\hat{\mathbf{a}}_\wedge\mathbf{r}$, where $\hat{\mathbf{a}}$ is the unit vector in $\overline{OC}$ so chosen that a right-handed screw advancing along it turns in the sense of ω.

$$\therefore\ \mathbf{v} = (\omega\, r \sin\theta)\hat{\mathbf{t}}\ ,$$

where $\hat{\mathbf{t}}$ is the unit vector in the sense $\hat{\mathbf{a}}_\wedge\mathbf{r}$.

$$\therefore\ \mathbf{v} = \omega\, \hat{\mathbf{a}}_\wedge\mathbf{r} = \boldsymbol{\omega}_\wedge\mathbf{r}\ ,$$

where $\boldsymbol{\omega} = \omega\hat{\mathbf{a}}$. The quantity $\boldsymbol{\omega}$ so defined is termed the (*instantaneous*) *vector angular velocity* of the rigid body about O. Its magnitude ω is the physical angular velocity about the instantaneous axis OC and $\hat{\mathbf{a}}$ specifies the direction of this instantaneous axis.

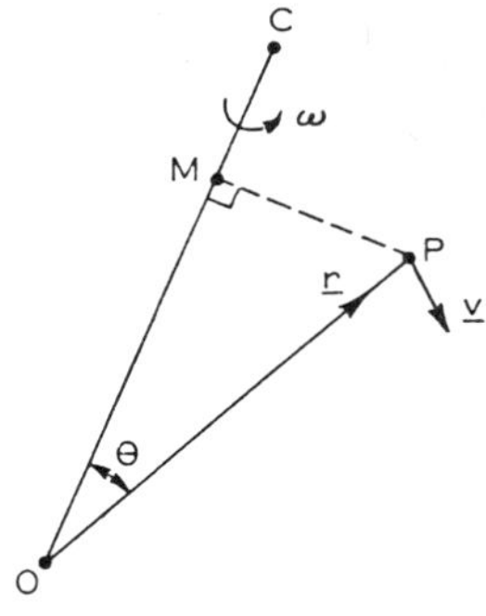

Fig. 2.9

If we write the vectors **ω**, **r** in Cartesian components in the forms

$$\boldsymbol{\omega} = [\omega_1, \omega_2, \omega_3] \ , \quad \mathbf{r} = [x, y, z] \ ,$$

referred to co-ordinate axes through O fixed in space, then

$$\mathbf{v} = \boldsymbol{\omega} \wedge \mathbf{r} = (\omega_2 z - \omega_3 y)\mathbf{i} + (\omega_3 x - \omega_1 z)\mathbf{j} + (\omega_1 y - \omega_2 x)\mathbf{k} \ .$$

Hence

$$\text{curl } \mathbf{v} = 2\omega_1 \mathbf{i} + 2\omega_2 \mathbf{j} + 2\omega_3 \mathbf{k} \ ,$$

i.e. $\boldsymbol{\omega} = \frac{1}{2}$ curl **v** for a rigid body rotating about a fixed point O.

2.6 GENERAL MOTION OF A RIGID BODY

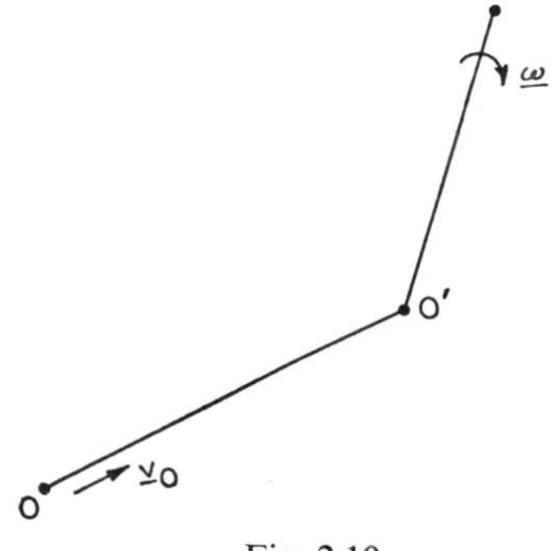

Fig. 2.10

In the general case where no point of the rigid body is fixed, we may regard the body's motion as being the displacement of one point O of it to a new position O', followed by an appropriate turning about a certain axis through O' until the body takes up the required orientation. Thus the general motion may be described as a translation and a rotation.

When we take the limit as $O' \to O$, the motion is specified by a vector $\mathbf{v}_o$ giving the instantaneous velocity of O and an angular velocity $\boldsymbol{\omega}$, being the instantaneous vector angular velocity about O.

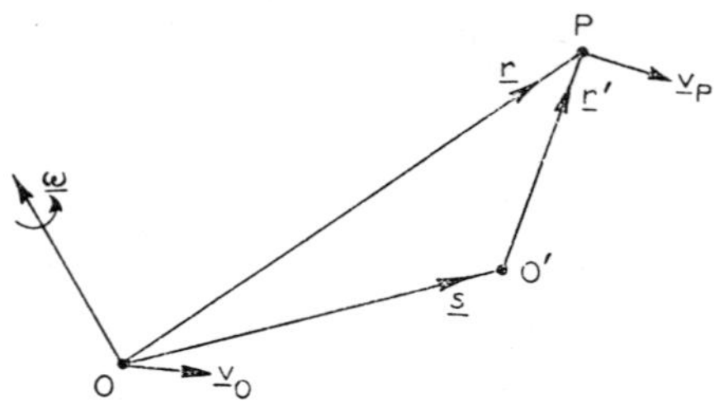

Fig. 2.11

Suppose now that P is any other point of the body and that its velocity is $\mathbf{v}_P$ (Fig. 2.11). If $\overline{OP} \equiv \mathbf{r}$, then the velocity of P relative to O is $\boldsymbol{\omega}_\wedge\mathbf{r}$. Hence, since $\mathbf{v}_o$ is the velocity of O,

$$\underline{\mathbf{v}_P = \mathbf{v}_o + \boldsymbol{\omega}_\wedge\mathbf{r}} \ .$$

We now prove that $\boldsymbol{\omega}$ is independent of the position of O.

Proof

In Fig. 2.11, choose O' so that $\overline{OO'} \equiv \mathbf{s}$, $\overline{O'P} \equiv \mathbf{r}'$. Then $\mathbf{r} = \mathbf{s} + \mathbf{r}'$. Also,

$$\mathbf{v}_P = \mathbf{v}_o + \boldsymbol{\omega}_\wedge\mathbf{r} \ ,$$
$$\mathbf{v}_o' = \mathbf{v}_o + \boldsymbol{\omega}_\wedge\mathbf{s} \ ,$$

and so by subtraction,

$$\mathbf{v}_P = \mathbf{v}_o' + \boldsymbol{\omega}_\wedge\mathbf{r}' \ .$$

This establishes the required result.

2.7 GENERAL RIGID BODY MOTION AS A SCREW MOTION*

Let the point O, fixed in the rigid body, have velocity $\mathbf{v}$, and let $\boldsymbol{\omega}$ be the instantaneous angular velocity of the body. Then if O' be another point of the body such that $\overline{OO'} \equiv \mathbf{s}$ and $\mathbf{v}'$ be the velocity of O',

$$\mathbf{v}' = \mathbf{v} + \boldsymbol{\omega}_\wedge\mathbf{s} \ . \tag{1}$$

Let us try to find a point O' such that $\mathbf{v}'$ and $\boldsymbol{\omega}$ are parallel vectors: then $\boldsymbol{\omega}_\wedge\mathbf{v}' = \mathbf{0}$. Hence (1) gives

$$\mathbf{0} = \boldsymbol{\omega}_\wedge\mathbf{v} + \boldsymbol{\omega}_\wedge(\boldsymbol{\omega}_\wedge\mathbf{s})$$
$$= \boldsymbol{\omega}_\wedge\mathbf{v} + (\boldsymbol{\omega}.\mathbf{s})\boldsymbol{\omega} - \omega^2\mathbf{s} \ .$$

* Numbers 16 and 17 in Exercise 2 extend the methods developed here to the problem of reducing a system of forces to a wrench.

Then, if $\omega \neq 0$,

$$\mathbf{s} = (\boldsymbol{\omega}_\wedge \mathbf{v})/\omega^2 + \lambda\boldsymbol{\omega} \tag{2}$$

where $\lambda = (\boldsymbol{\omega}.\mathbf{s})/\omega^2$. As λ varies, (2) shows that the point O' having position vector $\mathbf{s}$ traces out a straight line through the point represented by $\mathbf{s} = (\boldsymbol{\omega}_\wedge \mathbf{v})/\omega^2$ and having the same direction as $\boldsymbol{\omega}$ (Fig. 2.12).

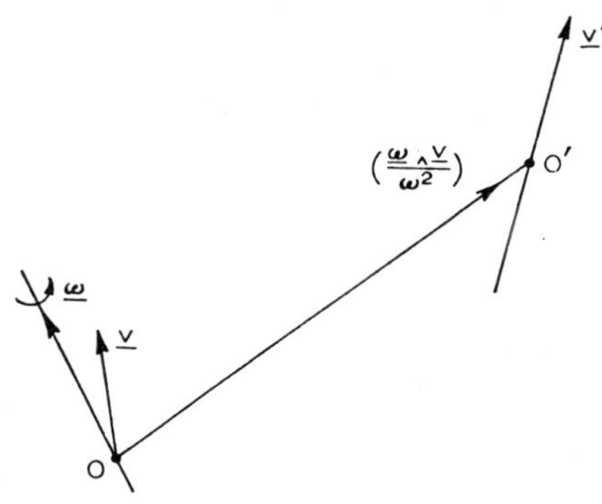

Fig. 2.12

At the particular instant considered, every point on the line (2) has a velocity parallel to $\boldsymbol{\omega}$. Hence the instantaneous motion of the body is a *screw motion* about this line. This line is called the *axis of the screw* or *central axis.* Every point on the axis moves along it and the body turns about the axis.

The velocity $\mathbf{V}$ at any point P where $\overline{OP} \equiv \mathbf{r}$ is

$$\mathbf{V} = \mathbf{v} + \boldsymbol{\omega}_\wedge \mathbf{r} \ .$$

Hence,

$$\mathbf{V}.\boldsymbol{\omega} = \mathbf{v}.\boldsymbol{\omega} \ .$$

This shows that the scalar quantity $\Gamma = \mathbf{v}.\boldsymbol{\omega}$ is invariant for the system at any instant. Also ω^2 is invariant. Write

$$p = \Gamma/\omega^2 \ .$$

p is, then, another invariant. At any point on the axis of the screw with position vector $\mathbf{v}'$, since $\mathbf{v}'$ and $\boldsymbol{\omega}$ are parallel, we see that $p = \mathbf{v}'.\boldsymbol{\omega}/\omega^2 = v'/\omega$. This latter ratio is seen to be the distance the body advances along the axis of the screw per unit angle of turn. Hence p is termed the *pitch* of the screw.

2.8 COMPOSITION OF ANGULAR VELOCITIES

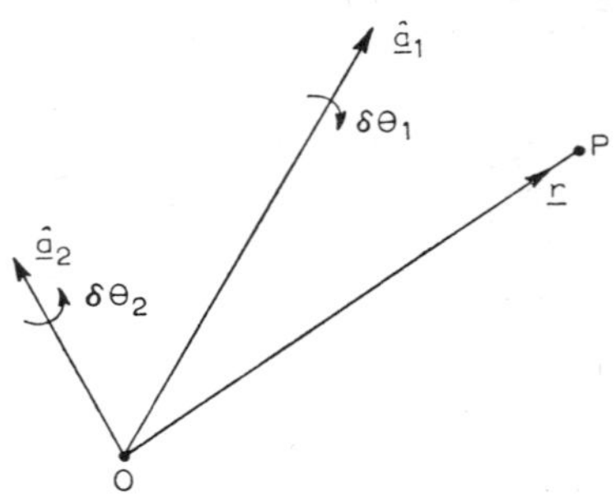

Fig. 2.13

In Fig. 2.13, O and P are points fixed in a rigid body and $\overline{OP} \equiv \mathbf{r}$. Let the rigid body turn through a small angle $\delta\theta_1$ in the positive sense about an axis through O specified by the unit vector $\hat{\mathbf{a}}_1$. Then the movement of P is $(\hat{\mathbf{a}}_1\delta\theta_1)_\wedge\mathbf{r}$; and so the new position of P relative to O is now

$$\mathbf{r}_1 = \mathbf{r} + (\hat{\mathbf{a}}_1\delta\theta_1)_\wedge\mathbf{r} \; .$$

Now suppose the body rotates through another small angle $\delta\theta_2$ in the positive sense about an axis through O specified by the unit vector $\hat{\mathbf{a}}_2$. Then, since $(\hat{\mathbf{a}}_2\delta\theta_2)_\wedge\mathbf{r}_1$ is the movement of P, the new position of P relative to O is now

$$\begin{aligned}\mathbf{r}_{12} &= \mathbf{r}_1 + (\hat{\mathbf{a}}_2\delta\theta_2)_\wedge\mathbf{r}_1\\ &= \mathbf{r} + \delta\theta_1\hat{\mathbf{a}}_{1\wedge}\mathbf{r} + (\hat{\mathbf{a}}_2\delta\theta_2)_\wedge(\mathbf{r} + \delta\theta_1\hat{\mathbf{a}}_{1\wedge}\mathbf{r})\\ &= \mathbf{r} + (\delta\theta_1\hat{\mathbf{a}}_1 + \delta\theta_2\hat{\mathbf{a}}_2)_\wedge\mathbf{r}, \text{ to first order.}\end{aligned}$$

If the operations were performed in the reverse order, we should obtain for the displacement of P relative to O,

$$\mathbf{r}_{21} = \mathbf{r} + (\delta\theta_2\hat{\mathbf{a}}_2 + \delta\theta_1\hat{\mathbf{a}}_1)_\wedge\mathbf{r}, \text{ to first order.}$$

Thus, to first order, $\mathbf{r}_{12} = \mathbf{r}_{21}$, *but this result would not hold for finite rotations.*

If δt be the time interval in which these operations take place, the velocity of P relative to O at time t is

$$\begin{aligned}&\lim_{\delta t \to 0}\{(\mathbf{r}_{12} - \mathbf{r})/\delta t\} = (\dot{\theta}_1\hat{\mathbf{a}}_1 + \dot{\theta}_2\hat{\mathbf{a}}_2)_\wedge\mathbf{r}\\ &= (\boldsymbol{\omega}_1 + \boldsymbol{\omega}_2)_\wedge\mathbf{r} \; ,\end{aligned}$$

where $\boldsymbol{\omega}_1 = \dot{\theta}_1\hat{\mathbf{a}}_1$ = vector angular velocity about $\hat{\mathbf{a}}_1$; $\boldsymbol{\omega}_2 = \dot{\theta}_2\hat{\mathbf{a}}_2$ = vector angular velocity about $\hat{\mathbf{a}}_2$.

This result confirms that vector angular velocities about a point of a rigid body are additive in the same way as other vector quantities. It is, therefore, permissible to resolve such quantities into components as illustrated in the following example.

Example 1

A rigid body S has a spin $\boldsymbol{\omega}$ and a particle A of S has velocity $\mathbf{v}$. Show that every particle P of S with velocity vector parallel to $\boldsymbol{\omega}$ lies on the line $\overline{AP} \equiv (\boldsymbol{\omega}_\wedge\mathbf{v})\omega^2 + \mu\boldsymbol{\omega}$, μ being an arbitrary scalar.

The instantaneous velocities of particles at points $(a, 0, 0)$, $(0, a/\sqrt{3}, 0)$, $(0, 0, 2a)$ of a rigid body are $[u, 0, 0]$, $[u, 0, v]$, $[u+v, -v\sqrt{3}, v/2]$, respectively, referred to a rectangular Cartesian frame. Find the magnitude and direction of spin of the body and the point at which the central axis cuts the xz-plane. (L.U.)

Let A, B, C be the points $(a, 0, 0)$, $(0, a/\sqrt{3}, 0)$, $(0, 0, 2a)$, and let $\boldsymbol{\omega} = [\omega_1, \omega_2, \omega_3]$ be the vector angular velocity of the body.

Denoting $\overline{AB} \equiv \mathbf{r}_1 = [-a, a\sqrt{3}, 0]$, $\overline{AC} \equiv \mathbf{r}_2 = [-a, 0, 2a]$, the velocities of B and C relative to A are given by

$$\mathbf{v}_1 = [0, 0, v] = \boldsymbol{\omega}_\wedge\mathbf{r}_1 \ ,$$
$$\mathbf{v}_2 = [v, -v\sqrt{3}, v/2] = \boldsymbol{\omega}_\wedge\mathbf{r}_2 \ .$$

Hence

$$\begin{vmatrix} \mathbf{i} & \mathbf{j} & \mathbf{k} \\ \omega_1 & \omega_2 & \omega_3 \\ -a & a/\sqrt{3} & 0 \end{vmatrix} = [0, 0, v] \ , \tag{1}$$

$$\begin{vmatrix} \mathbf{i} & \mathbf{j} & \mathbf{k} \\ \omega_1 & \omega_2 & \omega_3 \\ -a & 0 & 2a \end{vmatrix} = [v, -v\sqrt{3}, v/2] \ . \tag{2}$$

We thus obtain the consistent equations

$$\omega_3 = 0 \ ; \qquad (\omega_1/\sqrt{3} + \omega_2)a = v \ ; \qquad 2a\omega_2 = v \ ;$$
$$-(\omega_3 + 2\omega_1)a = -v\sqrt{3} \ ; \qquad a\omega_2 = v/2 \ .$$
$$\therefore \ \underline{\boldsymbol{\omega} = (v/a)[\sqrt{3}/2, 1/2, 0] \ ; \qquad \omega = v/a \ .}$$

Now

$$\boldsymbol{\omega}_\wedge\mathbf{v}_A = \frac{v}{a}\begin{vmatrix} \mathbf{i} & \mathbf{j} & \mathbf{k} \\ \sqrt{3}/2 & 1/2 & 0 \\ u & 0 & 0 \end{vmatrix} = -(uv/2a)\mathbf{k} \ ,$$

and so the equation of the C.A. is

$$\overline{AP} \equiv (\boldsymbol{\omega}_\wedge\mathbf{v}_A)/\omega^2 + \mu\boldsymbol{\omega} = (\mu v\sqrt{3}/2a)\mathbf{i} + (\mu v/2a)\mathbf{j} - (ua/2v)\mathbf{k} \ .$$
$$\therefore \ \overline{OP} = \overline{OA} + \overline{AP} \equiv (a + \mu v\sqrt{3}/2a)\mathbf{i} + (\mu v/2a)\mathbf{j} - (ua/2v)\mathbf{k} \ .$$

The C.A. cuts the xz-plane where $\mu v/2a = 0$, i.e. $\mu = 0$.

$\therefore$ <u>The required intersection is $(a, 0, -ua/2v)$.</u>

Example 2

Show that two equal and opposite rotations of a rigid body about distinct parallel axes are equivalent to a translation of the body.

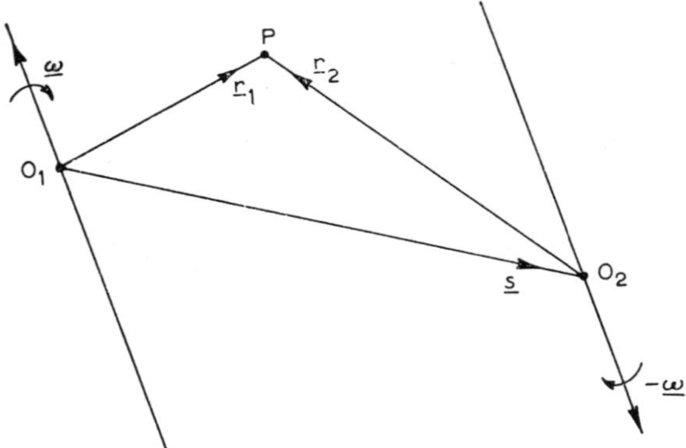

Fig. 2.14

Let O_1, O_2 be any points chosen on the axes of rotation about which the angular velocities are $\boldsymbol{\omega}$, $-\boldsymbol{\omega}$ respectively and let $\overline{O_1O_2} \equiv \mathbf{s}$, $\overline{O_1P} \equiv \mathbf{r}_1$, $\overline{O_2P} \equiv \mathbf{r}_2$, where P is any point of the rigid body (Fig. 2.14). The velocity of P is then

$$\mathbf{v}_P = (+\boldsymbol{\omega})_\wedge \mathbf{r}_1 + (-\boldsymbol{\omega})_\wedge \mathbf{r}_2$$
$$= \boldsymbol{\omega}_\wedge (\mathbf{r}_1 - \mathbf{r}_2) = \boldsymbol{\omega}_\wedge \mathbf{s} \ .$$

This is independent of the position of P and perpendicular to the plane of $+\boldsymbol{\omega}$, $-\boldsymbol{\omega}$, $\mathbf{s}$. Thus every point of the rigid body moves instantaneously with this velocity, and so the total velocity is one of translation only.

Because of the essentially vectorial character of both linear and angular velocities, we can generalize the results of rigid body motion as follows.

If a rigid body moves in such a way that a velocity $\mathbf{v}_1$ *of a point O fixed in it together with a spin* $\boldsymbol{\omega}_1$ *is to be combined with a velocity* $\mathbf{v}_2$ *of O and a spin* $\boldsymbol{\omega}_2$ *of the body, then the net motion is a velocity* $\mathbf{v}_1 + \mathbf{v}_2$ *of O and a spin* $\boldsymbol{\omega}_1 + \boldsymbol{\omega}_2$ *of the body.*

The proof is left as an exercise to the reader.

2.9 MOVING AXES

Let OX, OY, OZ form a tri-rectangular frame, the point O being fixed, for the moment, and the frame rotating instantaneously about an axis through O with angular velocity

$$\boldsymbol{\omega} = \omega_1 \mathbf{i} + \omega_2 \mathbf{j} + \omega_3 \mathbf{k} \ ,$$

where $\mathbf{i}$, $\mathbf{j}$, $\mathbf{k}$ are the unit vectors in $\overline{OX}$, $\overline{OY}$, $\overline{OZ}$ (Fig. 2.15). Let $P\ (x, y, z)$ be any point in space which can move relative to the frame and let $\overline{OP} \equiv \mathbf{r} = x\mathbf{i} + y\mathbf{j} + z\mathbf{k}$. Then the velocity of P is $\mathbf{v}$ where

$$\mathbf{v} = d\mathbf{r}/dt = \dot{x}\mathbf{i} + \dot{y}\mathbf{j} + \dot{z}\mathbf{k} + x d\mathbf{i}/dt + y d\mathbf{j}/dt + z d\mathbf{k}/dt \ .$$

Since the axes are not fixed in this case, $d\mathbf{i}/dt$, etc., are non-zero. Now $d\mathbf{i}/dt$

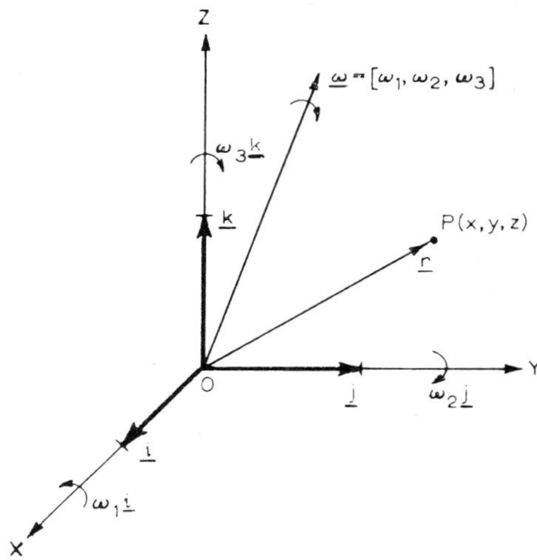

Fig. 2.15

is the velocity of the point (1, 0, 0). This is due to the spins ω_2 about OY and ω_3 about OZ. Thus, since ω_2 about OY produces a velocity $-\omega_2\mathbf{k}$ of this point and ω_3 about OZ a velocity $+\omega_3\mathbf{j}$, we infer

$$d\mathbf{i}/dt = \omega_3\mathbf{j} - \omega_2\mathbf{k} \ ,$$

with similar expressions for $d\mathbf{j}/dt$, $d\mathbf{k}/dt$. Hence

$$\mathbf{v} = (\dot{x}\mathbf{i} + \dot{y}\mathbf{j} + \dot{z}\mathbf{k}) + x(\omega_3\mathbf{j} - \omega_2\mathbf{k}) + y(\omega_1\mathbf{k} - \omega_3\mathbf{i}) + z(\omega_2\mathbf{i} - \omega_1\mathbf{j}) \ .$$

The term $\dot{x}\mathbf{i} + \dot{y}\mathbf{j} + \dot{z}\mathbf{k}$ represents the velocity of P *relative to the frame*: were the frame at rest ($\boldsymbol{\omega} = \mathbf{0}$), $\mathbf{v}$ would be equal to this. Let this quantity be denoted by $\partial\mathbf{r}/\partial t$, where the partial differentiations mean the motion is *relative to the frame*. The reader will easily verify the remaining terms on the R.H.S. are equal to $\boldsymbol{\omega}_\wedge\mathbf{r}$. Thus, finally,

$$\underline{d\mathbf{r}/dt = \partial\mathbf{r}/\partial t + \boldsymbol{\omega}_\wedge\mathbf{r} \ .} \tag{1}$$

The reader can further show that for any differentiable vector function $\mathbf{F}(x, y, z)$,

$$\underline{d\mathbf{F}/dt = \partial\mathbf{F}/\partial t + \boldsymbol{\omega}_\wedge\mathbf{F} \ ,} \tag{2}$$

where the partial derivative is evaluated relative to the frame.

If now we allow the origin to move with velocity $\mathbf{v}_0$, the total velocity of P is simply

$$\underline{\mathbf{v}_P = \mathbf{v}_o + \partial\mathbf{r}/\partial t + \boldsymbol{\omega}_\wedge\mathbf{r} \ ,} \tag{3}$$

since the last two terms on the R.H.S. of (3) give the velocity of P relative to O.

Now suppose O is again fixed, $\mathbf{v}$ is the velocity of P and $\mathbf{f}$ the acceleration of P. Then, using (2)

$$\mathbf{f} = d\mathbf{v}/dt = \partial\mathbf{v}/\partial t + \boldsymbol{\omega}_\wedge\mathbf{v} \ .$$

From (1),

$$\begin{aligned}\partial\mathbf{v}/\partial t &= \partial^2\mathbf{r}/\partial t^2 + \boldsymbol{\omega}_\wedge(\partial\mathbf{r}/\partial t) + (\partial\boldsymbol{\omega}/\partial t)_\wedge\mathbf{r} \\ &= \partial^2\mathbf{r}/\partial t^2 + \boldsymbol{\omega}_\wedge(\partial\mathbf{r}/\partial t) + \dot{\boldsymbol{\omega}}_\wedge\mathbf{r} \ .\end{aligned}$$

Also, $\boldsymbol{\omega}_\wedge\mathbf{v} = \boldsymbol{\omega}_\wedge(\partial\mathbf{r}/\partial t) + \boldsymbol{\omega}_\wedge(\boldsymbol{\omega}_\wedge\mathbf{r})$, and so finally

$$\underline{\mathbf{f} = \partial^2\mathbf{r}/\partial t^2 + 2\boldsymbol{\omega}_\wedge(\partial\mathbf{r}/\partial t) + \dot{\boldsymbol{\omega}}_\wedge\mathbf{r} + \boldsymbol{\omega}_\wedge(\boldsymbol{\omega}_\wedge\mathbf{r}) \ .} \tag{4}$$

If, however, O moves with acceleration $\mathbf{f}_0$, then the total acceleration of P is now

$$\underline{\mathbf{f} = \mathbf{f}_o + \partial^2\mathbf{r}/\partial t^2 + 2\boldsymbol{\omega}_\wedge(\partial\mathbf{r}/\partial t) + \dot{\boldsymbol{\omega}}_\wedge\mathbf{r} + \boldsymbol{\omega}_\wedge(\boldsymbol{\omega}_\wedge\mathbf{r}) \ .} \tag{5}$$

We now apply these results to develop expressions for velocity and acceleration components in terms of cylindrical and spherical polar co-ordinates.

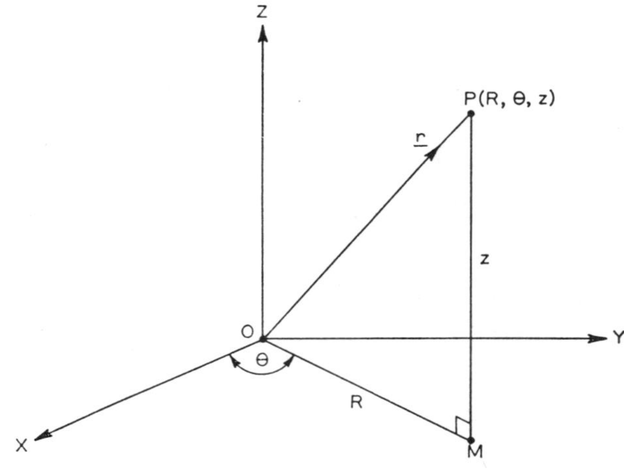

Fig. 2.16

In Fig. 2.16, OX, OY, OZ are *fixed* axes, mutually perpendicular. PM is the perpendicular from P on the plane XOY, $\angle MOX = \theta$, $OM = R$, $MP = z$. Then (R, θ, z) are called the *cylindrical polar co-ordinates of P*.

Let $\overline{OP} \equiv \mathbf{r}$ and let P vary with time. Then we have

$$\mathbf{r} = R\hat{\mathbf{R}} + 0\hat{\boldsymbol{\theta}} + z\hat{\mathbf{z}} = R\hat{\mathbf{R}} + z\hat{\mathbf{z}} \ ,$$

where $\hat{\mathbf{R}}$, $\hat{\boldsymbol{\theta}}$, $\hat{\mathbf{z}}$ denote unit vectors in the directions of R, θ, z increasing. These constitute a right-handed frame for which $\hat{\mathbf{R}}_\wedge\hat{\boldsymbol{\theta}} = \hat{\mathbf{z}}$, etc. Then since P is in a frame specified by these unit vectors and moving with angular velocity $\boldsymbol{\omega} = \dot{\theta}\hat{\mathbf{z}}$, the velocity of P is

$$\begin{aligned}\mathbf{v} &= \partial\mathbf{r}/\partial t + \boldsymbol{\omega}_\wedge\mathbf{r}\\ &= \dot{R}\hat{\mathbf{R}} + \dot{z}\hat{\mathbf{z}} + \dot{\theta}\hat{\mathbf{z}}_\wedge(R\hat{\mathbf{R}} + z\hat{\mathbf{z}})\\ &= \dot{R}\hat{\mathbf{R}} + \dot{z}\hat{\mathbf{z}} + R\dot{\theta}\,\hat{\boldsymbol{\theta}}\ ,\end{aligned}$$

i.e.

$$\underline{\mathbf{v} = \dot{R}\hat{\mathbf{R}} + R\dot{\theta}\,\hat{\boldsymbol{\theta}} + \dot{z}\hat{\mathbf{z}}\ .}$$

The acceleration of P is

$$\begin{aligned}\mathbf{f} &= \partial\mathbf{v}/\partial t + \boldsymbol{\omega}_\wedge\mathbf{v}\\ &= \ddot{R}\hat{\mathbf{R}} + (R\ddot{\theta} + \dot{R}\dot{\theta})\hat{\boldsymbol{\theta}} + \ddot{z}\hat{\mathbf{z}}\\ &\qquad + \dot{\theta}\hat{\mathbf{z}}_\wedge(\dot{R}\hat{\mathbf{R}} + R\dot{\theta}\,\hat{\boldsymbol{\theta}} + \dot{z}\hat{\mathbf{z}})\ .\end{aligned}$$

On simplification,

$$\underline{\mathbf{f} = (\ddot{R} - R\dot{\theta}^2)\hat{\mathbf{R}} + (R\ddot{\theta} + 2\dot{R}\dot{\theta})\hat{\boldsymbol{\theta}} + \ddot{z}\hat{\mathbf{z}}\ .}$$

The case of $z = \text{const.}$ recovers formulae for radial and transverse plane motion obtained previously.

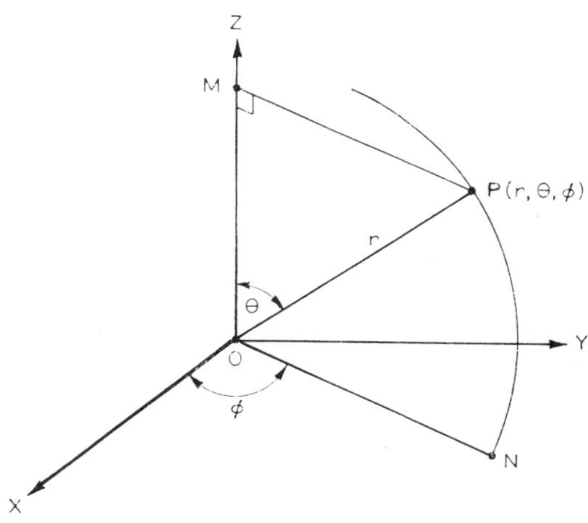

Fig. 2.17

In Fig. 2.17, OX, OY, OZ are *fixed* axes, P is variable. $OP = r$, $\angle POZ = \theta$. The circle in the plane POZ having centre O and radius OP, meets the plane XOY in N and $\angle NOX = \phi$. Then $(r,\ \theta,\ \phi)$ are termed the *spherical polar co-ordinates* of P.

Let $\hat{\mathbf{r}}$, $\hat{\boldsymbol{\theta}}$, $\hat{\boldsymbol{\phi}}$ denote the unit vectors in the directions of r, θ, ϕ increasing, respectively. These constitute a right-handed frame for which $\hat{\mathbf{r}}_\wedge\hat{\boldsymbol{\theta}} = \hat{\boldsymbol{\phi}}$, etc. This frame has angular velocity

$$\begin{aligned}\boldsymbol{\omega} &= \dot{\theta}\,\hat{\boldsymbol{\phi}} + \dot{\phi}\,\mathbf{k} = \dot{\theta}\hat{\boldsymbol{\phi}} + \dot{\phi}\,(\cos\theta\hat{\mathbf{r}} - \sin\theta\,\hat{\boldsymbol{\theta}})\\ &= \dot{\phi}\cos\theta\,\hat{\mathbf{r}} - \dot{\phi}\sin\theta\,\hat{\boldsymbol{\theta}} + \dot{\theta}\,\hat{\boldsymbol{\phi}}\ .\end{aligned}$$

Hence the velocity of P is

$$\begin{aligned}\mathbf{v} &= d\hat{\mathbf{r}}/dt = \partial\mathbf{r}/\partial t + \boldsymbol{\omega}_\wedge\mathbf{r}\\ &= \dot{r}\hat{\mathbf{r}} + (\dot{\phi}\cos\theta\,\hat{\mathbf{r}} - \dot{\phi}\sin\theta\,\hat{\boldsymbol{\theta}} + \dot{\theta}\,\hat{\boldsymbol{\phi}})_\wedge\mathbf{r}\\ &= \underline{\dot{r}\hat{\mathbf{r}} + r\dot{\theta}\,\hat{\boldsymbol{\theta}} + r\dot{\phi}\sin\theta\,\hat{\boldsymbol{\phi}}\ .}\end{aligned}$$

The reader will readily show that the acceleration is

$$\begin{aligned}\mathbf{f} = {} & (\ddot{r} - r\dot{\theta}^2 - r\dot{\phi}^2\sin^2\theta)\hat{\mathbf{r}} + (r\ddot{\theta} + 2\dot{r}\dot{\theta} - r\dot{\phi}^2\sin\theta\cos\theta)\hat{\boldsymbol{\theta}}\\ & \underline{+ [d(r\dot{\phi}\sin\theta)/dt + r\dot{\theta}\dot{\phi}\cos\theta + \dot{r}\dot{\phi}\sin\theta]\hat{\boldsymbol{\phi}}\ .}\end{aligned}$$

2.10 INSTANTANEOUS AXIS OF ROTATION AND INSTANTANEOUS CENTRE OF ROTATION

We have seen in Section 2.7 that if a point O fixed in a rigid body has velocity $\mathbf{v}_o$ and if $\boldsymbol{\omega}$ is the angular velocity of the body, then the equation of the central axis is

$$\mathbf{s} = (\boldsymbol{\omega}_\wedge\mathbf{v}_o)/\omega^2 + \lambda\boldsymbol{\omega} \tag{1}$$

where λ is a variable scalar parameter and $\mathbf{s}$ is the position vector of any point P on the central axis with respect to O. The velocity of P is thus

$$\begin{aligned}\mathbf{v}_P &= \mathbf{v}_o + \boldsymbol{\omega}_\wedge\mathbf{s}\\ &= \mathbf{v}_o + [(\boldsymbol{\omega}.\mathbf{v}_o)\boldsymbol{\omega} - \omega^2\mathbf{v}_o]/\omega^2 = (\boldsymbol{\omega}.\mathbf{v}_o)\boldsymbol{\omega}/\omega^2\ .\end{aligned}$$

This shows that, if $\boldsymbol{\omega}$ and $\mathbf{v}_o$ are non-zero, $\mathbf{v}_P$ can only be zero if the vectors $\boldsymbol{\omega}$, $\mathbf{v}_o$ are perpendicular. In these circumstances, every point on the central axis is instantaneously at rest and it is then termed the *instantaneous axis of rotation*, since the body as a whole is at that instant rotating about the central axis. If the two vectors are not perpendicular, such a line cannot be determined.

The above notion is especially important in two-dimensional dynamics when a rigid lamina is moving in its own plane. In these circumstances the velocity $\mathbf{v}_o$ of every point O of the lamina is at right angles to the vector angular velocity $\boldsymbol{\omega}$, since the direction of the latter is normal to the plane. Thus for such a case, the instantaneous axis of rotation will exist and its point of intersection with the plane of the lamina is termed the *instantaneous centre of rotation* of the lamina. As the motion of the lamina proceeds, this centre, I, will trace out a locus fixed in space termed the *space centrode* and one fixed in the body called the *body centrode*.

In Fig. 2.18, C, C' represent the space and body centrodes for a lamina. At a particular instant of time t, the body rotates about I, which lies on both curves. At time δt later, there will be a new instantaneous centre, common to both curves. Thus rotation about I brings I_1 of C and I_1' of C' into coincidence. After interval $2\delta t$, I_2 of C and I_2' of C' are brought into coincidence. Thus there are points $I, I_1, I_2, \ldots$ fixed in space and $I, I_1', I_2', \ldots$ fixed in the body which are brought into coincidence in pairs

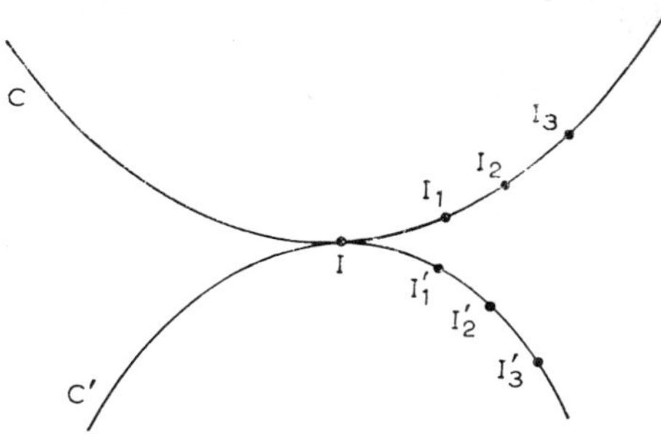

Fig. 2.18

at times t, $t+\delta t$, $t+2\delta t$, . . . The first set of points trace out the curve C: the second the curve C'. We thus see that the *motion of the lamina is equivalent to the rolling of the space centrode C on the body centrode C'.*

Example 1

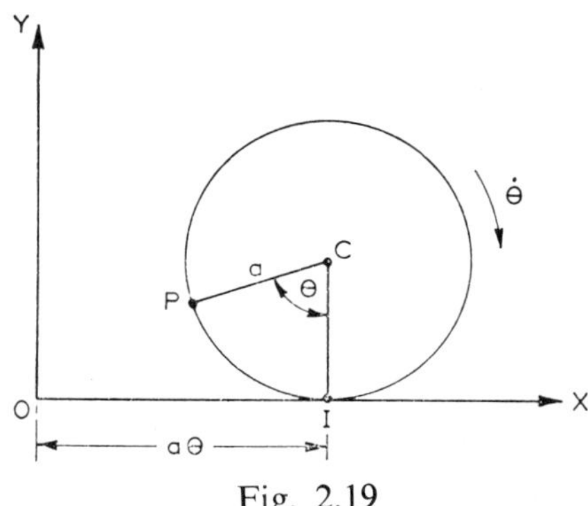

Fig. 2.19

A circular disc of radius a and centre C rolls without slipping along OX and in the plane of YOX (Fig. 2.19). I is the point of contact at any time t, P is a fixed point on the rim of the disc, originally coinciding with O. Then, since $OI=\text{arc } PI=a\theta$,

$$x_P=a(\theta-\sin\theta), \qquad y_P=a(1-\cos\theta)\ , \tag{1}$$

and

$$\dot{x}_P=a\dot{\theta}(1-\cos\theta), \qquad \dot{y}_P=a\dot{\theta}\sin\theta\ . \tag{2}$$

These show that $\dot{x}_P=0$, $\dot{y}_P=0$ when $\theta=0$, i.e. that the point I is instantaneously at rest. Thus I is the instantaneous centre. The body centrode is the rim of the circle, the space centrode is the line OX. Since I is at rest, the velocity of any point P on the rim is $\dot{\theta}(PI)$ in the direction perpendicular to PI in the sense of θ increasing. The reader will also readily confirm this from the expressions (2) for $\dot{x}_P$, $\dot{y}_P$.

Example 2

The ends P, Q of a rod PQ are constrained to move along straight lines OA, OB (Fig. 2.20).

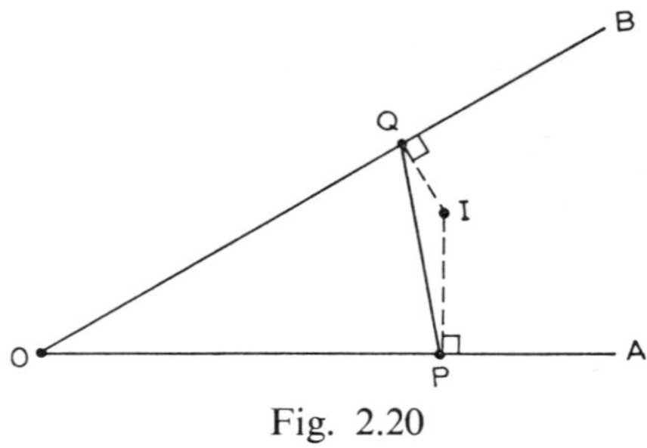

Fig. 2.20

The velocity of P is along OA. Hence the instantaneous centre I is such that PI is perpendicular to OA. Similarly, QI is perpendicular to OB. Thus O, P, I, Q are concyclic, OI being the diameter of their circumcircle. Since the angle $\angle POQ$ is fixed and PQ is of fixed length, the diameter OI is of fixed length. Thus the space centrode of I is a circle of centre O. The body centrode is the circle $OPIQ$, having OI as diameter. The two circles touch at I.

Example 3

A rigid lamina is moving in its own plane and one point A in it has velocity $\mathbf{u}$ relative to a fixed origin O. If $\boldsymbol{\omega}$ is the angular velocity of the lamina, show that the point P in it has velocity $\mathbf{u}+\boldsymbol{\omega}_\wedge\mathbf{r}$ relative to O, where $\mathbf{r} \equiv \overline{AP}$. Hence or otherwise prove that the position vector $\mathbf{r}'$ of the instantaneous centre I of the lamina relative to A is given by $\mathbf{r}' = (\boldsymbol{\omega}_\wedge\mathbf{u})/\omega^2$.

Show also that the acceleration of the particle in the lamina which instantaneously coincides with I has a component $du/dt-(u/\omega)(d\omega/dt)$ parallel to $\mathbf{u}$. (L.U.)

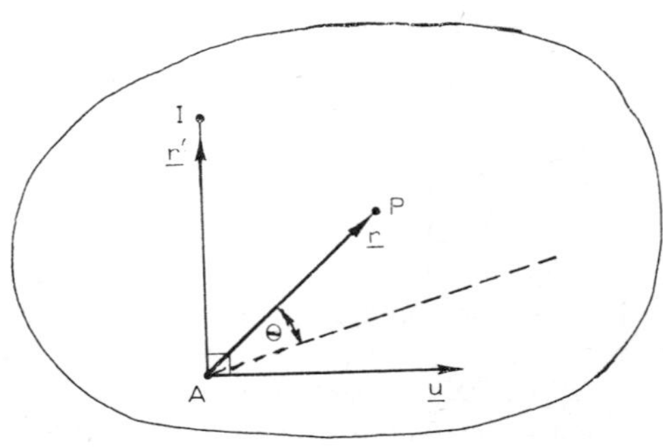

Fig. 2.21

The velocity of I is $\mathbf{u}+\boldsymbol{\omega}_\wedge\mathbf{r}'$. Since I is instantaneously at rest,

$$\mathbf{u}+\boldsymbol{\omega}_\wedge\mathbf{r}'=\mathbf{0} \;;$$

$$\therefore\; \boldsymbol{\omega}_\wedge\mathbf{u}+\boldsymbol{\omega}_\wedge(\boldsymbol{\omega}_\wedge\mathbf{r}')=\mathbf{0} \;;$$

$$\therefore\; \boldsymbol{\omega}_\wedge\mathbf{u}+(\boldsymbol{\omega}.\mathbf{r}')\boldsymbol{\omega}-\omega^2\mathbf{r}'=\mathbf{0} \;.$$

Since $\boldsymbol{\omega}$, $\mathbf{r}'$ are at right angles, $\boldsymbol{\omega}.\mathbf{r}'=0$ and so

$$\underline{\mathbf{r}'=(\boldsymbol{\omega}_\wedge\mathbf{u})/\omega^2 \;.}$$

Relative to A, the acceleration of P has radial and transverse components $\ddot{r}-r\dot{\theta}^2$ and $r\ddot{\theta}+2\dot{r}\dot{\theta}$, where θ is the angle AP makes with a direction AX chosen to be fixed in space. These components are $\ddot{r}-r\omega^2$ and $r\dot{\omega}+2\dot{r}\omega$. When $P\equiv I$, $\mathbf{r}=\mathbf{r}'$ and so the component of acceleration of I relative to A is the transverse component $r'\dot{\omega}$ in the direction of $-\mathbf{u}$, since $\dot{r}'=0$, the velocity of I relative to A being perpendicular to AI.

Now

$$u^2=\mathbf{u}^2 \;,$$

so that

$$2u\, du/dt=2\mathbf{u}.\, d\mathbf{u}/dt \;,$$

i.e.

$$\hat{\mathbf{u}}.\, d\mathbf{u}/dt=du/dt \;.$$

But $d\mathbf{u}/dt$ is the acceleration of A in space and $\hat{\mathbf{u}}.d\mathbf{u}/dt$ the component in the direction of $\mathbf{u}$. Hence the total component of acceleration of I in the direction of $\mathbf{u}$ is

$$du/dt-r'd\omega/dt=du/dt-(u/\omega)(d\omega/dt) \;.$$

Example 4—Kinematics of rigid lithospheric plate motions on a rotating earth.
We assume the earth to be a sphere of radius a, its centre O moving with velocity $\mathbf{v}_0$ at a considered instant of time. This measurement is made relatively to an inertial frame. According to the modern theory of *plate tectonics*, the continents and oceans rest upon rigid lithospheric plates which float over the molten asthenosphere that forms part of the earth's upper mantle. We consider the relative motions of two such plates which will be regarded as shells, each of radius a.

Suppose the outer layers of the asthenosphere on which the plates reside have an angular velocity $\boldsymbol{\Omega}$ due to the earth's rotation. Let $\boldsymbol{\omega}_1$, $\boldsymbol{\omega}_2$ be the angular velocities of the two plates, measured relatively to the asthenosphere. If $\mathbf{R}$ is the position vector from O to a point P which is common to the two plates, then the instantaneous velocities $\mathbf{V}_1$, $\mathbf{V}_2$ of P on the plates are given by the equations

$$\mathbf{V}_i=\mathbf{v}_0+(\boldsymbol{\omega}_i+\boldsymbol{\Omega})_\wedge\mathbf{R} \qquad (i=1,2) \;,$$

since $\boldsymbol{\omega}_i+\boldsymbol{\Omega}$ are the angular velocities of the plates in the inertial frame. Let us seek points at which $\mathbf{V}_1=\mathbf{V}_2$, i.e.

$$(\boldsymbol{\omega}_1-\boldsymbol{\omega}_2)_\wedge\mathbf{R}=\mathbf{0} \;, \qquad |\mathbf{R}|=a \;.$$

As $\boldsymbol{\omega}_1\neq\boldsymbol{\omega}_2$ and $\mathbf{R}\neq\mathbf{0}$, we can find a scalar K such that

$$\mathbf{R}=K(\omega_1-\omega_2) \ .$$

Then, taking moduli,

$$a=\pm K|\boldsymbol{\omega}_1-\boldsymbol{\omega}_2|$$

leading to the two values of $\mathbf{R}$:

$$\mathbf{R}=\frac{\pm a(\boldsymbol{\omega}_1-\boldsymbol{\omega}_2)}{|\boldsymbol{\omega}_1-\boldsymbol{\omega}_2|} \ .$$

These determine two antipodal points on the surface of the sphere and their positions are quite independent of $\mathbf{v}_o$ and $\boldsymbol{\Omega}$. If $\boldsymbol{\omega}_1$ and $\boldsymbol{\omega}_2$ remain constant, the positions of these points remain fixed on the sphere and they determine the axis about which either plate is rotating relatively to the other. The two points are called *Euler poles*, after Euler who first discovered the result, and they are of great importance in modern geophysics.

EXERCISE 2

1. P is the point specified by polar co-ordinates (r, θ) in a plane and its position vector $\overline{OP}$ at any time t with respect to a fixed origin O is represented by the complex number $z=re^{i\theta}$. Show that

$$\dot{z}=\dot{r}\,e^{i\theta}+r\dot{\theta}(ie^{i\theta}) \ ;$$
$$\ddot{z}=(\ddot{r}-r\dot{\theta}^2)e^{i\theta}+(r\ddot{\theta}+2\dot{r}\dot{\theta})(ie^{i\theta}) \ ,$$

and hence deduce the radial and transverse velocity and acceleration components of P.

2. Obtain the radial and transverse components of acceleration of a particle, which describes the plane curve $r=f(\theta)$, in the form

$$\ddot{r}-r\dot{\theta}^2, \qquad (1/r)d(r^2\dot{\theta})/dt \ .$$

If the curve is the equiangular spiral $r=a\exp(\theta\cot\alpha)$, and if the radius vector to the particle has a constant angular velocity, show that the resultant acceleration of the particle makes an angle 2α with the radius vector and is of magnitude v^2/r, where v is the speed of the particle.

(L.U.)

3. A car of width b is moving with constant velocity V close to the edge of a straight road. If a pedestrian steps on to the road at a point distant d in front of the car, what is the least uniform velocity at which he must be able to walk in order to cross the road in safety?

If the car is at rest, but moves off with constant acceleration f at the same instant that the pedestrian starts to cross the road from a point on the edge of road distant d in front of the car, show that the least uniform velocity at which he must be able to walk is

$$[f\{(d^2+b^2)^{\frac{1}{2}}-d\}]^{\frac{1}{2}} \ ,$$

and that if he walks at this velocity his direction must be inclined to the edge of the road at an angle

$$\cot^{-1}[\{(d^2+b^2)^{\frac{1}{2}}-d\}/b] \ .$$

[Throughout this question it is assumed that the pedestrian's position on the road is located by a single point and that the distance between the car and the edge of the road is negligible.] (L.U.)

4. An aeroplane travels at a maximum speed V. It has to travel from A to B and back in the shortest possible time in a cross-wind blowing at constant velocity v in a direction making an angle ϕ with $\overline{BA}$. Find the time taken and determine the directions in which the aircraft must steer on the outward and return journeys.

5. A wheel is rolling without slipping along a plane with angular velocity ω. A and B are taken at different distances from the centre on two different spokes. Show that at any time the velocity of A relative to B is $\omega \times AB$ in a direction perpendicular to AB.

6. A point P moves on the surface of the anchor ring

$$x=(c+a\sin\theta)\cos\phi,\ y=(c+a\sin\theta)\sin\phi,\ z=a\cos\theta.$$

Denoting by $\hat{\mathbf{a}}$, $\hat{\mathbf{b}}$, $\hat{\mathbf{c}}$ the unit vectors in the normal, meridian and transverse directions at P, so that $\hat{\mathbf{a}}\wedge\hat{\mathbf{b}}=\hat{\mathbf{c}}$, etc., show that the position vector $\mathbf{r}$ of $\overline{OP}$ is

$$\mathbf{r}=(a+c\sin\theta)\hat{\mathbf{a}}+c\cos\theta\,\hat{\mathbf{b}} \ ,$$

and that the vector angular velocity $\boldsymbol{\omega}$ of the ($\hat{\mathbf{a}}$, $\hat{\mathbf{b}}$, $\hat{\mathbf{c}}$) frame is

$$\boldsymbol{\omega}=\dot{\phi}\cos\theta\,\hat{\mathbf{a}}-\dot{\phi}\sin\theta\,\hat{\mathbf{b}}+\dot{\theta}\,\hat{\mathbf{c}} \ .$$

Hence or otherwise deduce that the acceleration components of P are

$$[-\dot{\phi}^2\sin\theta(c+a\sin\theta)-a\dot{\theta}^2,$$
$$a\ddot{\theta}-\dot{\phi}^2(c+a\sin\theta)\cos\theta,\ 2a\dot{\theta}\dot{\phi}\cos\theta+(c+a\sin\theta)\ddot{\phi}].$$

7. The points $(a, 2a, -a)$, $(-a, -a, a)$, (a, a, a) of a rigid body have instantaneous velocities $[\sqrt{3}v/2, 0, \sqrt{3}v/2]$, $[-v/\sqrt{3}, 0, -v/\sqrt{3}]$, $[0, -v/\sqrt{3}, v/\sqrt{3}]$. Show that the body has the line through the origin having direction cosines $[1/\sqrt{3}, -1/\sqrt{3}, -1/\sqrt{3}]$ as instantaneous axis of rotation and that the magnitude of the angular velocity is $v/(2a)$.

8. S is a circle, centre C and radius a and t is a fixed tangent to S. A rod AB is constrained to move so that its end A slides along t whilst AB touches S at a variable point P. Prove that CP and the perpendicular to t at A meet at the instantaneous centre of rotation I. Show that $IA=IC$ and hence deduce that the space centrode is a parabola having C as focus and t as directrix. Taking $AI=r$, $\angle IAP=\theta$, show that $a/r=1-\sin\theta$ and hence determine the body centrode.

9. A bar AB moves in the plane of the rectangular axes Ox, Oy so that it always touches the parabola $y^2=4ax$ and the end A of the bar moves along the x-axis. Find the co-ordinates of the instantaneous centre in terms of θ, the angle between AB and Oy, and prove that the equation of the

space centrode is $ay^2+4x(a-x)^2=0$. Prove also that, referred to A as pole, and AB as initial line, the equation of the body centrode is $r=2a\sin\theta\sec^3\theta$. (L.U.)

10. A circular cylinder of radius a and axis OQ touches two planes hinged together along the line AB, $OQBA$ being a rectangle. One plane is fixed horizontally and the other inclined at an obtuse angle $\pi-\theta$ to it. Contact between the cylinder and the inclined plane is perfectly rough, but the horizontal plane is smooth. If contact is maintained between the cylinder and the planes as θ varies, show that the spin of the cylinder is $\dot{\theta}\{1-\frac{1}{2}\sec^2(\theta/2)\}$.

Taking A as origin, Ax in the horizontal plane perpendicular to AB and Ay vertically downwards, show that the space centrode is given by $y(a^2-x^2)=2ax^2$. (L.U.)

11. Assuming the formula $\mathbf{v}_P=\mathbf{v}_A+\boldsymbol{\omega}_\wedge\overline{AP}$ relating the velocities of particles, A, P of a lamina L moving in its own plane with spin $\boldsymbol{\omega}$, and that $\dot{\overline{AP}}=\boldsymbol{\omega}_\wedge\overline{AP}$, show that the acceleration vectors $\mathbf{f}_A$, $\mathbf{f}_P$ are related by the equation

$$\mathbf{f}_P=\mathbf{f}_A+\dot{\boldsymbol{\omega}}_\wedge\overline{AP}-\omega^2\overline{AP}\ .$$

Prove that if J is the particle of L given by

$$\overline{AJ}\equiv(\dot{\boldsymbol{\omega}}_\wedge\mathbf{f}_A+\omega^2\mathbf{f}_A)/(\omega^4+\dot{\omega}^2)\ ,$$

then the acceleration of J is instantaneously zero (instantaneous centre of acceleration).

If the spin of L is constant and $\mathbf{f}_A=a(\cos pt\,\mathbf{i}+\sin pt\,\mathbf{j})$, where a, p are constants and $\mathbf{i}$, $\mathbf{j}$ are constant orthogonal unit vectors, show that the instantaneous centre of acceleration J moves relative to A uniformly round a circle of radius a/ω^2 with period $2\pi/p$. (L.U.)

12. A point P moves so that it always has a velocity compounded of V_1 parallel to a fixed line and V_2 in the transverse direction perpendicular to OP, where O is a fixed point. Prove that its orbit is a conic having O as focus and discuss the cases $V_1>V_2$, $V_1=V_2$, $V_1<V_2$.

13. P is a point moving along a plane curve C and distant r from a fixed origin O at time t. $OM=p$ is the length of the perpendicular from O on the tangent at P and $PM=T$. PM makes an angle ψ with the fixed line OX. Show that P has velocity components $\dot{T}-p\dot{\psi}$, $\dot{p}+T\dot{\psi}$ along $\overline{MP}$, $\overline{OM}$ and hence that $\dot{p}+T\dot{\psi}=0$, $\dot{T}-p\dot{\psi}=\dot{s}$, s being the length of arc along C from some fixed point on it to position P. Show further that $r\dot{r}=p\dot{p}+T\dot{T}$, and hence derive the formula $\rho=|rdr/dp|$ for the radius of curvature $\rho=|ds/d\psi|$ of the curve C at P.

14. *Moving axes in two dimensions.*

(i) At any instant t, the Cartesian co-ordinates of a point P moving in a plane are (x, y) relative to perpendicular axes OX, OY in the plane and rotating with angular velocity $\omega\mathbf{k}$. Show that the components of velocity and acceleration along $\overline{OX}$, $\overline{OY}$ are respectively

$$\dot{x} - \omega y,\ \dot{y} + \omega x\ ;$$
$$\ddot{x} - \dot{\omega} y - 2\omega\dot{y} - \omega^2 x,\ \ddot{y} + \dot{\omega} x + 2\omega\dot{x} - \omega^2 y\ .$$

(ii) If ω be constant and if the x- and y- acceleration components be constant and equal to f_1, f_2, by forming $f_1 + if_2 (i = \sqrt{-1})$, show that the equation of the trajectory of P is of the form

$$\begin{cases} x = (A + Bt)\cos\omega t + (C + Dt)\sin\omega t - f_1/\omega^2\ , \\ y = -(A + Bt)\sin\omega t + (C + Dt)\cos\omega t - f_2/\omega^2\ , \end{cases}$$

where A, B, C, D are constants.

15. A cyclist rides along a straight road with uniform speed V. At a given instant, the cyclist is at a point O on the road and a dog is at A, a point distant a from the road and such that OA is perpendicular to the road. The dog runs with speed V towards the cyclist. Taking OA and the road as axes of x and y respectively, and the co-ordinates of the dog as x, y, show that

$$x\, d^2y/dx^2 = \{1 + (dy/dx)^2\}^{\frac{1}{2}}\ ,$$

and hence that the equation of the dog's path is

$$y = \tfrac{1}{2}\{(x^2/2a) - a\log(x/a) - \tfrac{1}{2}a\}\ .$$

Show further that the distance between the cyclist and the dog tends to $\frac{1}{2}a$. (L.U.)

16. *Poinsot's reduction of a system of forces* (*cf. Section 2.7*). The n forces $\mathbf{F}_k$ act at points P_k of space. $\overline{OP_k} \equiv \mathbf{r}_k$ $(k = 1, \ldots, n)$ and O is a chosen origin. O' is another origin such that $\overline{OO'} \equiv \mathbf{s}$. Denoting by $\mathbf{R}$ the resultant of the forces $(= \sum_{k=1}^{n} \mathbf{F}_k)$ and by $\mathbf{G}, \mathbf{G}'$ the vector moments of the system about O, O' respectively, prove that

$$\mathbf{G}' = \mathbf{G} - \mathbf{s}_\wedge\mathbf{R}$$

and hence that $\mathbf{G}.\mathbf{R}$ is an *invariant* Γ for any origin O.

Now choose the origin O', if possible, so that $\mathbf{G}'$ and $\mathbf{R}$ are parallel, i.e. such that $\mathbf{R}_\wedge\mathbf{G}' = \mathbf{0}$. Show that the locus of O', when O is considered fixed, is the straight line

$$\mathbf{s} = (\mathbf{R}_\wedge\mathbf{G})/R^2 + t\mathbf{R}\ ,$$

where t is the variable parameter given by $t = (\mathbf{R}.\mathbf{s})/R^2$. This line is called the *central axis* of the system. The analysis shows that in general such *a given system of forces may be reduced to a single resultant force* $\mathbf{R}$ *acting through the point O' whose position vector is* $(\mathbf{R}_\wedge\mathbf{G})/R^2$, *together with a couple of vector moment* $\mathbf{G}'$ *about O' and in the line of action of* $\mathbf{R}$. Such a reduction is termed a *wrench*.

Show further that for such a couple $\mathbf{G}'$ (parallel to $\mathbf{R}$),

$$\mathbf{G}' = \Gamma\,\mathbf{R}/R^2\ .$$

Further, taking $\mathbf{G}'=p\mathbf{R}$, where p is a scalar, show that $p=\Gamma/R^2$. p is called the *pitch of the wrench.*

17. F_1 and F_2 are two forces whose lines of action are at right angles and distant d apart. They are combined to form a wrench. Prove that its pitch is $F_1F_2d/(F_1{}^2+F_2{}^2)$.

Chapter 3

Newton's Laws of Motion

3.1 MASS, MOMENTUM, FORCE. NEWTON'S LAWS OF MOTION

The science of dynamics, the preoccupation of which is the study of motion, starts with the simplest of bodies—the *particle*. This latter may be defined as a quantity of matter having negligible linear dimensions. The *mass* of the particle is a measure of the quantity of matter: it is measured variously in lb, kg, tons, etc. Let m denote the mass of the particle in some chosen system of units. Suppose at a particular instant the particle has a velocity $\mathbf{v}$. Then the vector

$$\mathbf{p} = m\mathbf{v}$$

defines the (*linear*) *momentum* of the particle at the instant.

For convenience we now summarize and discuss briefly *Newton's Laws of Motion* and the notion of *force*. This latter is the cause of momentum change. The statements of the laws are as follow:

(1) Every particle continues in its state of rest or of uniform velocity in a straight line unless compelled to do otherwise by a force acting on it.

(2) The rate of change of momentum is proportional to the impressed force and takes place in the direction of action of the force.

(3) To every action there is an equal and opposite reaction.

(1) defines force as the causal agent of motion change; (2) gives its measure as rate of change of momentum; and (3) describes the equal and opposite force interactions of two bodies. However, (2) needs careful interpretation. When the mass m varies, the force is not necessarily $d\mathbf{p}/dt$: we shall return to this point when changing mass systems are discussed. If, however, m is constant, the force $\mathbf{F}$ is certainly

$$\mathbf{F} = m\, d\mathbf{v}/dt = m\mathbf{f},$$

where $\mathbf{f}$ is the acceleration of the particle.

3.2 WORK, ENERGY AND POWER

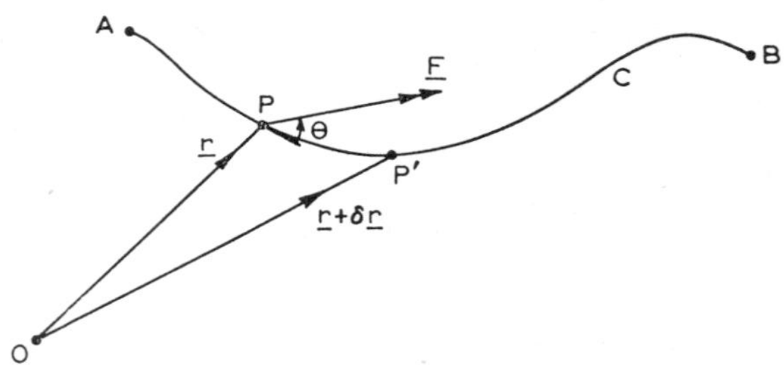

Fig. 3.1

Suppose a force **F** acts on a particle constrained to move along a curve C joining points A and B (Fig. 3.1). Let P, P' be neighbouring points on C such that $\overline{OP} \equiv \mathbf{r}$, $\overline{OP'} \equiv \mathbf{r} + \delta\mathbf{r}$: then $\overline{PP'} \equiv \delta\mathbf{r}$, or $\delta\mathbf{s}$, if we choose to work in terms of the arc length of C. The *work* done by **F** in moving its point of application from P to P' is defined to the scalar product of **F** and $\delta\mathbf{r}$, i.e. $\mathbf{F}.\delta\mathbf{r} = \mathbf{F}.\delta\mathbf{s}$: it is the resolved part of **F** in the tangent at P directed towards P' multiplied by the element of arc length PP'. Thus the total work done in moving the particle from A to B along C is the limiting sum of these elemental expressions and is represented by any of the expressions

$$W = \int_A^B \mathbf{F}.d\mathbf{r} = \int_A^B \mathbf{F}.d\mathbf{s} = \int_A^B F \cos\theta \, ds,$$

θ being the angle **F** makes with $\delta\mathbf{r}$ at P. These expressions are termed *line integrals* of **F**: they are all evaluated along the curve C and may depend on C.

If the particle P has constant mass m and velocity **v** at P, then

$$\mathbf{F}.d\mathbf{r} = m(d\mathbf{v}/dt).(\mathbf{v}\, dt) = m\mathbf{v}.d\mathbf{v} = md(\tfrac{1}{2}\mathbf{v}^2) = md(\tfrac{1}{2}v^2).$$

Hence the total work done on the particle is

$$\int_A^B \mathbf{F}.d\mathbf{r} = [\tfrac{1}{2}mv^2]_A^B = \tfrac{1}{2}mv_B{}^2 - \tfrac{1}{2}mv_A{}^2.$$

Thus the quantity $T = \frac{1}{2}mv^2$ has the same dimensions as work and is termed the *kinetic energy* of the particle. We have shown that the total work done by the external force **F** in carrying the particle from A to B along C is equal to the kinetic energy gained in the process

Since $\mathbf{F}.\delta\mathbf{r}$ is the work done by the force in moving the particle from P to P', if the motion takes place in time δt, the rate of working of the force at P is

$$\lim_{\delta t \to 0} (\mathbf{F}.\delta\mathbf{r}/\delta t) = \mathbf{F}.d\mathbf{r}/dt = \mathbf{F}.\mathbf{v}.$$

This is known as the *power* or the *activity* of the force. Denoting it by A, we see that the work done is

$$W = \int_{t_A}^{t_B} A\,dt.$$

We further note that

$$dT/dt = d(\tfrac{1}{2}m\mathbf{v}^2)/dt = m\mathbf{v}.(d\mathbf{v}/dt) = \mathbf{F}.\mathbf{v} = A.$$

3.3 CONSERVATIVE FORCES—POTENTIAL ENERGY

In the case considered in the last section, in general **F** will vary along C and even if the points A and B are kept fixed, we generally expect W to depend on the shape of the path C joining them. However, it frequently happens that there are forces for which W is dependent on the positions of A and B and not at all on the curve C: such forces are termed *conservative*. In such cases, if C' be a closed curve, the total line integral of **F** round C' is zero. This is conventionally denoted by

$$\oint_{C'} \mathbf{F}.d\mathbf{r} = 0\,.$$

If **F** be uniquely defined at every point of a region of space, then the totality of all such vector forces **F** throughout the space is termed a *field of force*. If at every point of the space **F** is conservative, the whole field of force is said to be conservative.

We now give examples of conservative and non-conservative fields.

Example 1—The earth's gravitational field.

Except over large distances, the acceleration **g** due to gravity is constant. Thus the force on a particle falling freely under gravity, i.e. its weight **W**, is

$$\mathbf{W} = m\mathbf{g}\,.$$

The work done by gravity in moving the particle from position A to position B along a curve C is thus

$$\int_A^B \mathbf{W}.d\mathbf{r} = m\int_A^B \mathbf{g}.d\mathbf{r} = m\int_A^B d(\mathbf{g}.\mathbf{r})$$
$$= m[\mathbf{g}.\mathbf{r}]_A^B = m[\mathbf{g}.\mathbf{r}_B - \mathbf{g}.\mathbf{r}_A]$$
$$= m\mathbf{g}.[\mathbf{r}_B - \mathbf{r}_A]\,.$$

This last expression only involves the position vectors of A, B. Thus the earth's gravitational field is conservative. The reader will easily confirm that this last expression reduces to

$$mg \times \text{Depth of } B \text{ below } A\,.$$

Example 2—Central force of the type $\mathbf{F} = F(r)\hat{\mathbf{r}}$.

The work done by **F** is

$$W = \int_A^B \mathbf{F}.d\mathbf{r} = \int_A^B F(r)\hat{\mathbf{r}}.d\mathbf{r}\,.$$

Now

$$\mathbf{r}^2 = r^2 ,$$

and so

$$2\mathbf{r}.d\mathbf{r} = 2r\, dr .$$

$$\therefore dr = \hat{\mathbf{r}}.d\mathbf{r} ;$$

$$\therefore W = \int_A^B F(r)dr .$$

This expression is dependent only on the functional form of $F(r)$ and on the end points A and B. Hence the field of $\mathbf{F}$ is conservative.

Example 3—$\mathbf{F} = k\hat{\mathbf{s}}$, where $k = \text{const.}$

$$W = k\int_A^B \hat{\mathbf{s}}.d\mathbf{s} = k\int_A^B ds = k(s_B - s_A) .$$

But $s_B - s_A$ is the length of the arc C and this will vary as C is varied. Hence the field of $\mathbf{F}$ is non-conservative.

In the case of a conservative field of force, we define the *potential energy* V to be the work done by the force in moving the particle from its existing position to some standard position. Thus if the existing position be specified by the vector $\mathbf{r} \equiv \overline{OP}$ and the standard position by $\mathbf{r}_0 \equiv \overline{OP}_0$, the P.E. is

$$V = \int_{\mathbf{r}}^{\mathbf{r}_0} \mathbf{F}.d\mathbf{r} .$$

If $\mathbf{r} \equiv x\mathbf{i} + y\mathbf{j} + z\mathbf{k}$, then $V = V(x, y, z)$. Let $\mathbf{F} = F_1\mathbf{i} + F_2\mathbf{j} + F_3\mathbf{k}$. Since $d\mathbf{r} = dx\mathbf{i} + dy\mathbf{j} + dz\mathbf{k}$,

$$V = -\int_{P_0}^{P} (F_1dx + F_2dy + F_3dz) .$$

$$\therefore F_1 = -\partial V/\partial x, \qquad F_2 = -\partial V/\partial y, \qquad F_3 = -\partial V/\partial z .$$

$$\therefore \mathbf{F} = -(\partial V/\partial x)\mathbf{i} - (\partial V/\partial y)\mathbf{j} - (\partial V/\partial z)\mathbf{k} = -\nabla V .$$

It thus follows that for such a conservative field of force, curl $\mathbf{F} = \mathbf{0}$ since curl grad $V \equiv \mathbf{0}$. It is further shown in works on the vector calculus that the condition curl $\mathbf{F} = \mathbf{0}$ implies the existence of a potential function V satisfying the relation $\mathbf{F} = -\text{grad}\ V$. The criterion curl $\mathbf{F} = \mathbf{0}$ is thus a necessary and a sufficient condition for a conservative field of force $\mathbf{F}$.

Also,

$$\mathbf{F}.\delta\mathbf{r} = -(\partial V/\partial x)\delta x - (\partial V/\partial y)\delta y - (\partial V/\partial z)\delta z = -\delta V.$$

$$\therefore \int_A^B \mathbf{F}.d\mathbf{r} = -\int_A^B dV = V_A - V_B .$$

But we have shown in Section 3.2 that

$$\int_A^B \mathbf{F}.d\mathbf{r} = \tfrac{1}{2}mv_B^2 - \tfrac{1}{2}mv_A^2 .$$

$$\therefore V_A + \tfrac{1}{2}mv_A^2 = V_B + \tfrac{1}{2}mv_B^2 ;$$

i.e. *for a single particle in a conservative field of force the sum of the kinetic and potential energies is constant.* This is the *Principle of Conservation of Energy* established for a single particle.

Clearly, for a system of particles moving in a conservative field of force, the Principle of Conservation of Energy would still hold, since it holds for each individual particle.

Example 4

A smooth wire bent in the form of a parabola is fixed with its axis vertical and vertex downwards. A particle of mass m oscillates on the wire coming to rest at the extremities of the latus rectum. Show that the reaction of the wire on the particle when passing through the vertex is $2mg$. (L.U.)

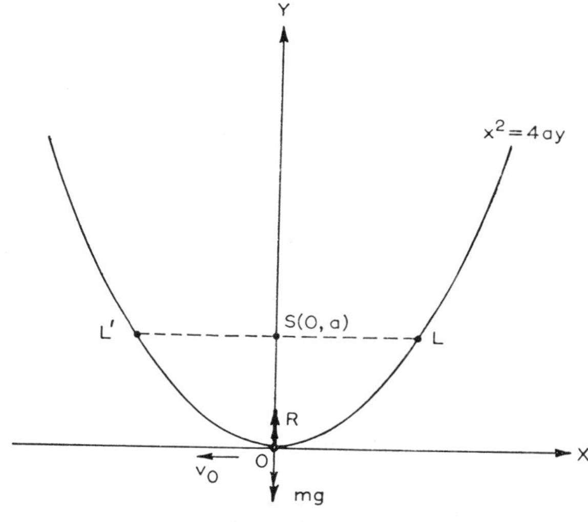

Fig. 3.2

Let the equation of the wire be $x^2 = 4ay$. The focus is $(0, a)$, so that in falling from an extremity L of the latus rectum to the vertex O, the particle loses P.E. $= mga$. Let v_o be the speed at O, so that the K.E. gained in the fall from L to O is $\frac{1}{2}mv_o^2$.

$$\therefore \tfrac{1}{2}mv_o^2 = mga ,$$

or

$$\underline{v_o^2 = 2ga} .$$

Let R_o be the inward normal reaction on the particle at O and let ρ_o be the radius of curvature at O. The net inward force towards the centre of curvature at O is $R_o - mg$ and the acceleration inwards is v_o^2/ρ_o.

$$\therefore R_o - mg = mv_o^2/\rho_o \,.$$

$$\therefore R_o = mg + 2mga/\rho_o \,.$$

Now $y = x^2/(4a)$, $y' = x/(2a)$, $y'' = 1/(2a)$.

$$\therefore \rho_o = [(1+y'^2)^{3/2}/y'']_o = 2a \,.$$

$$\therefore R_o = mg + 2mga/(2a) = \underline{2mg} \,.$$

3.4 IMPULSIVE FORCES

The *impulse of a force* acting on a particle in any interval of time is defined to be the momentum change produced. Thus if a particle of constant mass m has a velocity changed from $\mathbf{v}_1$ to $\mathbf{v}_2$ in a time t by a force $\mathbf{F}$ acting, then the impulse $\mathbf{I}$ is

$$\mathbf{I} = m\mathbf{v}_2 - m\mathbf{v}_1 \,.$$

$$\therefore \mathbf{I} = m\int_{t_1}^{t_2} (d\mathbf{v}/dt)dt = \int_{t_1}^{t_2} \mathbf{F}\, dt \,,$$

since $\mathbf{F} = m\, d\mathbf{v}/dt$. Thus the impulse of the force $\mathbf{F}$ is the time-integral of the force.

Now suppose that $\mathbf{F}$ grows very large, the interval $(t_2 - t_1)$ very small, but in such a way as to ensure that the time integral $\mathbf{I}$ stays finite. Such forces are called *impulsive forces*: their measurement as forces is impracticable, but one can measure the momentum change they produce.

Example

An impulse $\mathbf{I}$ changes the velocity of a particle of mass m from $\mathbf{v}_1$ to $\mathbf{v}_2$. Show that the K.E. gained is $\frac{1}{2}\mathbf{I}.(\mathbf{v}_1 + \mathbf{v}_2)$.

We have $\mathbf{I} = m(\mathbf{v}_2 - \mathbf{v}_1)$.

K.E. gained

$$= \tfrac{1}{2}m(\mathbf{v}_2{}^2 - \mathbf{v}_1{}^2)$$

$$= \tfrac{1}{2}m(\mathbf{v}_2 - \mathbf{v}_1).(\mathbf{v}_2 + \mathbf{v}_1)$$

$$= \underline{\tfrac{1}{2}\mathbf{I}.(\mathbf{v}_2 + \mathbf{v}_1)} \,.$$

3.5 RECTILINEAR PARTICLE MOTION

We now consider applications of the laws of motion to a particle moving in a straight line. For such motion, the vectorial character of displacement, velocity, acceleration, force, etc., need not be emphasized.

Let P be a particle of constant mass m moving on a straight line, O being a fixed point on the line and determining the position of the particle at zero time. Suppose P is the position of the particle at time t and that $OP = x$. Further, let v, f denote the velocity and acceleration of the particle in the direction of $\overline{OP}$ increasing at time t (Fig. 3.3).

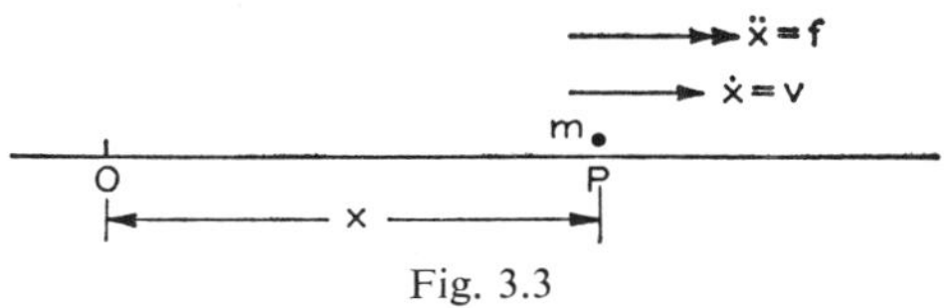

Fig. 3.3

We have at once $v = \dot{x}, f = \ddot{x}$. Also, we note that

$$f = dv/dt = (dv/dx)(dx/dt) = v dv/dx ,$$

the latter form being especially useful when connecting together velocity and distance. We now consider particular cases of rectilinear motion.

(i) *Uniformly accelerated motion*

In the above case, f is constant and the reader will easily obtain the results

$$v = u + ft, \ x = ut + \tfrac{1}{2}ft^2, \ v^2 = u^2 + 2fx ,$$

where u is the particle velocity at O.

(ii) *Resisted motion*

Suppose in Fig. 3.3 a force F acts on P in the direction $\overline{OP}$. Also suppose that there is a resistance R acting in the direction $\overline{PO}$. Often, R will be a function of velocity. Then the resultant force on the particle in the direction $\overline{OP}$ is $F - R$, and the equation of motion becomes

$$F - R = mf .$$

Example 1

A small stone of mass m is thrown vertically upwards with initial speed V. If the air resistance at speed v is mkv^2, where k is a positive constant, show that the stone returns to its starting point with speed $V(1 + kV^2/g)^{1/2}$.

If the stone has the same mass and initial speed, but if the air resistance is mkv, show that the stone returns to its starting point with speed U given by the equation

$$g - kU = (g + kV)\exp\{-k(U + V)/g\} .$$

(L.U.)

(*a*) Resistance of mkv^2 .

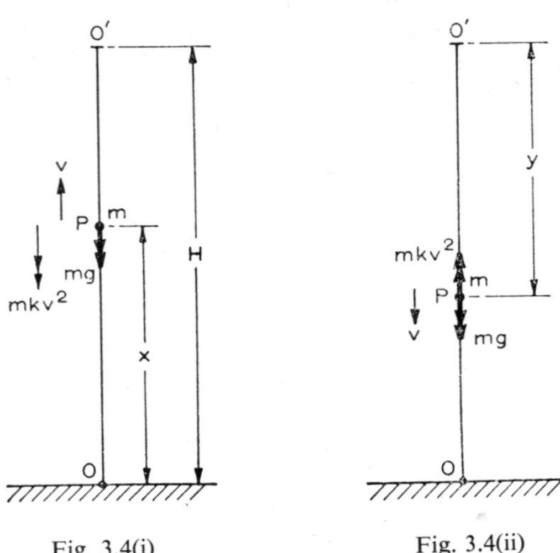

Fig. 3.4(i) Fig. 3.4(ii)

Fig. 3.4(i) depicts the upward motion, $OP = x \leqslant H$, the maximum height attained. Fig. 3.4(ii) depicts the downward motion, $O'P = y \leqslant H$.

In Fig. 3.4(i), the acceleration of P in the direction of x increasing is $v dv/dx$, where v is the upward velocity of P. The forces acting on the particle at P are (i) the weight mg downwards; (ii) the resistance mkv^2, also downwards. Thus the total force in the *direction of the acceleration* is $-(mg + mkv^2)$, so that the equation of motion is

$$-(mg + mkv^2) = mv\,dv/dx$$

$$\therefore (D + 2k)v^2 = -2g \qquad (D \equiv d/dx)\,,$$

and so $$v^2 = Ae^{-2kx} - g/k.$$

Since $v = V$ when $x = 0$, we find $A = V^2 + g/k$.

$$\therefore \underline{v^2 = (V^2 + g/k)e^{-2kx} - g/k}\,.$$

$\therefore$ Taking $v = 0$ when $x = H$, we find

$$\underline{H = (1/2k) \log (1 + kV^2/g)}\,.$$

For the downward motion depicted in Fig. 3.4(ii), since the downward acceleration is $v dv/dy$ and the net downward force $mg - mkv^2$,

$$mg - mkv^2 = mv\,dv/dy\,,$$

and so $$(D + 2k)v^2 = 2g \qquad (D \equiv d/dy)\,.$$

$$\therefore v^2 = Be^{-2ky} + g/k\,.$$

When $y = 0$, $v = 0$, so $B = -g/k$ and

$$v^2 = (g/k)(1 - e^{-2ky})\,.$$

Taking $y = H$ in this gives the stated result.

(*b*) The second part is attacked in the same way.

N.B. In (*a*), the time to the highest point from ground level would best be found using the upward acceleration in the form dv/dt, so that

$$-(mg+mkv^2)=m\,dv/dt$$

is the appropriate equation of motion. Then the time from O to O' is

$$T=-\int_V^0 \frac{dv}{g+kv^2}=(1/kg)^{\frac{1}{2}}\tan^{-1}(V\sqrt{k/g})\;.$$

Example 2

A car of mass M moves on a level road, the engine working at a constant rate R. There is a constant frictional resistance and the maximum speed attainable is W. If the car starts from rest and S, T are respectively the distance traversed and the time taken, before the car attains a speed V, show that

$$S=(MW^3/R)\{\log[W/(W-V)]-(V/W)-V^2/(2W^2)\}\;,$$
$$T=(S/W)+(MV^2/2R)\;. \qquad \text{(L.U.)}$$

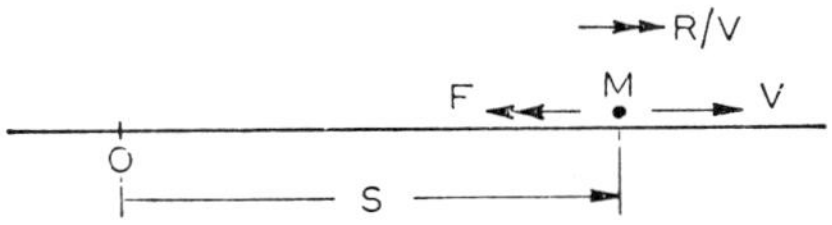

Fig. 3.5

The tractive force is R/V in the direction of S increasing and the resistance is the constant force F in the opposite direction (Fig. 3.5).

Then the equation of motion is

$$R/V-F=MVdV/dS=MdV/dT\;.$$

When $V=W$, $dV/dT=0$, and so $F=R/W$.

$$\therefore dS=\frac{MV\,dV}{R(1/V-1/W)}=\frac{MW}{R}\left[\frac{W^2}{W-V}-V-W\right]dV\;.$$

Since $V=0$ when $S=0$,

$$S=(MW/R)[-W^2\log(W-V)-\tfrac{1}{2}V^2-WV]+(MW^3/R)\log W,$$

i.e. $S=(MW^3/R)\{\log[W/(W-V)]-(V/W)-V^2/(2W^2)\}$. (1)

Also we have

$$R(1/V-1/W)=M\,dV/dT\;,$$

and so $dT=-(WM/R)[1+W/(V-W)]dV\;.$

Integrating and using $V=0$ when $T=0$,

$$T=(WM/R)\{W\log[W/(W-V)]-V\}\;. \qquad (2)$$

From (1), $\log[W/(W-V)] - V/W = V^2/(2W^2) + RS/(MW^3)$. Hence (2) gives

$$\underline{T = S/W + MV^2/(2R)}\ . \tag{3}$$

(iii) Simple harmonic motion

A *rectilinear simple harmonic motion* is one in which a particle moves along a straight line under a force which is always directed towards a point O in that line, the magnitude being proportional to the particle's displacement from O.

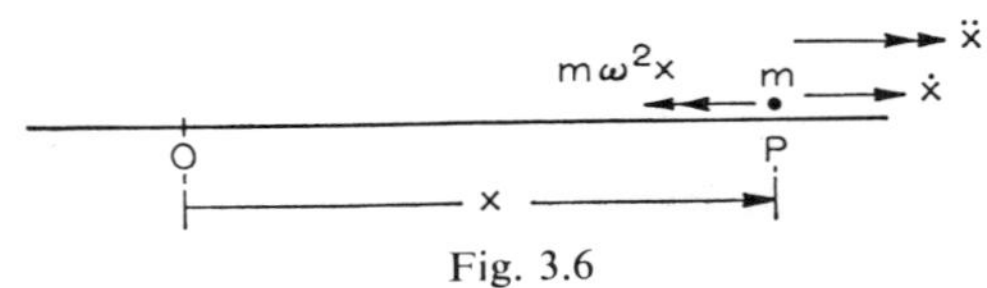

Fig. 3.6

Fig. 3.6 depicts a particle P of mass m moving in a straight line. At time t, $OP = x$, so that $\dot{x}$, $\ddot{x}$ are the velocity and acceleration of P measured in the direction of x increasing. The force on the particle is $m\omega^2 x$ towards O, i.e. $-m\omega^2 x$ in the direction of x increasing. Thus the equation of motion is

$$-m\omega^2 x = m\ddot{x}\ ,$$

$$\therefore\ (D^2 + \omega^2)x = 0 \qquad (D \equiv d/dt)\ .$$

The general solution of this differential equation may be expressed in two alternative forms:

$$x = A\cos\omega t + B\sin\omega t = a\sin(\omega t + x)\ ,$$

where A, B, a, α are arbitrary constants. Thus x has period $2\pi/\omega$. In the latter form, a is termed the *amplitude* of the motion and $(\omega t + \alpha)$ the *phase* at time t. The *frequency* is $\omega/(2\pi)$: the *angular frequency* is ω.

We shall often encounter non-rectilinear motions leading to equations of this type, e.g. $\ddot{\theta} = -\omega^2\theta$, where θ is an angle. Since the solution follows the same pattern, such a motion will be described as simple harmonic.

Example

A particle of mass m is constrained to execute a rectilinear S.H.M. under a force towards O of magnitude $m\omega^2 x$, x being the particle's displacement from O. When passing through O its velocity is V and when its velocity has become $\frac{1}{2}V$ in the same direction, an impulse I is applied to the particle in the direction of its motion. Assuming the same law of force, find the time and total distance travelled from O to the first position of instantaneous rest.

Throughout the motion, we have

$$-m\omega^2 x = m\ddot{x}$$

or

$$\ddot{x} = -\omega^2 x\ .$$

The general solution is $x = A\cos\omega t + B\sin\omega t$. Taking $x = 0$ when $t = 0$ gives $x = B\sin\omega t$, since $A = 0$.

$\therefore\ \dot{x} = B\omega \cos \omega t$, before the application of the impulse.

Since $\dot{x} = V$ when $t = 0$, we have $B = V/\omega$, and so

$$x = (V/\omega) \sin \omega t,\ \dot{x} = V \cos \omega t \ .$$

When $\dot{x} = \frac{1}{2}V$, $\cos \omega t = \frac{1}{2}$, and so

$$t = \pi/(3\omega),\ x = (\sqrt{3}V/2\omega) \ .$$

Let V_o be the velocity of the particle after applying the impulse. Then, from momentum considerations,

$$I = m(V_o - \tfrac{1}{2}V) \ ,$$

i.e.

$$V_o = \tfrac{1}{2}V + I/m \ .$$

During subsequent motion, take, for $t > \pi/(3\omega)$,

$$x = A' \cos \omega t + B' \sin \omega t \ .$$

$$\therefore\ \dot{x} = -A'\omega \sin \omega t + B'\omega \cos \omega t \ .$$

When $t = \pi/(3\omega)$, $x = (\sqrt{3}\,V/2\omega)$, $\dot{x} = V_o$. Thus

$$\begin{cases} \tfrac{1}{2}A' + (\sqrt{3}/2)B' = \sqrt{3}V/2\omega \ , \\ -(\sqrt{3}\omega/2)A' + \tfrac{1}{2}\omega B' = \tfrac{1}{2}V + I/m \ . \end{cases}$$

Solving for A', B', we find

$$x = -(\sqrt{3}I/2m\omega) \cos \omega t + [V/\omega + I/(2m\omega)] \sin \omega t \ ,$$

$$\dot{x} = (\sqrt{3}I/2m) \sin \omega t + [V + (I/2m)] \cos \omega t \ ,$$

for $t > \pi/(3\omega)$. Thus $\dot{x} = 0$ when

$$\tan \omega t = -[2mV/(\sqrt{3}I) + (1/\sqrt{3})] \ .$$

Hence we find t, x.

(iv) *Damped and forced vibrations*

Now suppose we have a particle of mass m moving in a straight line under a force $m\omega^2 x$ towards O fixed in the line, where x is the particle's distance from O at time t, and in addition suppose a force $mk\dot{x}$ acts in the direction opposing the velocity $\dot{x}$, where $k > 0$; then the resultant force acting on m in the sense of x increasing is $-(m\omega^2 x + mk\dot{x})$. Hence the equation of motion is

$$-(m\omega^2 x + mk\dot{x}) = m\ddot{x}$$

or

$$(D^2 + kD + \omega^2)x = 0 \qquad (D \equiv d/dt) \ .$$

The case of $k = 0$, $\omega \neq 0$ gives the S.H.M. which has been considered. When $k \neq 0$, various cases arise.

Case I: $k^2 < 4\omega^2$.

$$\therefore\ [(D + \tfrac{1}{2}k)^2 + n^2]x = 0 \ ,$$

where $n^2 = \omega^2 - \frac{1}{4}k^2 > 0$. The general solution is

$$x = e^{-kt/2}(A \cos nt + B \sin nt) .$$

This may be put in the equivalent form

$$x = ae^{-kt/2} \sin(nt + \alpha) ,$$

where A, B, a, α are all constants. The motion is *periodic*, the period being $2\pi/n$, but the amplitude $ae^{-kt/2}$ is *damped*, since it decays with time. This type of motion is called *damped oscillatory motion*. This case corresponds to *small damping*.

Case II: $k^2 > 4\omega^2$.

Putting $\lambda^2 = \frac{1}{4}k^2 - \omega^2 > 0$, the differential equation is

$$(D + \tfrac{1}{2}k + \lambda)(D + \tfrac{1}{2}k - \lambda)x = 0 .$$

$$\therefore\ x = A \exp[-(\lambda + \tfrac{1}{2}k)t] + B \exp[-(\tfrac{1}{2}k - \lambda)t] .$$

The form of this solution is *aperiodic* or *dead beat*, both exponentials decaying with time. This case corresponds to *large damping*.

Case III: $k^2 = 4\omega^2$.

$$\therefore\ (D + \omega)^2 x = 0 ;$$

$$\therefore\ x = e^{-\omega t}(A + Bt) .$$

Again the motion is aperiodic and the system is now said to be *critically damped*.

Now suppose that, in addition to the forces which act, the particle has an applied periodic force $mF \cos pt$ in the direction of x increasing. Then the total force in this direction is $m(F \cos pt - \omega^2 x - k\dot{x})$, so that the equation of motion becomes

$$m(F \cos pt - \omega^2 x - k\dot{x}) = m\ddot{x} ,$$

or

$$(D^2 + kD + \omega^2)x = F \cos pt .$$

The general solution of this second order equation consists of a complementary function and a particular integral superposed. The nature of the former, corresponding to zero applied force, has already been discussed. This portion of the solution tends to zero as t tends to infinity whenever $k \neq 0$: it is consequently called the *transient*.

The particular integral is

$$x = \frac{F}{D^2 + kD + \omega^2}(\cos pt) = \frac{F}{(\omega^2 - p^2) + kD}(\cos pt)$$

$$= F\frac{(\omega^2 - p^2) - kD}{(\omega^2 - p^2)^2 - k^2D^2}(\cos pt)$$

$$= F\frac{(\omega^2 - p^2)\cos pt + kp \sin pt}{(\omega^2 - p^2)^2 + k^2p^2} .$$

This does not vanish permanently at any time, however large: it is called the *steady state* and in practice is the more important part of the general solution. The steady state solution can also be expressed in the form

$$x = F\sin(pt+\alpha)/\{(\omega^2-p^2)^2+k^2p^2\}^{\frac{1}{2}},$$

where $\cos\alpha = kp/R$, $\sin\alpha = (\omega^2-p^2)/R$, $R=\{(\omega^2-p^2)^2+k^2p^2\}^{\frac{1}{2}}$. Thus the amplitude of the steady state is F/R. Now $R^2 = p^4-2(\omega^2-\frac{1}{2}k^2)p^2+\omega^4 = [p^2-(\omega^2-\frac{1}{2}k^2)]^2+\omega^4-(\omega^2-\frac{1}{2}k^2)^2$. Thus, if $\omega > k/\sqrt{2}$, R^2 is a minimum when

$$p = (\omega^2-\tfrac{1}{2}k^2)^{\frac{1}{2}} \ .$$

The amplitude of the steady state is then a maximum for this value of p, and the corresponding value of $p/(2\pi)$ is called the *resonance frequency* of the applied periodic force.

3.6 ELASTIC STRINGS AND SPRINGS

When an elastic string is subjected to tensile forces, it is found experimentally that, over an appreciable range, the extension produced is proportional to the tension. The same is also true of springs, but in addition compressive forces will produce contractions for which the proportionality law still holds. This law is known as *Hooke's law*. If l_o be the natural length of the string (or spring) and ϵ the extension produced by a tensile force T, then $T=\lambda\epsilon/l_o$, where λ is a constant known as the *modulus* of the string (or spring). λ may also be interpreted as the force required to double the length of the string (or spring) provided Hooke's law still holds for such a loading. Essentially λ is a constant of the string (or spring) but it depends on the geometry as well as on the material of the string (or spring). In consequence it is sometimes useful to state Hooke's law in a form which involves quantities separately dependent on the geometry and on the material of the string. Thus, if we have a steel wire of natural length l_o, tensile force T, uniform cross-sectional area A and extension ϵ, then it is found that

$$T/A = E(\epsilon/l_o) \ ,$$

where E is a constant dependent only on the material and termed *Young's modulus*, a concept of greater importance in engineering.

When an elastic string is stretched, work is done by the tensile force: this work is stored as elastic potential energy in the string. Thus, if a tensile force T applied to a string of natural length l_o and modulus λ produces an extension ϵ, the work done in producing a further extension $\delta\epsilon$ is $(\lambda\epsilon/l_o)\delta\epsilon$, so that the total energy stored in the string is

$$W = \int_0^{\varepsilon} (\lambda\epsilon/l_o)d\epsilon = \tfrac{1}{2}\,\lambda\epsilon^2/l_o \ .$$

The form of W indicates that elastic forces are conservative.

Consider now the motion of a particle P of mass m hanging from the lower end of an elastic string suspended vertically from a fixed point O (Fig. 3.7).

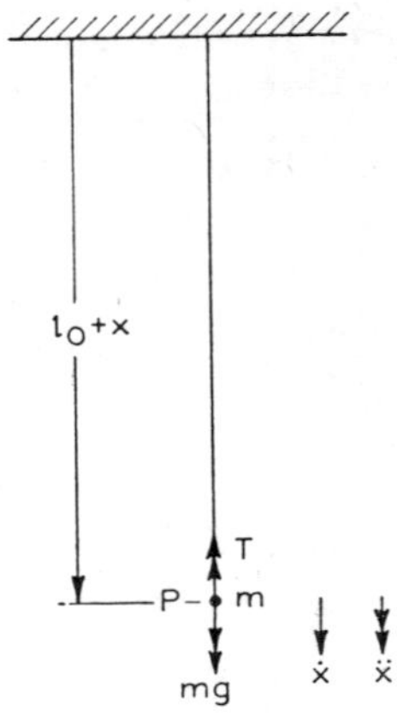

Fig. 3.7

Let l_o+x be the length OP at time t, l_o being the natural length. The weight of the particle is mg and the tensile force in the string is $T=\lambda x/l_o$, so that the net downward force is $mg-\lambda x/l_o$. Since the downward acceleration is $d^2(l_o+x)/dt^2=\ddot{x}$, the equation of motion is

$$mg-\lambda x/l_o=m\ddot{x}\ .$$

The static equilibrium position is given by $x=mgl_o/\lambda$, and so putting $z=x-mgl_o/\lambda$ in the above equation gives

$$\ddot{z}=-(\lambda/ml_o)z\ .$$

This equation shows that, so long as the string does not slacken, the particle executes a rectilinear S.H.M. about the equilibrium position, the period being $2\pi(ml_o/\lambda)^{\frac{1}{2}}$. This requires that the depth z below the equilibrium position should never exceed mgl_o/λ or that $x\leqslant 2mgl_o/\lambda$.

Example 1

An elastic string has natural length l_o and is held vertically with the upper end fixed. A heavy particle attached to its lower end produces a static deflection d of the string. If the particle is pulled down a further distance c below equilibrium and released show that in the case when $d(d+4l_o)>c^2>d^2$, the time taken for the particle to return to its original position of release is $2(d/g)^{\frac{1}{2}}(\pi-\theta+\tan\theta)$, where $\cos\theta=d/c$, and θ is acute.

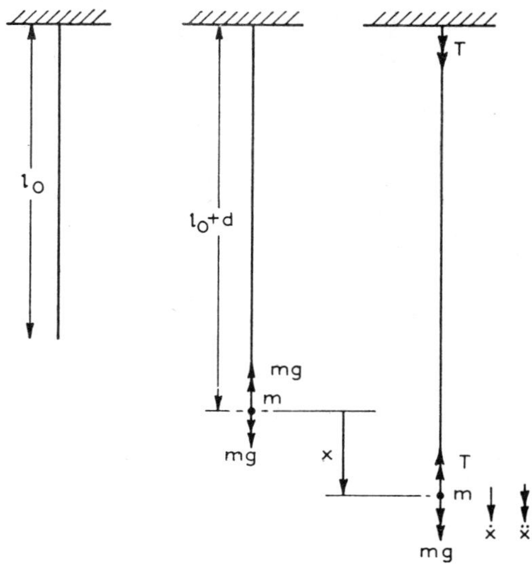

Fig. 3.8(i) Fig. 3.8(ii) Fig. 3.8(iii)

Fig. 3.8(i) shows the unstretched string, Fig. 3.8(ii) the equilibrium position, and Fig. 3.8(iii) the string of length $(l_o + d + x)$ at time t after release, the string still being taut.

Taking m to be the mass of the particle in Fig. 3.8(ii) we have

$$mg = \lambda(d/l_0), \text{ or } \lambda = mgl_0/d \ .$$

Thus in Fig. 3.8(iii) the tension is

$$T = \lambda(x+d)/l_0 = mg(x+d)/d \ .$$

Since the weight of the particle is mg, the net downward force in Fig. 3.8(iii) is $mg - T = -mgx/d$, and since the downward acceleration of m is $\ddot{x}$, we have

$$-mgx/d = m\ddot{x}$$

or

$$\ddot{x} = -(g/d)x \ . \tag{1}$$

(1) holds only so long as the string is taut. The general solution of (1) is

$$x = A\cos(g/d)^{\frac{1}{2}}t + B\sin(g/d)^{\frac{1}{2}}t \ .$$

Since $x = c$, $\dot{x} = 0$ when $t = 0$, we find $B = 0$, $A = c$ and so the solution of (1) is

$$x = c\cos(g/d)^{\frac{1}{2}}t \ . \tag{2}$$

(2) holds only as long as the string remains taut. But when $x = -d$, the string has attained its natural length. If T_1 be the time taken for this to happen,

$$-d = c\cos(g/d)^{\frac{1}{2}}T_1 \ ,$$

i.e.

$$\underline{T_1 = (d/g)^{\frac{1}{2}}(\pi - \theta)} \ . \tag{3}$$

At this instant, $\dot{x}=-c(g/d)^{\frac{1}{2}}\sin(g/d)^{\frac{1}{2}}T_1=-c(g/d)^{\frac{1}{2}}\sin\theta$. Thus at time $t=T_1$, the particle is rising against gravity with an upward velocity of $V=c(g/d)^{\frac{1}{2}}\sin\theta$. Let T_2 be the further time taken for the particle to attain its greatest height. The greatest height the particle would attain under free gravitational flight would be

$$V^2/(2g)=\tfrac{1}{2}(c^2/d)\sin^2\theta=\tfrac{1}{2}(c^2/d)[1-(d/c)^2] \ .$$

If the string does not tighten in this phase of the motion, then this quantity $<2l_o$,

i.e. $$c^2-d^2<4dl_0$$

or $$d(d+4l_0)>c^2 \ .$$

As this condition holds, it follows that in the subsequent upward flight of the particle, the string does not tighten. Hence

$$T_2=V/g=(c/g)(g/d)^{\frac{1}{2}}\sin\theta=(d/g)^{\frac{1}{2}}\tan\theta \ . \tag{4}$$

Hence the total time from release to the highest point attained is

$$T_1+T_2=(d/g)^{\frac{1}{2}}(\pi-\theta+\tan\theta) \ .$$

Thus the total time required is twice this.

Example 2

A and B are two points in a horizontal plane distant l apart and are joined by a light elastic string of natural length l_o and modulus λ, where $l_o<l$. The string has a particle of mass m attached to its midpoint. This particle is drawn aside a small distance and released. Show that the ensuing motion of the particle is approximately simple harmonic and find the period.

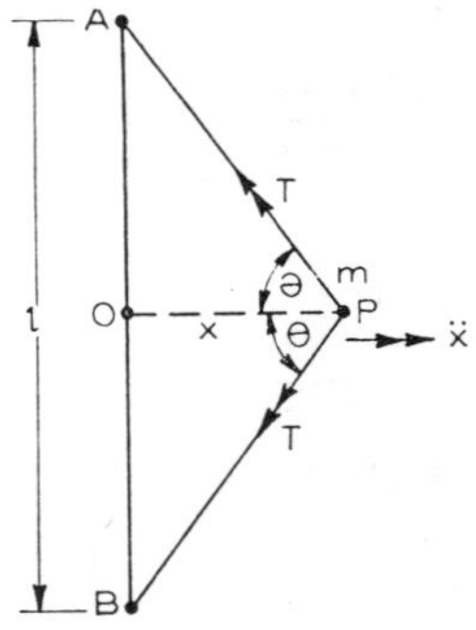

Fig. 3.9

Fig. 3.9 shows the configuration at time t, T being the tension in the strings AP, BP. Then $AP=BP=(\frac{1}{4}l^2+x^2)^{\frac{1}{2}}$.

$\therefore$ Stretch of $AP=(\frac{1}{4}l^2+x^2)^{\frac{1}{2}}-\frac{1}{2}l_0\simeq\frac{1}{2}(l-l_0)$, to 1st order.

$\therefore$ $T\simeq\frac{1}{2}\lambda(l-l_0)\div\frac{1}{2}l_0=\lambda(l/l_0-1)$, to 1st order.

If x denote the displacement of the particle O at time t, the equation of motion is

$$-2T\cos\theta = m\ddot{x} .$$

Now $\cos\theta = x/(\frac{1}{4}l^2 + x^2)^{\frac{1}{2}} \simeq 2x/l$, to 1st order. Hence to first order, the equation of motion is

$$m\ddot{x} = -4\lambda(1/l_o - 1/l)x ,$$

and so the motion is S.H.M. of period $\pi\left[\dfrac{ml l_o}{\lambda(l-l_o)}\right]^{\frac{1}{2}}$.

Example 3

A heavy particle of mass m hangs at the lower end of a light vertical elastic spring of natural length l_o and modulus $2mg$. At zero time the system is in equilibrium and the upper end of the spring is made to execute vertical oscillations so that its downward displacement at time t later is $a\sin pt$, where $p^2 = 2g/l_o$. Find the displacement of the particle at this instant.

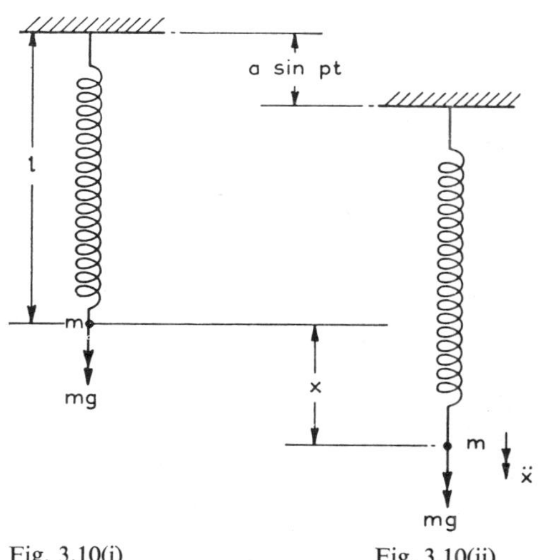

Fig. 3.10(i) Fig. 3.10(ii)

Fig. 3.10(i) shows the system in equilibrium at time $t=0$. Then

$$mg = 2mg(l - l_o)/l_o, \text{ so } l = 3l_o/2 .$$

In Fig. 3.10(ii), the situation at time t is depicted, the particle having been displaced a distance x. Let T be the tension in the spring. The total stretch of the spring is $(l_o/2 + x - a\sin pt)$ and so

$$T = 2mg(\tfrac{1}{2}l_o + x - a\sin pt)/l_o .$$

Hence the equation of motion of m is

$$mg - T = m\ddot{x} \ ,$$

i.e. $$(D^2 + p^2)x = (2ga/l_o)\sin pt \qquad (D \equiv d/dt) \ .$$

This equation exhibits the case of *resonance*. The C.F. is

$$x = A\cos pt + B\sin pt \ ,$$

and the particular integral is found from

$$(2ga/l_o)\frac{1}{D^2+p^2}(\sin pt) \ .$$

Now $$\frac{1}{D^2+p^2}(e^{ipt}) = \frac{1}{(D+ip)(D-ip)}(e^{ipt})\cdot$$

$$= \frac{1}{2ip}\cdot\frac{1}{D-ip}(e^{ipt}) = \frac{e^{ipt}}{2ip}\cdot\frac{1}{D}(1)$$

$$= -it\,e^{ip}/(2p) = -t(i\cos pt - \sin pt)/(2p) \ .$$

$$\therefore \ \frac{1}{D^2+p^2}(\sin pt) = -(t\cos pt)/(2p) \ .$$

Thus the complete primitive is

$$x = A\cos pt + B\sin pt - (gat\cos pt/pl_o) \ .$$

When $t = 0$, $x = 0$, $\dot{x} = 0$, so $A = 0$, $B = (ga/p^2 l_o)$.

$$\therefore \ \underline{x = (ga/p^2 l_o)(\sin pt - pt\cos pt)} \ .$$

3.7 CHANGING MASS PROBLEMS

When the mass of a body varies, the technique is to find the momentum gain of the system over an interval of time from t to $t + \delta t$, then obtain the rate of momentum gain at time t and equate to the resultant force which acts at that instant.

Example 1

Establish the formula $\mathbf{F}(t) = m(t)d\mathbf{v}/dt + \lambda\mathbf{u}$ for the motion of a particle of varying mass $m(t)$ with velocity $\mathbf{v}$ under a force $\mathbf{F}(t)$, matter being emitted at a constant rate λ with velocity $\mathbf{u}$ relative to the particle.

A rocket whose mass at time t is $m_o(1 - \alpha t)$, where m_o and α are constants travels vertically upwards from rest at $t = 0$. The matter emitted has constant backward speed $4g/\alpha$ relative to the rocket. Assuming that the gravitational field $\mathbf{g}$ is constant and that the resistance of the atmosphere is $2m_o v\alpha$, where v is the speed of the rocket, show that half of the original mass is left when the rocket reaches a height $(g/3\alpha^2)$.

(L.U.)

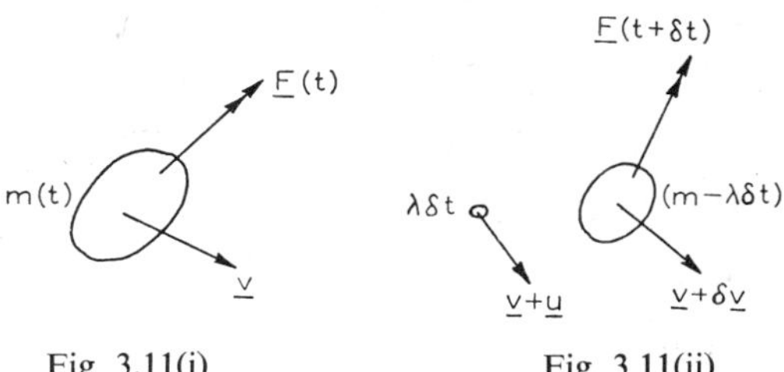

Fig. 3.11(i) Fig. 3.11(ii)

(i) Fig. 3.11(i) depicts the particle of mass $m(t)$ at time t moving with velocity $\mathbf{v}$ under a resultant force $\mathbf{F}(t)$. Fig. 3.11(ii) shows the particle at time $t+\delta t$, its mass now being $m-\lambda\delta t$, together with the discarded matter of mass $\lambda\delta t$ moving with velocity $\mathbf{v}+\mathbf{u}$. The particle of mass $m-\lambda\delta t$ moves with velocity $\mathbf{v}+\delta\mathbf{v}$ and the force now acting is $\mathbf{F}(t+\delta t)$.

The momentum of the system in Fig. 3.11(i) is $m\mathbf{v}$: that in Fig. 3.11(ii) is $(m-\lambda\delta t)(\mathbf{v}+\delta\mathbf{v})+\lambda\delta t(\mathbf{v}+\mathbf{u})$ and so the gain in time δt is

$$(m-\lambda\delta t)(\mathbf{v}+\delta\mathbf{v})+\lambda\delta t(\mathbf{v}+\mathbf{u})-m\mathbf{v}=m\delta\mathbf{v}+\lambda\mathbf{u}\,\delta t-\lambda\delta t\,\delta\mathbf{v}\ .$$

Thus the mean rate of increase of linear momentum over this interval is $m\delta\mathbf{v}/\delta t+\lambda\mathbf{u}-\lambda\delta\mathbf{v}$ and so at time t the rate of increase in linear momentum is

$$\lim_{\delta t\to 0}[m\delta\mathbf{v}/\delta t+\lambda\mathbf{u}-\lambda\delta\mathbf{v}]=md\mathbf{v}/dt+\lambda\mathbf{u}\ .$$

This is equal to the resultant force on the particle in Fig. 3.11(i).

$$\therefore\ \underline{\mathbf{F}=md\mathbf{v}/dt+\lambda\mathbf{u}}\ .$$

(ii)

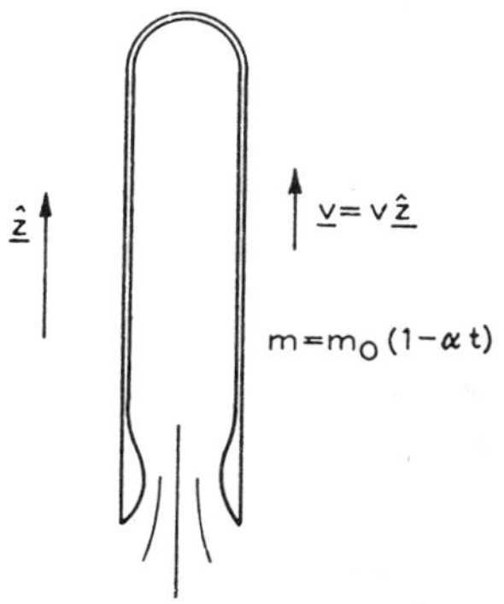

Fig. 3.12

Take $\hat{\mathbf{z}}$ to be the unit vector in the upward vertical direction.

$$m = m_o(1-\alpha t), \qquad \mathbf{v} = v\hat{\mathbf{z}}, \qquad \mathbf{u} = -(4g/\alpha)\hat{\mathbf{z}} \ ,$$
$$\mathbf{F} = -m_o(1-\alpha t)g\hat{\mathbf{z}} - 2m_8 v\alpha\hat{\mathbf{z}} \ ,$$
$$\lambda = m_o\alpha \ .$$

Substituting in the previous result gives

$$-2m_o v\alpha\hat{\mathbf{z}} - m_o(1-\alpha t)g\hat{\mathbf{z}} = m_o(1-\alpha t)(dv/dt)\hat{\mathbf{z}} + m_o\alpha(-4g/\alpha)\,\hat{\mathbf{z}} \ ,$$

and so, on simplification,

$$dv/dt + 2\alpha v/(1-\alpha t) = (3+\alpha t)g/(1-\alpha t) \ .$$

The integrating factor is $\exp \int 2\alpha(1-\alpha t)^{-1}dt = (1-\alpha t)^{-2}$.

$$\therefore \ d[v(1-\alpha t)^{-2}]/dt = (3+\alpha t)g/(1-\alpha t)^3 = 4g/(1-\alpha t)^3 - g/(1-\alpha t)^2 \ .$$

Integrating and taking $v=0$ when $t=0$,

$$v(1-\alpha t)^{-2} = [2g/\alpha(1-\alpha t)^2] - [g/\alpha(1-\alpha t)] - g/\alpha \ .$$

Taking s to be the rise at time t,

$$ds/dt = 2g/\alpha - g(1-\alpha t)/\alpha - g(1-\alpha t)^2/\alpha \ .$$

Integrating and remembering $s=0$ when $t=0$,

$$s = 2gt/\alpha + g(1-\alpha t)^2/(2\alpha^2) + g(1-\alpha t)^3/(3\alpha^2) - (5g/6\alpha^2) \ .$$

Taking $\alpha t = \frac{1}{2}$, gives $s = (g/3\alpha^2)$.

Example 2

A machine-gun of mass M_o stands on a horizontal plane and contains a shot of total mass m, which is fired horizontally at a uniform rate with constant velocity u relative to the gun. If the coefficient of sliding friction

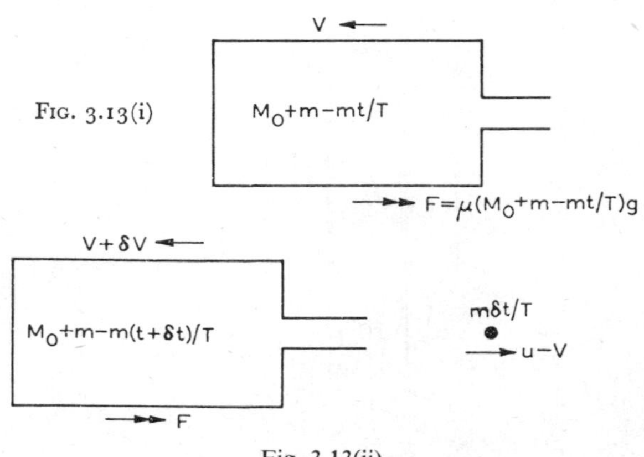

FIG. 3.13(i)

Fig. 3.13(ii)

between the gun and the plane is μ, and the shot is all expended in time T, show that the velocity of the gun is then

$$u \log(1+m/M_o) - \mu g T \ . \qquad \text{(I.C.S.)}$$

Let V be the velocity of the gun at time t when the total mass is $M_o + m - (mt/T)$ and the friction is $F = \mu(M_o + m - mt/T)g$ (Fig. 3.13(i)). At time $t + \delta t$, the mass of the gun and contents is $M_o + m - m(t+\delta t)/T$, the mass of emitted shot in time δt being $m\delta t/T$ and the velocities of the gun and shot being $V + \delta V$, $u - V$ in the senses indicated in Fig. 3.13(ii). Then in time δt, the linear momentum gained in the direction of F is

$$\begin{aligned}(m\delta t/T)(u-V) &- \{M_o + m - m(t+\delta t)/T\}(V+\delta V) + (M_o + m - mt/T)V \\ &= (m\delta t/T)(u-V) + (mV\delta t/T) - (M_o + m - mt/T)\delta V + O(\delta t \delta V) \ .\end{aligned}$$

$\therefore$ Rate of change of linear momentum in direction of F at time t

$$= (mu/T) - (M_o + m - mt/T)dV/dt \ .$$

$$\therefore \ (mu/T) - (M_o + m - mt/T)dV/dt = \mu g(M_o + m - mt/T) \ .$$

$$\therefore \ dV = -\mu g \, dt + (mu/T)dt/(M_o + m - mt/T) \ .$$

Hence the final velocity after time T is

$$-\mu g T + u\left[-\log(M_o + m - mt/T\right]_{t=0}^{t=T}$$

$$= \underline{-\mu g T + u \log(1 + m/M_o)} \ .$$

3.8 DETAILED TREATMENT OF ROCKET MOTION

(i) *Performance in flight of a single-stage rocket*

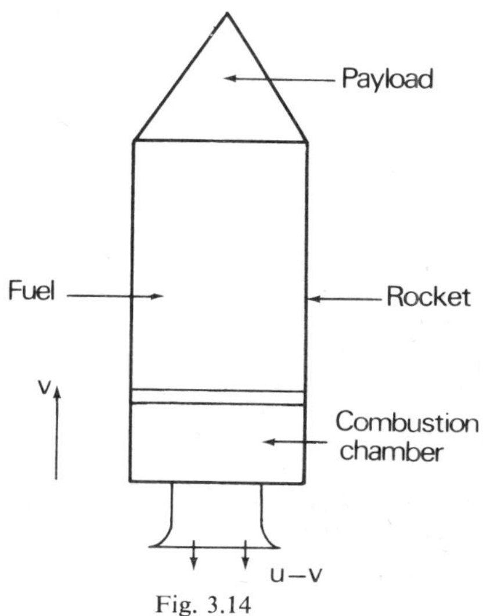

Fig. 3.14

Fig. 3.14 shows a single-stage rocket moving vertically upwards. The rocket has an initial mass of M_0 (rocket casing and fuel), together with a payload of mass P. Suppose the fuel burns at a uniform rate of mass m per unit time, so that the mass of the rocket at time t where $0 \leqslant t \leqslant \tau$, and τ is the time to the stage of 'all burnt', is

$$M(t) = P + M_0 - mt \ .$$

The upward velocity of the rocket at such time t is v and we suppose that the velocity of emission of the fuel relative to the rocket is u, giving the total efflux velocity of the fuel as $u - v$ downwards. The upward linear momentum of the system at time t is

$$(P + M_0 - mt)v \ .$$

At time $t + \delta t$, let $v + \delta v$ be the upward velocity of the rocket which now has a linear momentum (upwards) of

$$[P + M_0 - m(t + \delta t)]\,(v + \delta v) \ ,$$

whilst the burnt fuel emitted during the interval $(t, t + \delta t)$ has mass $m\,\delta t$ and upward linear momentum and $m\delta t(v - u)$. Then, at the instant of time t, the rate of change of linear momentum of the system, measured vertically upwards is

$$\lim_{\delta t \to 0} \left\{ \frac{[P + M_0 - m(t + \delta t)]\,(v + \delta v) + m(v - u)\delta t - (P + M_0 - mt)v}{\delta t} \right\}$$

$$= (P + M_0 - mt)\frac{dv}{dt} - mu \ .$$

As the upward force on the rocket (presumed to be due solely to gravity, air resistance being neglected) at time t is $-(P + M_0 - mt)g$, the equation of motion is

$$(P + M_0 - mt)\frac{dv}{dt} - mu = -(P + M_0 - mt)g \ .$$

Integrating this from $t = 0$ to $t = \tau = M_1/m$, where $M_1 (< M_0)$ is the initial mass of fuel, and taking $v = 0$ when $t = 0$, we have

$$v - 0 = \int_0^\tau \frac{mu}{P + M_0 - mt}\,dt - g\tau$$

i.e.

$$\underline{v = -u \log\left[1 - \frac{M_1}{P + M_0}\right] - g\tau \ .}$$

The last term on the right-hand side is due solely to gravity and with gravity neglected, it would be absent. It is easy to show that in practical cases the gravitational contribution to the velocity at 'all burnt' is often negligible. We take as typical values:

$$M_0/P = 100, \quad I = u/g = 300 \text{ s}, \quad M_1/M_0 = 0.8 \ ,$$
$$M_0 = 10^5 \text{ kg}, \quad \tau = 16 \text{ s} \ .$$

I is a quantity known as the specific impulse of the propellant and is often used by rocket ballisticians. Calculation shows that the first term in the expression for v is approximately 4.64 km s^{-1} and the second (gravitational) about 0.16 km s^{-1}. We shall, therefore, neglect the gravitational effect in what follows.

(ii) *Performance of a two-stage rocket*

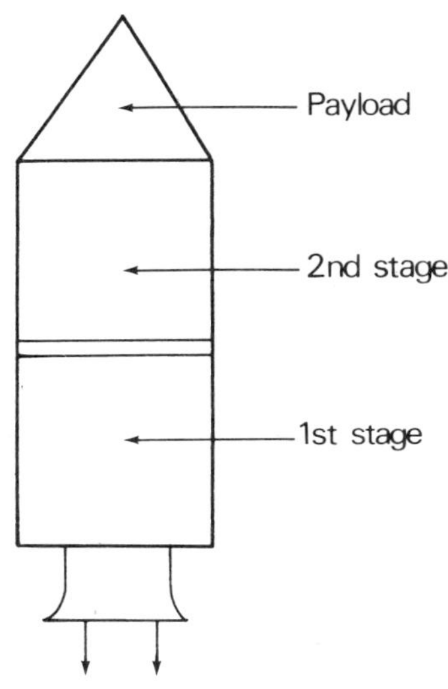

Fig. 3.15

Fig. 3.15 illustrates a two-stage rocket. The design is such that when all the fuel in the first stage has been consumed, the rocket casing and instruments of this stage are detached and ignition of the second-stage fuel takes effect. Let us suppose that P is the mass of the payload, M_1 the initial mass of the first stage of which ϵM_1 is the initial mass of the fuel, M_2 the initial mass of the second stage of which ϵM_2 is the initial fuel mass. Then if u is the ejection velocity of the fuel (relative to the casing) and m its rate of burning, neglecting gravity and air resistance, the analysis of (i) indicates that the velocity attained by the rocket in the first stage of motion is

$$v_1 = -u \log\left[1 - \frac{\epsilon M_1}{M_1 + M_2 + P}\right] \ .$$

The time required for the fuel to burn in the second stage of flight is $\epsilon M_2/m$. The differential equation governing the rocket's motion during the second stage is

$$(P+M_2-mt)\frac{dv}{dt}=mu$$

and so, if the velocity v increases from v_1 to v_2 in the second stage of flight, then

$$\int_{v_1}^{v_2} dv=\int_0^{\tau_2}\frac{mu\,dt}{P+M_2-mt}\ (\tau_2=\epsilon M_2/m)\ ,$$

i.e.

$$v_2-v_1=-u\log\left[1-\frac{\epsilon M_2}{P+M_2}\right]\ .$$

Thus the final velocity attained by the two-stage rocket is

$$v_2=-u\log\left[1-\frac{\epsilon M_1}{M_1+M_2+P}\right]-u\log\left[1-\frac{\epsilon M_2}{M_2+P}\right]\ .$$

With the values $M_1=M_2=50P$, $\epsilon=0.8$, $I=300\text{s}$,

$$v_2\simeq 6.0\,\text{km}\,\text{s}^{-1}\ ,$$

compared with the figure of $0.16\,\text{km}\,\text{s}^{-1}$ attained for single-staging.

If the greatest value of v_2 is required, as for example when it is required to put the payload into orbit round the Earth, we can proceed as follows. Let $M_0=M_1+M_2$ so that

$$v_2=-u\log\left[1-\frac{\epsilon(M_0-M_2)}{M_0+P}\right]-u\log\left[1-\frac{\epsilon M_2}{M_2+P}\right]\ ,$$

and, taking M_0 as constant, the stationary condition $dv_2/dM_2=0$ gives

$$\frac{\epsilon/(M_0+P)}{[1-\epsilon(M_0-M_2)/(M_0+P)]}=\frac{\epsilon P/(M_2+P)^2}{[1-\epsilon M_2/(M_2+P)]}$$

which on simplifying gives for the case $\epsilon\neq 1$,

$$M_2{}^2+2PM_2-PM_0=0\ .$$

The appropriate root M_2 may be expressed by the dimensionless formula

$$\frac{M_2}{M_0}=-\frac{P}{M_0}+\left(\frac{P^2}{M_0^2}+\frac{P}{M_0}\right)^{1/2}$$

When $P/M_0=\alpha$ and is small, this gives, on expansion,

$$M_2/M_0=\alpha^{1/2}-\alpha+\tfrac{1}{2}\alpha^{3/2}+\ldots=\alpha^{1/2}-\alpha+0(\alpha^{3/2})\ .$$

With $\alpha=1/100$, $M_2/M_0\simeq 1/10$ or $M_1:M_2\simeq 9:1$. This optimum choice of ratio $M_1:M_2$ shows that the first stage must be very much larger than the second.

Taking $P/M_0=\alpha$, $M_2/M_0=\alpha^{1/2}-\alpha+0(\alpha^{3/2})$, the velocity v_2 at the end of the second stage of propulsive flight, is

$$v_2=-u\log\left\{1-\frac{\epsilon[1-\alpha^{1/2}+\alpha+0(\alpha^{3/2})]}{1+\alpha}\right\}$$

$$-u\log\left\{1-\frac{\epsilon[\alpha^{1/2}-\alpha+0(\alpha^{3/2})]}{\alpha^{1/2}+0(\alpha^{3/2})}\right\}$$

$$=-u\log\{1-\epsilon[1-\alpha^{1/2}+0(\alpha^{3/2})]\}$$

$$-u\log\left\{1-\frac{\epsilon[1-\alpha^{1/2}+0(\alpha)]}{1+0(\alpha)}\right\}$$

$$=-u\log\{1-\epsilon[1-\alpha^{1/2}+0(\alpha^{3/2})]\}$$

$$-u\log\{1-\epsilon[1-\alpha^{1/2}+0(\alpha)]\}$$

$$\simeq -2u\log\{1-\epsilon(1-\alpha^{1/2})\}$$

With $P/M_0=1/100$, $\epsilon=0.8$, $I=300\,\mathrm{s}$, this gives

$$\underline{v_2\simeq 7.5\,\mathrm{km\,s^{-1}}}\ .$$

(iii) *Performance of a multi-stage rocket*

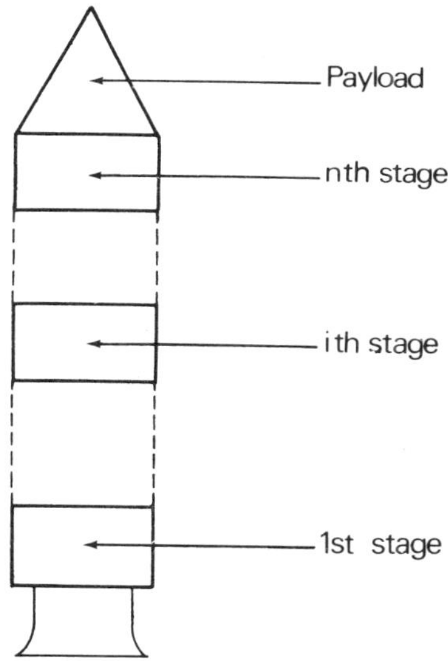

Fig. 3.16

Fig. 3.16 depicts an n-stage rocket with payload. The initial mass of the ith stage is M_i $(i=1, \ldots, n)$. Here we take the simple case in which m, the rate of mass burning of fuel per unit time, is constant throughout all stages as well as the efflux velocity, u and the structural factor ϵ expressing the initial ratio of the mass of fuel in a stage to the total mass (of fuel and casement) of that stage. The analysis of the two-stage case shows that the velocity increment during the entire time of combustion of the ith stage (neglecting gravity and air resistance) is

$$v_i = -u \log\left\{1 - \frac{\epsilon M_i}{M_i + M_{i+1} + \ldots + M_n + P}\right\}$$

$$= u \log\left\{\frac{M_i + M_{i+1} + \ldots + M_n + P}{(1-\epsilon)M_i + M_{i+1} + \ldots + M_n + P}\right\} .$$

Thus the velocity v attained when combustion of all n stages is just completed is

$$v = u \sum_{i=1}^{n} \log\left\{\frac{M_i + M_{i+1} + \ldots + M_n + P}{(1-\epsilon)M_i + M_{i+1} + \ldots + M_n + P}\right\}$$

The total initial mass of the n stages is M_0 and so

$$M_1 + M_2 + \ldots + M_n = M_0 .$$

We have to fix v and choose the values of $M_1, M_2, \ldots, M_n$ to ensure that M_0 is a minimum. For this purpose, we invoke the Lagrange multiplier technique, writing

$$f = M_1 + M_2 + \ldots + M_n + \lambda\left[v - u\sum_{i=1}^{n}\log\left\{\frac{M_i + M_{i+1} + \ldots + M_n + P}{(1-\epsilon)M_i + M_{i+1} + \ldots + M_n + P}\right\}\right]$$

Then the required minima of f are found from the n equations

$$\partial f / \partial M_i = 0 \qquad (i = 1, \ldots, n)$$

coupled with the equation of constraint. For simplicity, we write

$$\mu_i = \frac{M_i + M_{i+1} + \ldots + M_n + P}{(1-\epsilon)M_i + M_{i+1} + \ldots + M_n + P} \, (i = 1, \ldots, n) .$$

Now

$$\frac{M_i + M_{i+1} + \ldots + M_n + P}{M_{i+1} + \ldots + M_n + P}$$

$$= \frac{M_i + M_{i+1} + \ldots + M_n + P}{[(1-\epsilon)M_i + M_{i+1} + \ldots + M_n + P] - (1-\epsilon)[M_i + M_{i+1} + \ldots + M_n + P]} \times \epsilon$$

$$= \frac{\epsilon \mu_i}{1 - (1-\epsilon)\mu_i}$$

Hence

$$\frac{M_0 + P}{P} = \left[\frac{M_1 + M_2 + \ldots + M_n + P}{M_2 + \ldots + M_n + P}\right]\left[\frac{M_2 + \ldots + M_n + P}{M_3 + \ldots + M_n + P}\right]$$

$$\ldots \left[\frac{M_{n-1} + M_n + P}{M_n + P}\right]\left[\frac{M_n + P}{P}\right]$$

$$= \prod_{i=1}^{n} \frac{\epsilon \mu_i}{1 - (1-\epsilon)\mu_i}$$

$$= \epsilon^n \prod_{i=1}^{n} \frac{\mu_i}{1 - (1-\epsilon)\mu_i} .$$

Minimizing M_0 is equivalent to minimizing $(M_0+P)/P$, since P is constant, and hence also to minimizing $\log[(M_0+P)/P]$. Now

$$\log[(M_0+P)/P]=n\log\epsilon+\sum_{i=1}^{n}\{\log\mu_i-\log[1-(1-\epsilon)\mu_i]\} .$$

This is subject to the constraint that

$$u\sum_{i=1}^{n}\log\mu_i=v=\text{const.}$$

Hence we form the function

$$\log[(M_0+P)/P]+\lambda u\sum_{i=1}^{n}\log\mu_i$$

and equate to zero its n partial derivatives with respect to $\mu_i(i=1,\ldots,n)$. This leads to the n equations

$$\frac{1}{\mu_i}+\frac{1-\epsilon}{1-(1-\epsilon)\mu_i}+\frac{\lambda u}{\mu_i}=0\ (i=1,\ldots,n)$$

together with the above equation of constraint. We obtain

$$\mu_i=\frac{1+\lambda u}{\lambda u(1-\epsilon)}\quad(i=1,\ldots,n),$$

and so $\mu_1=\mu_2=\ldots=\mu_n=\mu$, say.

Thus

$$v=u\log\sum_{i=1}^{n}\log\mu_i=nu\log\left[\frac{1+\lambda u}{\lambda u(1-\epsilon)}\right],$$

giving

$$\lambda=\frac{1}{u[(1-\epsilon)\exp(v/nu)-1]}$$

so that

$$\frac{M_0}{P}=\epsilon^n\prod_{i=1}^{n}\frac{\mu_i}{1-(1-\epsilon)\mu_i}-1$$

$$=\frac{\epsilon^n\mu^n}{[1-(1-\epsilon)\mu]^n}-1$$

$$=\frac{\epsilon^n\exp(v/u)}{[1-(1-\epsilon)\exp(v/nu)]^n}-1$$

A multi-stage rocket may be used to put a satellite into orbit. For a typical circular orbit, 1000 km above the Earth, the satellite must be given a speed of about 7.8 km s^{-1}. In the above formula for M_0/P let us take as typical values:

$$v=7.8\,\text{km s}^{-1},\quad I=u/g=300\,\text{s},\quad \epsilon=0.8 .$$

The payload is the artificial satellite to be placed. Then calculation gives the table:

n	1	2	3	4	5	∞
M_0/P	neg.	147	52	40	36	13.2

The table shows a very considerable decrease in rocket mass M_0 for a given payload P as the number of stages increases from 2 to 3. On the other hand, the decrease is much less as n changes from 3 to 4 or from 4 to 5. Consequently NASA uses 3-stage rockets for such orbits. Further, the cost of the design increases rapidly as the number of stages is increased.

To find the ratio of the masses of the various stages, we use the result (previously obtained)

$$\mu = \frac{M_i + M_{i+1} + \ldots + M_n + P}{(1-\epsilon)M_i + M_{i+1} + \ldots + M_n + P} \quad (i = 1, \ldots, n) .$$

Taking $i = n$,

$$\mu = (M_n + P)/[(1-\epsilon)M_n + P]$$

or

$$\frac{M_n}{P} = \frac{\mu - 1}{1 - (1-\epsilon)\mu} = \beta, \text{ say} .$$

Taking $i = n - 1$,

$$\mu = [M_{n-1} + (\beta + 1)P]/[(1-\epsilon)M_{n-1} + (\beta+1)P]$$

or

$$M_{n-1}/P = \beta(\beta + 1) .$$

Similarly,

$$M_{n-2}/P = \beta(\beta + 1) ,$$

$$\cdots\cdots$$

$$M_1/P = \beta(\beta+1)^{n-1} .$$

Thus

$$M_1/M_{i+1} = 1 + \beta = \frac{\epsilon\mu}{[1 - \mathrm{T}(1-\epsilon)]} .$$

Thus

(i) for $n = 2$, $M_1 : M_2 \simeq 12:1$;

(ii) for $n = 3$, $M_1 : M_2 : M_3 \simeq 13.7:3.7:1.0$.

EXERCISE 3

1. A particle of mass m is placed on a horizontal board which is made to execute vertical simple harmonic oscillations of period T and amplitude a. If $a < (gT^2/4\pi^2)$, show that the particle does not lose contact with the board at any time.

2. A, B, C, D, E, F are the vertices of a regular hexagon having O as centre. A particle of mass m, initially in equilibrium at O, is displaced to a

point P. If, for any position of P, the particle is subjected to forces represented completely by $\mu\overline{PA}$, $\mu\overline{PB}$, $\mu\overline{PC}$, $\mu\overline{PD}$, $\mu\overline{PE}$, $\mu\overline{PF}$, where $\mu > 0$, show that when released P executes a S.H.M. about O of period $2\pi(m/6\mu)^{\frac{1}{2}}$.

3. A body of mass M, travelling in a straight line, is supplied with constant power P and is subjected to a resistance Mkv^2, where v is its speed and k is a constant. Prove that the speed of the body cannot exceed a certain value and that, if it starts from rest, it acquires half the maximum speed after travelling a distance $(1/3k)\log(8/7)$. If the power is then cut off and an additional retarding force of constant value F is imposed, find the subsequent time which elapses before the body comes to rest. (L.U.)

4. A car of mass m starts with its engine exerting a constant pull mP. At any subsequent time it can be switched over to work at a constant rate mH. Show that if a possible speed $V(>H/P)$ is to be attained from rest against a resistance mkv^2, where v is the speed, then the distance covered will be a maximum if the engine is switched over when the speed is H/P. Show also that this maximum distance is

$$(1/6k)\log[P^3H^2/\{(P^3-kH^2)(H-kV^3)^2\}]\ .$$

(L.U.)

5. A particle P of mass m moves along a straight line through a point O and at any instant the distance OP is x. When $x > a$, the particle is attracted towards O by a force mk/x^2, and when $x < a$ the particle is repelled from O by a force mak/x^3. If the particle is released from rest at a distance $2a$ from O show that it will come to rest instantaneously when $x = a/\sqrt{2}$, and find the time the particle takes to travel from $x = a$ to $x = a/\sqrt{2}$. (L.U.)

6. A particle, moving in a straight line, is subject to a retardation of amount kv^n per unit mass, where v is the speed at time t. Show that, if $n < 1$, the particle will come to rest at a distance $u^{2-n}/k(2-n)$ from the point of projection at time $t = u^{1-n}/k(1-n)$, where u is the initial speed. What happens when (i) $1 < n < 2$, (ii) $n > 2$? (L.U.)

7. A particle moves under gravity in a medium which offers a resistance $kv^2/(a+y)$ per unit mass, where v is the speed of the particle, y is the height above a fixed point O and a and $k(\neq -\frac{1}{2})$ are constants. If the particle is projected vertically upwards from O with speed u, show that it will come to rest instantaneously when $y = h$, where

$$(a+h)^{2k+1} = a^{2k+1}\{1 + u^2(2k+1)/2ag\}\ .$$

Find also the speed with which the particle will return to O, and show that, if u^2 and ku^2 are both small, compared with ag, this speed will be approximately $u(1-ku^2/2ag)$. (L.U.)

8. A particle of mass m oscillates in a line with natural period $2\pi/n$. If an applied periodic force $F\cos pt$ now acts in the line so that the particle is instantaneously at rest at zero time at a distance d from the centre of oscillation, prove that the displacement of the particle at a subsequent time t is $d\cos nt + F(\cos pt - \cos nt)/(n^2-p^2)m$.

9. In the absence of external forces, a particle of mass m would execute a rectilinear S.H.M. When a disturbing force $F\cos pt$ acts, the maximum speed

attained is V. Show that the angular frequency of the free oscillations is $[(F+mpV)p/mV]$.

10. Define the impulse of a force over a finite time interval, and derive the equation of impulsive motion of a particle.

A cosmic-ray particle A of charge e_1 passes at a distance b from a particle B of charge e_2 and mass m initially at rest. A repulsive central force of magnitude e_1e_2/ρ^2 acts between the particles when their separation is ρ. The momentum of the particle A is so great that its path may be treated as rectilinear, its speed v as effectively constant, and the encounter so rapid that the displacement of B is negligible. Show that the impulse of the force acting on B is

$$\frac{2\,e_1e_2b}{v}\int_b^\infty \frac{d\rho}{\rho^2\sqrt{(\rho^2-b^2)}}$$

in the direction perpendicular to the path of A. Evaluate the integral and find the kinetic energy acquired by B during the encounter.

Calculate the maximum displacement of B during the subsequent motion, assuming that B is attracted towards its initial position O by a force equal to mk^2 times the distance from O, where k is a constant, and that its motion is resisted by a force equal to $2mk$ times the velocity. (L.U.)

11. A particle of unit mass moves in a straight line and is attracted to an origin O in the line by a force n^2x, where x is the displacement from O. There is also a damping force $2ku$, where u is the speed. Find the condition that the motion should be periodic, and show that if this condition is satisfied, the successive maximum displacements from the origin form a geometric progression with common ratio $e^{-k\pi/\lambda}$, where $\lambda=\sqrt{(n^2-k^2)}$.

If the particle is also acted on by a force $F\cos pt$ and $n^2>2k^2$, find for what value of p the amplitude A of the forced oscillation has a maximum value. Denoting these particular values of p and A by p_o and A_o, show that

$$4(A_o^2/A^2-1)=(p^2/p_o^2-1)^2(\lambda/k-k/\lambda)^2\ ,$$

where A is the amplitude of the forced oscillation for any value of p. (L.U.)

12. Prove that, if a particle moves in a straight line under the action of a force directed towards a fixed point P of that line and of magnitude proportional to the distance from the point P, it will oscillate with a period independent of the amplitude of the oscillations.

A particle of mass m is at rest on a smooth horizontal plane and is connected by three elastic strings, each of modulus λ and natural length a, to three points on the plane at the corners of an equilateral triangle of side $2\sqrt{3}a$. Prove that, if the particle is displaced in the direction of one of the strings and then released, it will perform oscillations of period $(4\pi/3)\sqrt{(am/\lambda)}$. (L.U.)

13. A particle of mass m is attached to the mid-point of an unstretched elastic string of natural length a and modulus mg. The string, with the mass attached, is then stretched between two points in the same vertical line, distant $2a$ apart. Find the position of equilibrium of the particle.

If the particle is now slightly displaced from its equilibrium position in *either* a vertical *or* a horizontal direction, show that in each case the ensuing motion is simple harmonic, but that the period of a horizontal oscillation is $(15/7)^{\frac{1}{2}}$ times the period of the vertical oscillation. (L.U.)

14. An elastic string of natural length a is fixed at its ends to points A, B on a smooth horizontal table distant $2a$ apart. A heavy particle is attached to the midpoint of the string and it is projected with velocity u along the table and at right angles to the string. The modulus of elasticity is equal to the weight of the particle. Show that when the particle first comes to instantaneous rest the length of the string is $2ka$, where k is the positive root of the equation

$$k^2 - k = u^2/4ag \ .$$

Show also that if $u^2/4ag$ is small, the motion is approximately simple harmonic of period $\pi\sqrt{(2a/g)}$.

15. A particle falls from rest under gravity through a stationary cloud. The mass of the particle increases by accretion from the cloud at a rate which at any time is mkv, where m is the mass, v the speed of the particle at that instant and k is a constant. Show that after the particle has fallen a distance x,

$$kv^2 = g(1 - e^{-2kx}) \ ,$$

and find the distance the particle has fallen after time t. (L.U.)

16. A particle moving along a straight line has mass m and speed v at time t. The particle ejects matter backwards at a constant rate λm_o with constant velocity u relative to the particle itself. Show that the equation of motion of the particle is

$$m\, dv/dt - \lambda m_o u = F \ ,$$

where F is the external force acting on the particle.

From a rocket which is free to move vertically upwards, matter is ejected downwards with constant relative velocity gT at a constant rate $2M/T$. Initially the rocket is at rest and has mass $2M$, half of which is available for ejection. Neglecting air resistance and variations in gravitational attraction, show that the greatest upward speed is attained when the mass of the rocket is reduced to M, and determine this speed.

Show that the rocket rises to a height

$$\tfrac{1}{2}gT^2\{1 - \log 2\}^2 \ .$$

(L.U.)

Chapter 4

Particle Dynamics

4.1 PROBLEMS IN TWO AND THREE DIMENSIONS

Most of the problems in the previous chapter were one-dimensional in kind. We now extend Newton's laws of motion to two- and three-dimensional cases for particle motion.

First suppose a particle of constant mass m is situated at the point $P(x, y, z)$ at time t and that $OP \equiv \mathbf{r} = x\mathbf{i} + y\mathbf{j} + z\mathbf{k}$, in the usual notation. Let $\mathbf{F} = X\mathbf{i} + Y\mathbf{j} + Z\mathbf{k}$ be the resultant force which acts on the particle. Then, by the second law of motion,

$$\mathbf{F} = m\ddot{\mathbf{r}}$$

or

$$(X\mathbf{i} + Y\mathbf{j} + Z\mathbf{k}) = m(\ddot{x}\mathbf{i} + \ddot{y}\mathbf{j} + \ddot{z}\mathbf{k}) .$$

Hence $X = m\ddot{x}$, $Y = m\ddot{y}$, $Z = m\ddot{z}$. Similarly if the force and acceleration be resolved in any three mutually perpendicular directions, components of force and mass-acceleration can be equated. We now treat a number of illustrative cases.

4.2 PROJECTILE MOTION UNDER GRAVITY

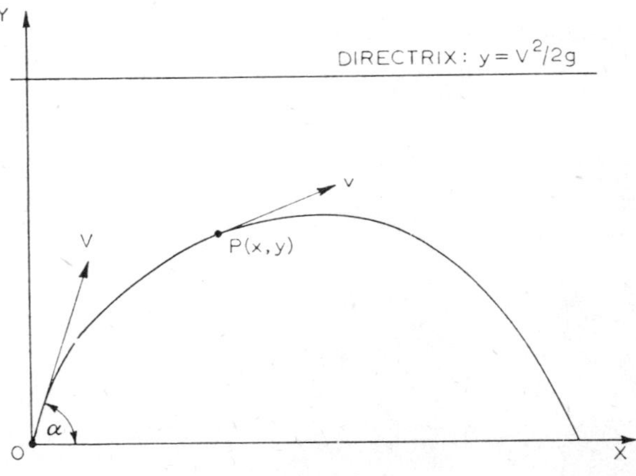

Fig. 4.1

Suppose a particle to be projected from a point O with velocity V in a direction making an angle of elevation α with the horizontal direction $\overline{OX}$ (Fig. 4.1). $\overline{OY}$ is the upward vertical through O. Let $P(x, y)$ be the position of the particle at time t after projection. Then, in the absence of air resistance, the particle is subjected throughout to uniform acceleration g in the downward vertical direction and so the equations of motion are

$$\ddot{x}=0\ , \qquad \ddot{y}=-g\ ,$$

subject to the initial conditions $\dot{x}=V\cos\alpha$, $\dot{y}=V\sin\alpha$; $x=0$, $y=0$ when $t=0$. Solving these equations gives

$$x=Vt\cos\alpha\ , \qquad y=Vt\sin\alpha-\tfrac{1}{2}gt^2\ ,$$

whence, eliminating t,

$$y=x\tan\alpha-(gx^2/2V^2\cos\alpha)\ . \tag{1}$$

The equation (1) shows that the trajectory is *parabolic*. Putting $y=0$ in (1) and taking $x\neq 0$ gives the *range* on the horizontal plane as $V^2\sin 2\alpha/g$. For a given range and a given V there are two possible values of the angle of elevation, viz. α and $(\frac{1}{2}\pi-\alpha)$. The range assumes its maximum value of V^2/g for $\alpha=\frac{1}{4}\pi$.

By completing the square of the quadratic expression in x, (1) may be written in the form

$$(x-V^2\sin\alpha\cos\alpha/g)^2=(-2V^2\cos^2\alpha/g)(y-V^2\sin^2\alpha/2g) \tag{2}$$

showing that the parabola has vertex at the point $(V^2\sin\alpha\cos\alpha/g,\ V^2\sin^2\alpha/2g)$, and latus rectum of length $2V^2\cos^2\alpha/g$. Thus the directrix is horizontal and at a height $V^2\cos^2\alpha/2g$ above the vertex, i.e. $V^2/2g$ above the line OX. Hence the equation of the directrix is $y=V^2/2g$.

At P the velocity v of the projectile is given by

$$\begin{aligned} v^2 &= \dot{x}^2+\dot{y}^2=V^2\cos^2\alpha+(V\sin\alpha-gt)^2 \\ &= V^2-2Vg\sin\alpha.\,t+g^2t^2=V^2-2gy\ , \end{aligned}$$

$$\therefore\ \tfrac{1}{2}mv^2=\tfrac{1}{2}mV^2-mgy=mgH-mgy\ ,$$

where $H=(V^2/2g)$. Thus we have shown that *at any point P of its flight in the parabolic trajectory, the kinetic energy of the particle is the same as that which would be acquired by it in falling freely under gravity to P from that point on the directrix vertically above P.*

Example 1—Shot putt analysis

A shot putter at the top A of a cliff of height h above sea level throws a shot at a speed V and angle of elevation α to the horizontal. The shot describes a parabolic trajectory under gravity, in the absence of air resistance, and enters the sea at B (Fig. 4.2). If the horizontal distance R of B from the foot of the cliff is to be a maximum, we find R, α in terms of V.

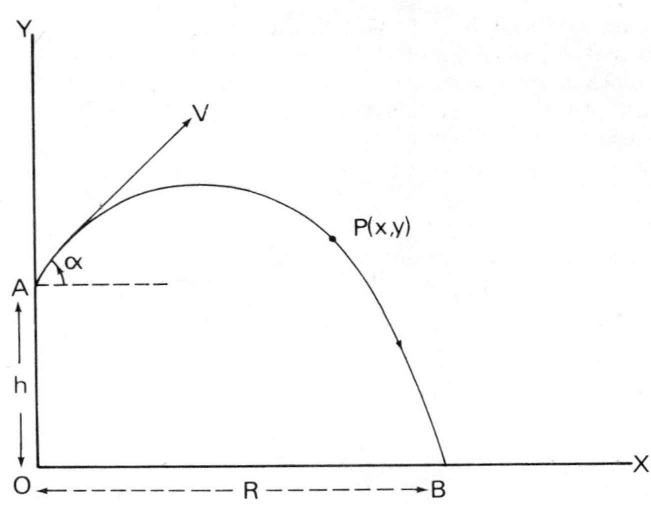

Fig. 4.2

At time t, suppose the shot has reached $P(x, y)$. Then, as above,

$$\ddot{x}=0\ , \qquad \ddot{y}=-g$$

subject to the initial conditions that, when $t=0$,

$$\dot{x}=V\cos\alpha, \qquad \dot{y}=V\sin\alpha\ , \qquad x=0\ , \qquad y=h\ .$$

Integrating the differential equations twice and using the initial conditions gives

$$\begin{aligned}\dot{x}&=V\cos\alpha\ ,\\ x&=Vt\cos\alpha\ ;\\ \dot{y}&=V\sin\alpha-gt\ ,\\ y&=h+Vt\sin\alpha-\tfrac{1}{2}gt^2\ .\end{aligned}$$

Eliminating t between the forms for x and y gives

$$y=h+x\tan\alpha-\frac{gx^2}{2V^2}\sec^2\alpha\ .$$

When $x=R$, $y=0$, giving

$$\frac{gR^2}{2V^2}\tan^2\alpha-R\tan\alpha+\left(\frac{gR^2}{2V^2}-h\right)=0\ .$$

Regarding this as a quadratic equation for $\tan\alpha$, the condition for its roots to be real is

$$R^2 \geqslant \frac{gR^2}{V^2}\left(\frac{gR^2}{V^2} - 2h\right)$$

or

$$R \leqslant \frac{V^2}{g}\left(1 + \frac{2hg}{V^2}\right)^{1/2} .$$

This last form shows that

$$\underline{R_{\max} = \frac{V^2}{g}\left(1 + \frac{2hg}{V^2}\right)^{1/2} .}$$

Writing $R = R_{\max}$ in the quadratic equation for $\tan \alpha$ we obtain for α the critical value α^* where

$$\underline{\tan \alpha^* = \frac{V^2}{gR_{\max}} = \left(1 + \frac{2hg}{V^2}\right)^{-1/2} ,}$$

which shows that $\tan \alpha^* < 1$ or $\alpha^* < \pi/4$, provided $h > 0$.

The special case of $h = 0$ gives

$$R_{\max} = V^2/g \; ; \qquad \alpha^* = \pi/4 .$$

It is instructive to see how the value of R as given by the quadratic equation

$$\frac{gR^2}{2V^2} \sec^2 \alpha - R \tan \alpha - h = 0$$

varies with R and α for a particular case. The appropriate solution is

$$\underline{R = \frac{V^2 \sin 2\alpha}{2g}\left[1 + \left(1 + \frac{2gh}{V^2 \sin^2 \alpha}\right)^{1/2}\right] .}$$

If we take the special values

$$h = 1.5\,\text{m} , \qquad V = 10\,\text{m}\,\text{s}^{-1} , \qquad g = 9.81\,\text{m}\,\text{s}^{-2} ,$$

then the critical value of α for maximum R is

$$\alpha^* = \tan^{-1}\left(1 + \frac{2hg}{V^2}\right)^{-1/2} \simeq 41°21' .$$

Also, in this case, $2gh < V^2 \sin^2 \alpha$ and so

$$R = \frac{V^2 \sin 2\alpha}{2g}\left[2 + \frac{gh}{V^2 \sin^2 \alpha} + 0\left(\frac{g^2h^2}{V^4 \sin^4 \alpha}\right)\right]$$

giving the approximation

$$\underline{R \simeq \frac{V^2}{g} \sin 2\alpha + h \cot \alpha .}$$

Clearly $(V^2 \sin 2\alpha)/g$ is the dominant term in this expression and so R will be much more sensitive to changes in V than to those in α.

Example 2

Obtain expressions for the tangential and normal resolutes of acceleration of a point moving along a plane path.

Show that in the free motion of an unresisted particle P, projected under gravity, its speed at any instant is the same as that due to a fall from a height SP, where S is the focus of the parabolic path.

A particle of mass m moves under gravity in a resisting medium, and S is the focus of the parabola which the particle would proceed to describe if at any instant the resistance then ceased. Show that S moves with a velocity which at any instant is along SP and of magnitude (Rv/mg), where v is the speed of P, and R the resistance. (L.U.)

(i) It is shown in Chapter 2 that a particle moving with velocity v along a plane curve has acceleration components $\dot{v}$ tangentially in the direction of arc length increasing and v^2/ρ normally inwards towards the centre of curvature.

(ii) Since by the focus-directrix property SP is equal to the depth of P below the directrix, the result follows from the foregoing analysis.

(iii)

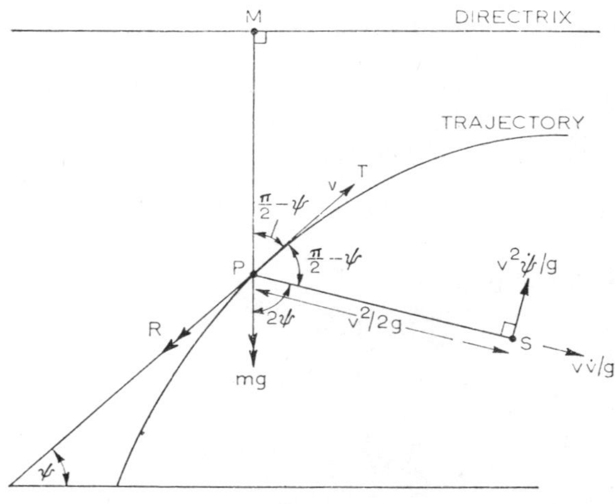

Fig. 4.3

In Fig. 4.3, S is the focus of the parabolic trajectory which would be described were $R=0$. PM is the perpendicular from P on to the directrix of the parabola. Then if v be the speed of the particle at P, $PS=(v^2/2g)$.

Let the tangent to the actual trajectory (for which $R \neq 0$) make angle ψ with the horizontal at P. Since PT instantaneously coincides with the tangent to the parabola at P, it bisects $\angle MPS$ and so $\angle MPS=\pi-2\psi$ and PS makes angle 2ψ with the downward vertical. It follows that the velocity components of S *relative* to P along $\overline{PS}$ and in the direction perpendicular to PS and in the sense of 2ψ increasing are respectively

$$d(v^2/2g)/dt = v\dot{v}/g \ ;$$
$$(v^2/2g) \times (2\dot{\psi}) = v^2\dot{\psi}/g \ .$$

Since P has tangential velocity v, the actual velocity components in the direction PS and at right angles to it in the sense of 2ψ increasing are V_1, V_2, where

$$\begin{cases} V_1 = (v\dot{v}/g) + v \sin \psi \ , \\ V_2 = (v^2\dot{\psi}/g) + v \cos \psi \ . \end{cases}$$

For the motion of the particle in the tangential and normal directions we have the equations

$$\begin{cases} -R - mg \sin \psi = m\dot{v} \ , \\ mg \cos \psi \qquad = mv^2/\rho \ . \end{cases}$$

From the last four equations one readily obtains

$$V_1 = -(Rv/mg) \ ,$$
$$V_2 = (v^2\dot{\psi}/g) + (v^3/\rho g) \ .$$

Since the curve is convex upwards, ψ decreases as s increases.

$$\therefore \ \rho = -ds/d\psi = -(ds/dt) \div (d\psi/dt) = -v/\dot{\psi} \ .$$
$$\therefore \ V_2 = (v^2\dot{\psi}/g) + (-v^2\dot{\psi}/g) = 0 \ .$$

Thus the motion of S is along $\overline{SP}$ and the velocity is $-Rv/mg$.

The equation (1) of this section may be written in the alternative form

$$(gx^2/2V^2)\tan^2\alpha - x \tan \alpha + [(gx^2/2V^2) + y] = 0 \ . \qquad (1')$$

Now suppose V be fixed: then as α varies, a family of trajectories is generated in the (x, y)-plane, each passing through $(0, 0)$. Since $(1')$ is a quadratic equation in $\tan \alpha$, the envelope of this family is obtained by equating the discriminant of the quadratic equation to zero giving

$$x^2 - (gx^2/V^2)[(gx^2/V^2) + 2y] = 0$$

whence, if $x \neq 0$,

$$x^2 = (2V^2/g)[(V^2/2g) - y] \ . \qquad (3)$$

(3) is a parabola and is known as the *parabola of safety*. When this parabola is rotated about the y-axis to form a paraboloid of revolution, it determines the entire region of space accessible from O for a given velocity of projection V.

(3) shows the vertex of the parabola of safety is $(0, V^2/2g)$. The reader will easily confirm that $(0, 0)$ is the focus and that the directrix is at height V^2/g above the point of projection O. Every trajectory, for given V, must touch the parabola of safety at some point T. The connection between the two curves is shown in Fig. 4.4.

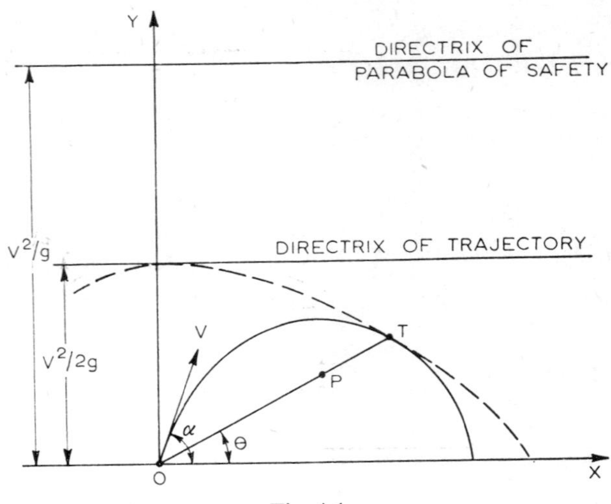

Fig. 4.4

We now prove that if P be any point on the line OT, the greatest value of OP for a given projection velocity V is attained when $P \equiv T$, T being the intersection of the line with the parabola of safety.

The focus S of the trajectory lies on a circle Σ_1 having centre O and radius $V^2/2g$, since $V^2/2g$ is the distance of O from the directrix. Since P is distant $(V^2/2g - y_P)$ from the directrix, S is on the circle Σ_2 centre P and radius $V^2/2g - y_P$. The circles Σ_1, Σ_2 may intersect in a pair of distinct real points, as shown in Fig. 4.5, a pair of imaginary points, or in a coincident pair. When there is no real intersection, P cannot be reached from O with the given projection velocity V. When there are two intersections, P may be attained via two parabolic paths. When the circles touch, as in Fig. 4.6, the distance OP is a maximum.

In Fig. 4.6, $OP = (V^2/2g) + (V^2/2g - y_P) = (V^2/g) - y_P$. This is the distance of P from the directrix of the parabola of safety. But O is the focus of the parabola of safety. Hence in Fig. 4.6, P lies on the parabola of safety and so the range OP is a maximum when $P \equiv T$.

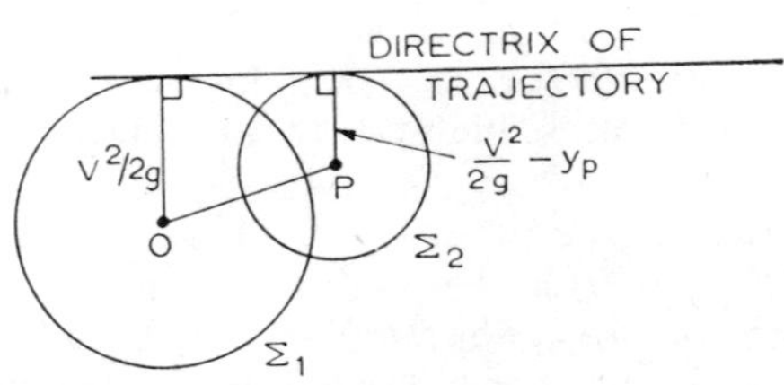

Fig. 4.5

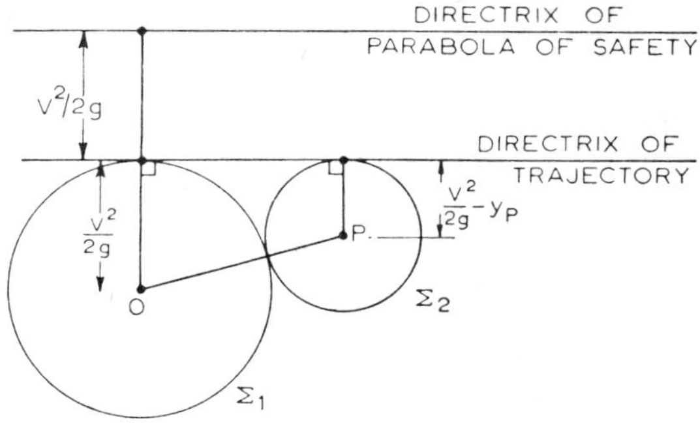

Fig. 4.6

The distance of T from the directrix of the parabola of safety is equal to OT. Hence

$$V^2/g - OT \sin\theta = OT$$
$$\therefore\ OT = V^2/\{g(1+\sin\theta)\}\ .$$

This gives the maximum range on a plane inclined at a fixed angle θ to the horizontal when the projection velocity V is prescribed. For an alternative treatment of this problem, see Example 1, Exercise 4.

Example 3—Projectile motion in a resisting medium.

A particle of unit mass is projected with velocity V and inclination α in a medium whose resistance is $k\times$ (velocity). Show that if k is small the equation of the path is approximately

$$y = x\tan\alpha - (gx^2/2V^2\cos^2\alpha) - (kgx^3/3V^3\cos^3\alpha)\ .$$

The particle is projected in the same medium from a point on a plane inclined at angle β to the horizontal. Prove that the range is approximately

$$\{2V^2\sin(\alpha-\beta)\cos\alpha/g\cos^2\beta\}\{1-[4kV\sin(\alpha-\beta)/3g\cos\beta]\}\ ,$$

where V is the velocity of projection and α the elevation. (L.U.)

(i) Equation of trajectory

Fig. 4.7 shows the particle at $P(x, y)$ at time t, the forces on it being g downwards and kv tangentially in the direction opposed to the velocity v. The equations of motion in the horizontal and vertical directions are:

$$\ddot{x} = -kv\cos\psi = -k\dot{x}\ , \qquad (1)$$
$$\ddot{y} = -g - kv\sin\psi = -g - k\dot{y}\ , \qquad (2)$$

ψ being the inclination of the tangent at P to the horizontal $\overline{OX}$.

Initially, $\dot{x} = V\cos\alpha$, $x = 0$, when $t = 0$ and so (1) gives

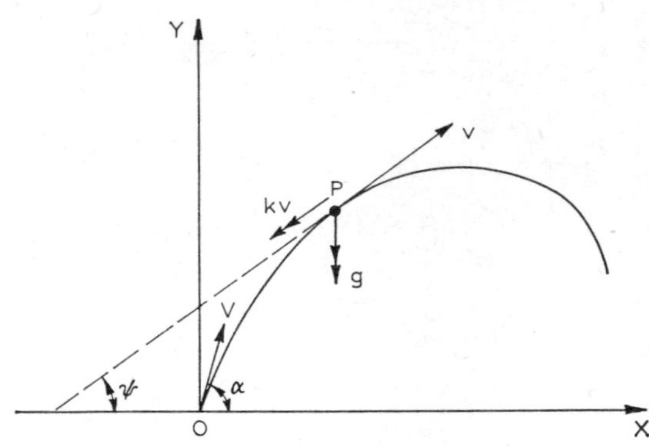

Fig. 4.7

$$(D+k)\dot{x}=0 \qquad (D \equiv d/dt) \ ,$$
$$\therefore \ \dot{x}=A_1e^{-kt}=V\cos\alpha\, e^{-kt} \ ,$$
$$\therefore \ x=A_2-(V/k)\cos\alpha\, e^{-kt} \ ,$$

i.e.
$$\underline{x=(V/k)\cos\alpha(1-e^{-kt})} \ . \tag{3}$$

Also, when $t=0$, $\dot{y}=V\sin\alpha$, $y=0$. Thus, from (2),

$$(D+k)\dot{y}=-g \ ,$$
$$\therefore \ \dot{y}=A_3e^{-kt}-g/k=(V\sin\alpha+g/k)e^{-kt}-g/k \ .$$
$$\therefore \ y=-(gt/k)-\frac{1}{k}(V\sin\alpha+g/k)e^{-kt}+A_4 \ ,$$

i.e.
$$\underline{y=(1/k)(V\sin\alpha+g/k)(1-e^{-kt})-gt/k} \ . \tag{4}$$

From (3), $-t=(1/k)\log[1-(kx/V\cos\alpha)]$ and so (4) becomes

$$\begin{aligned} y &= x\tan\alpha+(gx/kV\cos\alpha)+(g/k^2)\log[1-(kx/V\cos\alpha)] \\ &= x\tan\alpha+(gx/kV\cos\alpha)+(g/k^2)[-(kx/V\cos\alpha)- \\ &\qquad -(k^2x^2/2V^2\cos^2\alpha)-(k^3x^3/3V^3\cos^3\alpha)-\ldots] \end{aligned}$$

$$\therefore \ \underline{y \simeq x\tan\alpha-(gx^2/2V^2\cos^2\alpha)-(kgx^3/3V^3\cos^3\alpha)} \ . \tag{5}$$

(ii) Range on inclined plane

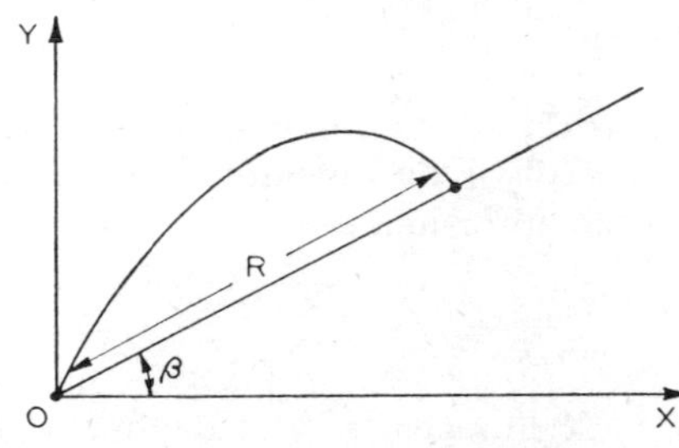

Fig. 4.8

Let R be the range on the plane inclined at β to $\overline{OX}$ (Fig. 4.8). Put $x = R\cos\beta$, $y = R\sin\beta$ in (5) to give

$$R\sin\beta = R\cos\beta\tan\alpha - (gR^2\cos^2\beta/2V^2\cos^2\alpha) - (kgR^3\cos^3\beta/3V^3\cos^3\alpha)\ .$$

$$\therefore\ (gR\cos^2\beta/2V^2\cos^2\alpha)[1 + (2kR\cos\beta/3V\cos\alpha)] = \sin(\alpha-\beta)/\cos\alpha\ .$$

$$\therefore\ R = [2V^2\cos\alpha\sin(\alpha-\beta)/g\cos^2\beta][1 + (2kR\cos\beta/3V\cos\alpha)]^{-1}$$

$$\simeq [2V^2\cos\alpha\sin(\alpha-\beta)/g\cos^2\beta][1 - (2kR\cos\beta/3V\cos\alpha)]\ .$$

$$\therefore\ R[1 + (4kV\sin(\alpha-\beta)/3g\cos\beta)] \simeq (2V^2\cos\alpha\sin(\alpha-\beta)/g\cos^2\beta]$$

$$\therefore\ R \simeq [2V^2\cos\alpha\sin(\alpha-\beta)/g\cos^2\beta][1 - (4kV\sin(\alpha-\beta)/3g\cos\beta)]$$

N.B. Equation (3) shows the trajectory, when continued, approaches the vertical asymptote $x = (V/k)\cos\alpha$.

4.3 CONSTRAINED PARTICLE MOTION

We now consider the motions of a single particle constrained to move on curves or surfaces. When the particle is smooth, there is no friction between it and the constraint: there is a normal reaction which does no work during the motion of the particle. In these circumstances, if the external forces be conservative, the principle of conservation of energy may be invoked as the worked Example 4 of Section 2, Chapter 3 shows.

Since the necessary dynamical principles of constrained motion have been established, it is adequate to illustrate this class of problem with worked examples.

Example 1—Simple pendulum.

Fig. 4.9 shows a simple pendulum of length l inclined at θ to the downward vertical. The mass of the bob P is m.

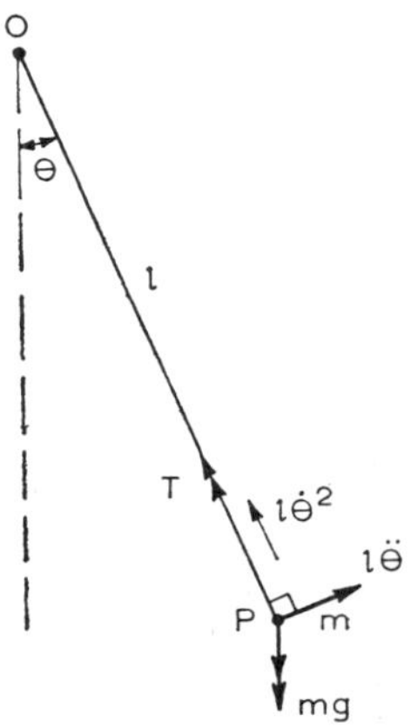

Fig. 4.9

The acceleration components of P are $l\dot{\theta}^2$ along $\overline{PO}$ and $l\ddot{\theta}$ in the transverse direction in the sense of θ increasing. The equations of motion for m in these two directions are

$$T - mg\cos\theta = ml\dot{\theta}^2 \; , \tag{1}$$

$$-mg\sin\theta = ml\ddot{\theta} \; . \tag{2}$$

In the case when θ is small, (2) gives to first order,

$$\ddot{\theta} \simeq -(g/l)\theta \; ,$$

which is seen to represent a S.H.M. of period $2\pi(l/g)^{\frac{1}{2}}$.

Example 2

Show that the acceleration of a particle P moving along a plane curve C is $\ddot{s}\,\mathbf{t} + (\dot{s}^2/\rho)\mathbf{n}$, where s denotes arc length along C, $\mathbf{t}$, $\mathbf{n}$ are unit vectors along the tangent and normal at P respectively and ρ is the radius of curvature at P.

A horizontal smooth fixed plane wire has the shape of the catenary $y = c\,\mathrm{ch}(x/c)$ and carries a small ring which is attracted towards the directrix with a force proportional to the distance y from the directrix. Show that the motion of the ring is S.H.M. and that the normal reaction varies inversely as y^2. (The formulae $s = c\tan\psi$, $y = c\sec\psi$ may be assumed). (L.U.)

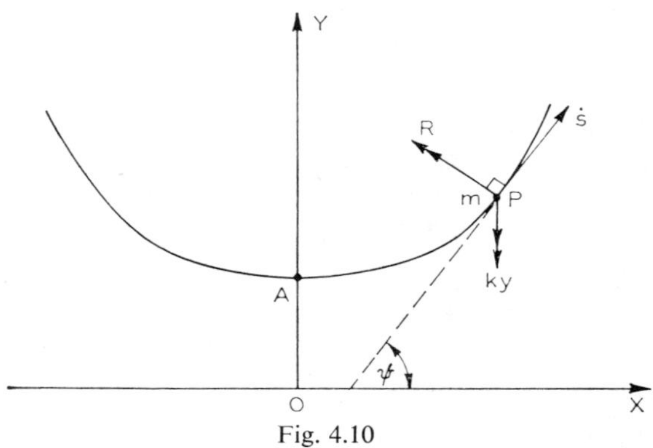

Fig. 4.10

Let s denote the arc length from the vertex A of the catenary to the ring at P (Fig. 4.10), and ky the attraction towards the directrix. R is the normal reaction of the wire on the ring at P.

Resolving tangentially,

$$-ky\sin\psi = m\ddot{s} \; . \tag{1}$$

Now $y\sin\psi = (c\sec\psi)\sin\psi = c\tan\psi = s$.

$$\therefore \; \ddot{s} = -(k/m)s \; . \tag{2}$$

Since $k>0$, this is a S.H.M. with period $2\pi(m/k)^{\frac{1}{2}}$.

Resolving normally, we have

$$R-ky\cos\psi=m\dot{s}^2/\rho\ . \tag{3}$$

Multiplying (2) through by $2\dot{s}$ and integrating gives

$$\dot{s}^2=A-(k/m)s^2\ ,$$

and so (3) gives

$$R=ky\cos\psi+(m/\rho)[A-(k/m)s^2]\ .$$

$$\therefore\ R=kc+(m/\rho)[A-(k/m)c^2\tan^2\psi]\ .$$

Now $\rho=ds/d\psi=c\sec^2\psi$.

$$\therefore\ R=kc+(mA/c)\cos^2\psi-kc\sin^2\psi$$
$$=[kc+(mA/c)]\cos^2\psi$$
$$=\underline{[kc+mA/c]c^2/y^2}\ .$$

Example 3

A fixed wire is in the shape of the cardioid $r=a(1+\cos\theta)$, the initial line being the downward vertical. A small ring of mass m can slide on the wire and is attached to the point $r=0$ of the cardioid by an elastic string of natural length a and modulus $4mg$. If the particle is released from rest when the string is horizontal show that

$$a\dot{\theta}^2(1+\cos\theta)-g\cos\theta(1-\cos\theta)=0\ .$$

Show further that $\theta=\pi/3$ is a position of stable equilibrium and that the period of small oscillations about this position is $2\pi(2a/g)^{\frac{1}{2}}$. (L.U.)

Fig. 4.11 shows the position of the ring at time t. The velocity components are then $[\dot{r},\ r\dot{\theta}]$ in the radial and transverse directions.

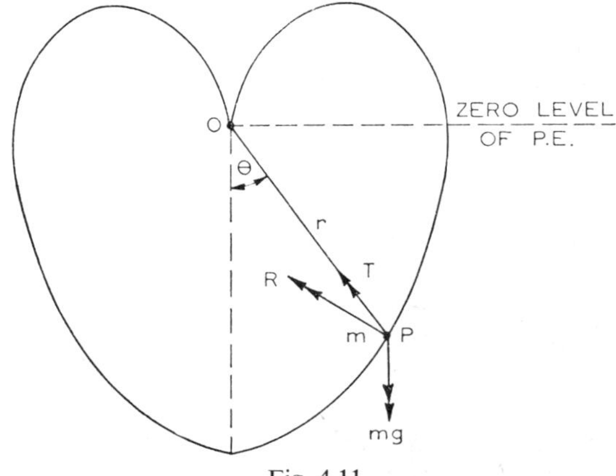

Fig. 4.11

$$\therefore\ \text{K.E. of bead} = \tfrac{1}{2}m(\dot{r}^2 + r^2\dot{\theta}^2)$$
$$= \tfrac{1}{2}ma[^2(\sin^2\theta\ \dot{\theta}^2 + (1+\cos\theta)^2\dot{\theta}^2]$$
$$= \underline{ma^2\dot{\theta}^2(1+\cos\theta)}\ .$$

The gravitational P.E. of the particle (referred to the horizontal through O as zero level) is

$$-mgr\cos\theta = -mga\cos\theta(1+\cos\theta)\ .$$

The elastic P.E. of the string is

$$\tfrac{1}{2}(4mg)(r-a)^2/a$$
$$= \underline{2mga\cos^2\theta}\ .$$

Since the forces are conservative,

$$\text{P.E.} + \text{K.E.} = \text{const.}$$

$\therefore\ mga(\cos^2\theta - \cos\theta) + ma^2\dot{\theta}^2(1+\cos\theta) = \text{const.} = 0$, since $\dot{\theta} = 0$ when $\theta = \pi/2$. Thus

$$\underline{a\dot{\theta}^2(1+\cos\theta) - g\cos\theta(1-\cos\theta) = 0}\ . \tag{1}$$

Differentiate (1) w.r.t. t. Then

$$-a\dot{\theta}^3\sin\theta + 2a\dot{\theta}\ddot{\theta}(1+\cos\theta) + g\dot{\theta}\sin\theta(1-\cos\theta) - g\dot{\theta}\cos\theta\sin\theta = 0\ .$$
$$\therefore\ 2a\ddot{\theta}(1+\cos\theta) - a\dot{\theta}^2\sin\theta + g\sin\theta(1-2\cos\theta) = 0\ . \tag{2}$$

For equilibrium, $\dot{\theta} = 0 = \ddot{\theta}$ and so $\theta = 0$ or $\pi/3$.
In (2), put $\theta = \pi/3 + \epsilon$. where ϵ is small.

$$\cos\theta = \cos(\ddot{\epsilon} + \pi/3) = \cos\epsilon\cos(\pi/3) - \sin\epsilon\sin(\pi/3) \simeq \tfrac{1}{2} - (\sqrt{3}/2)\epsilon\ .$$
$$\sin\theta = \sin(\ddot{\epsilon} + \pi/3) = \sin\epsilon\cos(\pi/3) + \cos\epsilon\sin(\pi/3) \simeq \tfrac{1}{2}\epsilon + (\sqrt{3}/2)\ .$$

Retaining terms of first order small quantity in ϵ, $\dot{\epsilon}$, $\ddot{\epsilon}$, we have from (2),

$$2a\ddot{\epsilon}[(3/2) - (\sqrt{3}/2)\epsilon] + g[\tfrac{1}{2}\epsilon + (\sqrt{3}/2)][\sqrt{3}\epsilon] = 0\ ,$$

i.e.
$$\ddot{\epsilon} + (g/2a)\epsilon = 0\ . \tag{3}$$

Equation (3) shows that for small perturbations about $\theta = \pi/3$, the motion is approximately S.H.M. Thus $\theta = \pi/3$ is a stable equilibrium position, the period of small oscillations being $2\pi(2a/g)^{\frac{1}{2}}$.

Example 4—Illustrating moving axes and vectorial methods.

A small bead slides on a smooth wire having the form of a plane curve. The wire is constrained to rotate about a vertical axis, which lies in its own plane, with constant angular velocity ω. The bead has a speed v along the wire when its distance from the axis is r and its height above a fixed horizontal plane is z. Show that $v^2 - r^2\omega^2 + 2gz = \text{const.}$

The wire has the form of a parabola of latus rectum $4a$ with its axis vertical and vertex downwards. The parabola rotates about its axis with constant angular velocity $(g/a)^{\frac{1}{2}}$. The bead is projected from the end of the latus rectum with a speed $2(ag)^{\frac{1}{2}}$ towards the vertex. Show that it will reach the vertex with speed $(2ag)^{\frac{1}{2}}$ after a time $(2a/g)^{\frac{1}{2}}$. (L.U.)

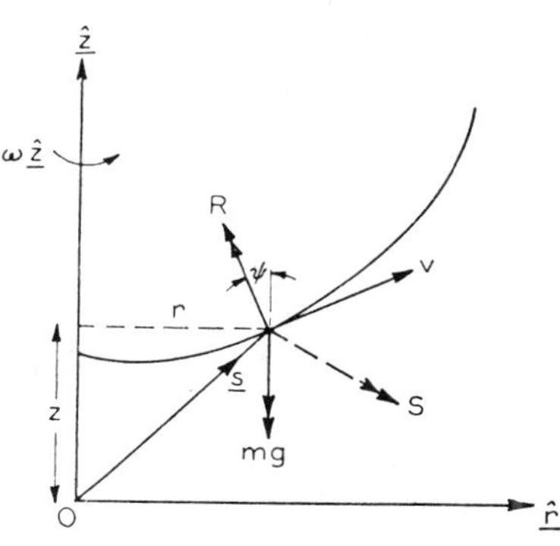

Fig. 4.12

Fig. 4.12 shows the position P of the bead at time t. Denote $\overline{OP} \equiv \mathbf{s}$, and work with cylindrical polar co-ordinates (r, θ, z), so that $\hat{\mathbf{r}}_\wedge\hat{\boldsymbol{\theta}} = \hat{\mathbf{z}}$, etc. Then

$$\mathbf{s} = r\hat{\mathbf{r}} + z\hat{\mathbf{z}} \ .$$
$$\therefore \ \mathbf{v} = d\mathbf{s}/dt = \partial\mathbf{s}/\partial t + \omega\hat{\mathbf{z}}_\wedge\mathbf{s} \ ,$$

using the results established in Chapter 2.

$$\begin{aligned} \therefore \ \mathbf{v} &= (\dot{r}\hat{\mathbf{r}} + \dot{z}\hat{\mathbf{z}}) + \omega\hat{\mathbf{z}}_\wedge(r\hat{\mathbf{r}} + z\hat{\mathbf{z}}) \\ &= \dot{r}\hat{\mathbf{r}} + \omega r\,\hat{\boldsymbol{\theta}} + \dot{z}\,\hat{\mathbf{z}} \ . \\ \therefore \ d\mathbf{v}/dt &= \partial\mathbf{v}/\partial t + \omega\hat{\mathbf{z}}_\wedge\mathbf{v} \\ &= \ddot{r}\hat{\mathbf{r}} + (\dot{\omega}r + \omega\dot{r})\hat{\boldsymbol{\theta}} + \ddot{z}\hat{\mathbf{z}} + \omega(\dot{r}\hat{\boldsymbol{\theta}} - \omega r\,\hat{\mathbf{r}}) \\ &= \underline{(\ddot{r} - \omega^2 r)\hat{\mathbf{r}} + (2\omega\dot{r} + \dot{\omega}r)\hat{\boldsymbol{\theta}} + \ddot{z}\,\hat{\mathbf{z}}} \ . \end{aligned}$$

The total force acting on the particle, taking R to be the normal reaction on it in the plane of the curve and ψ the inclination of the tangent to the horizontal at P, is

$$(R\cos\psi - mg)\hat{\mathbf{z}} - R\sin\psi\,\hat{\mathbf{r}} + S\,\hat{\boldsymbol{\theta}} \ ,$$

where S is the component of reaction normal to the plane of the curve. Thus the equation of motion of the particle is

$$(R\cos\psi - mg)\hat{\mathbf{z}} - R\sin\psi\,\hat{\mathbf{r}} + S\hat{\boldsymbol{\theta}} = m d\mathbf{v}/dt \ .$$

Equating coefficients of $\hat{\mathbf{r}}$, $\hat{\boldsymbol{\theta}}$, $\hat{\mathbf{z}}$, we have

$$m(\ddot{r} - \omega^2 r) = -R\sin\psi \ , \tag{1}$$
$$m(2\omega\dot{r} + \dot{\omega}r) = S \ , \tag{2}$$
$$m\,\ddot{z} = R\cos\psi - mg \ . \tag{3}$$

From (1), (3), we have

$$(\ddot{r}-\omega^2 r)\cos\psi+\ddot{z}\sin\psi+g\sin\psi=0\ ,$$

i.e. $$\ddot{r}-\omega^2 r+(\ddot{z}+g)\tan\psi=0\ .$$

Now $\tan\psi=\dot{z}/\dot{r}$. Hence

$$\dot{r}\ddot{r}-\omega^2 r\dot{r}+(\dot{z}\ddot{z}+g\dot{z})=0\ .$$

Integrating gives

$$\dot{r}^2-\omega^2 r^2+\dot{z}^2+2gz=\text{const., for constant } \omega\ .$$

i.e. $$\underline{v^2-r^2\omega^2+2gz=\text{const.}} \qquad (4)$$

Taking $\omega=(g/a)^{\frac{1}{2}}$, (4) gives

$$v^2-(g/a)r^2+2gz=A=\text{const.}$$

When $r=2a$, $z=a$, $v=2(ag)^{\frac{1}{2}}$, if $r^2=4az$ on the curve.

$$\therefore\ A=4ag-4ag+2ga=2ga\ .$$

$$\therefore\ v^2-(g/a)r^2+2gz=2ga\ .$$

When $r=0$, $z=0$, $v=v_o$, say.

$$\therefore\ \underline{v_o=(2ga)^{\frac{1}{2}}\ .} \qquad (5)$$

The equation of the curve is $r^2=4az$ and so

$$v^2=2g(a+z)\ .$$

Now $$r=2a^{\frac{1}{2}}z^{\frac{1}{2}},\ dr/dz=a^{\frac{1}{2}}z^{-\frac{1}{2}},\quad v^2=\dot{r}^2+\dot{z}^2\ .$$

$$\therefore\ v^2=(1+a/z)\dot{z}^2\ .$$

$$\therefore\ (1+a/z)\dot{z}^2=2g(a+z)\ .$$

$$\therefore\ \dot{z}=(2g)^{\frac{1}{2}}z^{\frac{1}{2}}\ ;$$

$$\therefore\ dt=(2g)^{-\frac{1}{2}}z^{-\frac{1}{2}}dz\ .$$

$$\therefore\ \text{Required time}=(2g)^{-\frac{1}{2}}\int_o^a z^{-\frac{1}{2}}dz=\underline{(2a/g)^{\frac{1}{2}}}\ .$$

[*N.B.* In this question, $v=(\dot{r}^2+\dot{z}^2)^{\frac{1}{2}}\neq|\mathbf{v}|$.]

4.4 ANGULAR MOMENTUM OF A PARTICLE

Fig. 4.13 shows the position of a particle P of mass m moving with velocity $\mathbf{v}$ at time t. For a fixed origin O, we take $\overline{OP}\equiv\mathbf{r}$ so that $\mathbf{v}=d\mathbf{r}/dt$. Then the (vector) *angular momentum* or *moment of momentum* of the particle about O is defined to be

$$\mathbf{H}=\mathbf{r}\wedge(m\mathbf{v})\ .$$

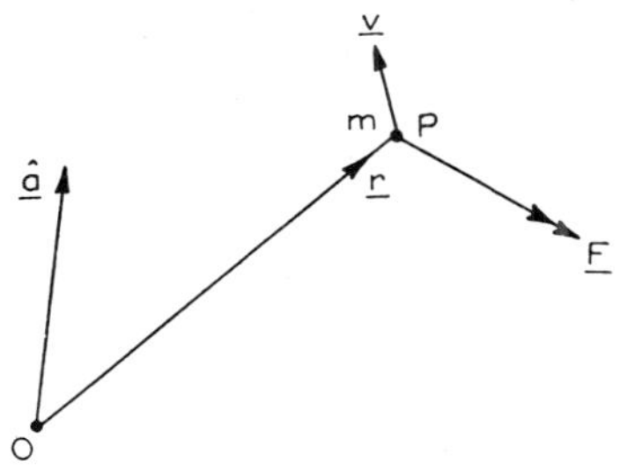

Fig. 4.13

Now let m be subjected to a force $\mathbf{F}$. Then, for m constant,

$$d\mathbf{H}/dt = \mathbf{r}_\wedge(md\mathbf{v}/dt) + \mathbf{v}_\wedge(m\mathbf{v})$$
$$= \mathbf{r}_\wedge\mathbf{F}, \text{ since } \mathbf{F} = md\mathbf{v}/dt \ .$$

Thus the rate of change of angular momentum of the particle is equal to the moment of the resultant force which acts on it.

It further follows that if $\mathbf{F} = \mathbf{0}$, then $\mathbf{H}$ is constant. *Thus the vector angular momentum about a fixed point of a particle moving under no forces is constant.* (Conservation of angular momentum).

If now $\hat{\mathbf{a}}$ specifies the unit vector in some axis drawn through O, we define the (physical) angular momentum about this axis to be the resolute of $\mathbf{H}$ about it, i.e. $\hat{\mathbf{a}}.\mathbf{H}$. Hence the rate of increase of angular momentum about the axis $\hat{\mathbf{a}}$ is

$$d(\hat{\mathbf{a}}.\mathbf{H})/dt = \hat{\mathbf{a}}.(d\mathbf{H}/dt) = \hat{\mathbf{a}}.(\mathbf{r}_\wedge\mathbf{F}) \ ,$$

i.e. *the rate of increase of angular momentum about the axis is equal to the moment of the resultant force which acts on the particle.* It further follows that *if the moment of the resultant force about the axis* $\hat{\mathbf{a}}$ *is zero, then the angular momentum* $\hat{\mathbf{a}}.\mathbf{H}$ *of the particle about the axis is constant.*

An important class of problems involving this last criterion is met in the case of *central forces* which will be dealt with more fully in the next chapter. Let a particle P of mass m be confined to move in a plane so that its polar co-ordinates at time t are (r, θ) and let the resultant force $\mathbf{F}$ acting on m be directed along $\overline{PO}$ (Fig. 4.14). Such a force $\mathbf{F}$ is termed a *central force* and the locus described by m under it is termed an *orbit.*

The radial and transverse components of velocity of P are $[\dot{r}, r\dot{\theta}]$. Hence the angular momentum about the normal to the plane through O is

$$r \times (mr\dot{\theta}) = mr^2\dot{\theta} \ .$$

Since $\mathbf{F}$ has no moment about this axis, the latter is constant, or

$$r^2\dot{\theta} = h = \text{constant} \ .$$

The reader can easily show that if $\mathbf{v}$ be the resultant velocity of the particle and p the length of the perpendicular from O on the direction of $\mathbf{v}$, then

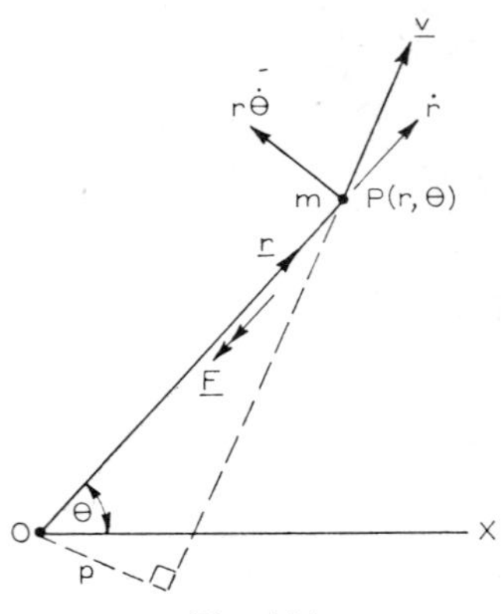

Fig. 4.14

$$h = pv = r^2\dot{\theta} \ .$$

h is known as the *angular momentum constant.*

(*N.B.* The constancy of h can also be inferred from the fact that the transverse acceleration component $(1/r)d(r^2\dot{\theta})/dt = 0$.)

Example 1

P, Q are two particles, each of mass m, connected by a light inextensible string of length $2l$, which passes through a small smooth hole O in a smooth horizontal table. P is free to slide on the table, Q hangs freely. Initially OQ is of length l, and P is projected from rest, at right angles to OP with velocity $(8gl/3)^{\frac{1}{2}}$. Show that in the ensuing motion Q will just reach the hole.

(L.U.)

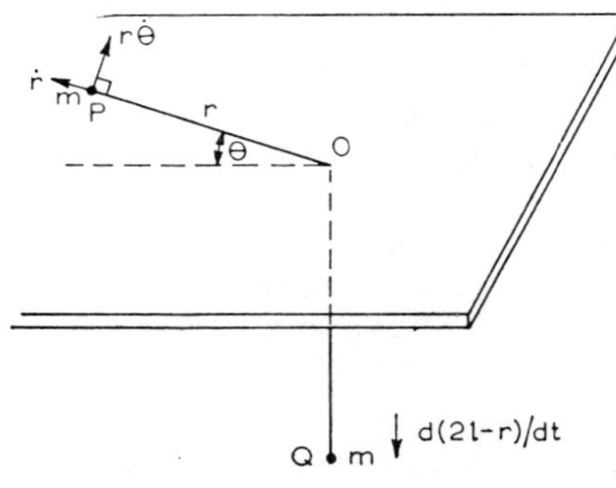

Fig. 4.15

Let (r, θ) be the polar co-ordinates of P referred to some initial line through O in the table. Then $2l - r$ is the distance of Q below the table (Fig. 4.15). P has radial and transverse velocity components $[\dot{r}, r\dot{\theta}]$ and Q has downward velocity $d(2l-r)/dt = -\dot{r}$. Hence the K.E. of the system is

$$\tfrac{1}{2}m(\dot{r}^2 + r^2\dot{\theta}^2) + \tfrac{1}{2}m(-\dot{r})^2 = m(\dot{r}^2 + \tfrac{1}{2}r^2\dot{\theta}^2) \ .$$

Referred to the table as zero level, the P.E. is $-mg(2l-r)$.

$$\therefore\ m(\dot{r}^2+\tfrac{1}{2}r^2\dot{\theta}^2)-mg(2l-r)=\text{const.}$$

i.e.
$$\dot{r}^2+\tfrac{1}{2}r^2\dot{\theta}^2+gr=\text{const.}$$

Initially when $t=0$, $\dot{r}=0$, $r=l$, $r\dot{\theta}=(8gl/3)^{\frac{1}{2}}$ and so

$$\dot{r}^2+\tfrac{1}{2}r^2\dot{\theta}^2+gr=7gl/3\ . \tag{1}$$

Further,

$$r^2\dot{\theta}=h=(8gl^3/3)^{\frac{1}{2}}\ . \tag{2}$$

Eliminating $\dot{\theta}$ from (1) and (2) gives

$$\dot{r}^2=(7gl/3)-(4gl^3/3r^2)-gr\ . \tag{3}$$

From (3), $\dot{r}=0$ when

$$3r^3-7lr^2+4l^3=0\ ,$$

i.e.
$$(r-l)(3r+2l)(r-2l)=0\ .$$

Thus the physically realizable values of r which make the system come to instantaneous rest are $r=l$, $r=2l$. When $r=2l$, the lower particle just reaches the hole.

[*N.B.* For this example, the principle of conservation of energy has been invoked for a two-particle system. This is justifiable since the tension in the string, an internal force, does not work. We shall extend the principle generally to systems of particles in Chapter 6.]

Example 2—Motion of a particle on a smooth surface of revolution.

A particle of mass m moves on the smooth inner surface of the paraboloid of revolution $x^2+y^2=4az$, whose axis is vertical and vertex downwards. Find the angular momentum of the particle about OZ if it describes a horizontal circle of radius $2a$ with constant speed. While the particle is describing this circle it receives an impulse $m(ag)^{\frac{1}{2}}$ along the surface of the paraboloid in a vertical plane through the axis. Show that in the subsequent motion the path of the particle lies between two horizontal planes.

(L.U.)

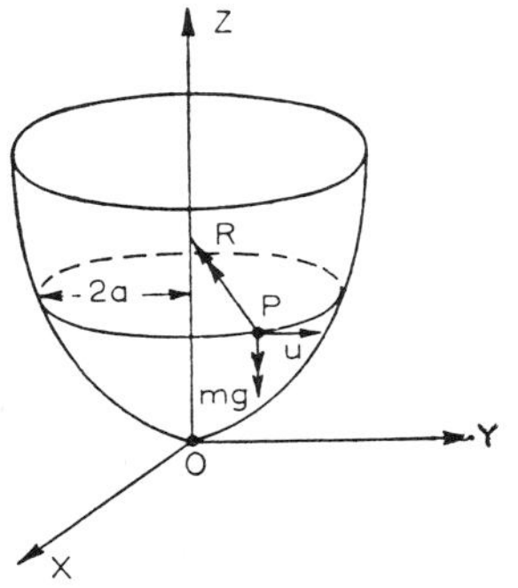

Fig. 4.16(i)

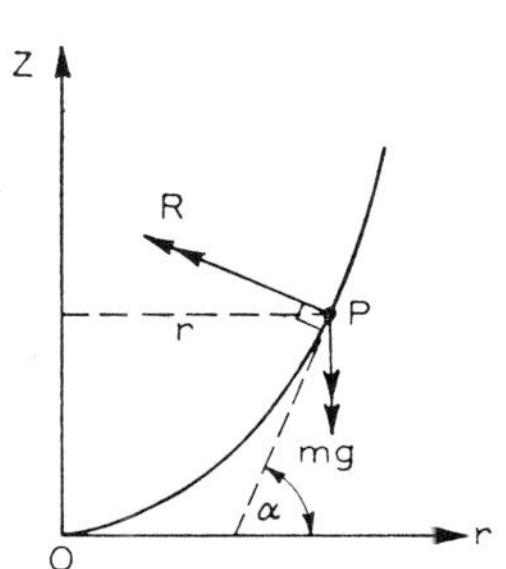

Fig. 4.16(ii)

Fig. 4.16(i) shows the particle moving in the circle $r=2a$ on the interior of the paraboloid of revolution. The forces acting on the particle are the weight mg downwards and the normal surface reaction R which intersects the line OZ. Hence angular momentum is conserved about OZ throughout the motion and its value H is

$$2amu=H=\text{const.} \tag{1}$$

The meridian section of the paraboloid through P is shown in Fig. 4.16(ii) and it has the equation $r^2=4az$, r being the distance of P from the axis OZ. Let α be the angle the tangent to the meridian curve makes with the horizontal. Then

$$\tan\alpha=dz/dr=(r/2a)\ .$$

When $z=a$, $r=2a$ and so $\alpha=\pi/4$.

Resolving along the direction OZ for the case $\alpha=\pi/4$,

$$R/\sqrt{2}=mg\ . \tag{2}$$

Resolving along the radius of the circle $r=2a$,

$$R/\sqrt{2}=m(u^2/2a)\ . \tag{3}$$

From (2), (3), $u^2=2ag$ and hence the A.M. about OZ is

$$H=2am(2ag)^{\frac{1}{2}}\ . \tag{4}$$

In general motion on the surface, the cylindrical polar velocity components of P are $[\dot{r},\ r\dot{\theta},\ \dot{z}]$ and so by conservation of energy

$$\tfrac{1}{2}m(\dot{r}^2+r^2\dot{\theta}^2+\dot{z}^2)+mgz=\text{const.}$$

or

$$\dot{r}^2+r^2\dot{\theta}^2+\dot{z}^2+2gz=\text{const.} \tag{5}$$

Now $H=2am(2ag)^{\frac{1}{2}}=mr^2\dot{\theta}$ and so

$$\dot{\theta}^2=8a^3g/r^4\ .$$

$\therefore$ (5) becomes

$$\dot{r}^2+8a^3g/r^2+\dot{z}^2+2gz=\text{const.}$$

Put $v^2=\dot{r}^2+\dot{z}^2$, so that v is the velocity component along the meridian curve. Then

$$v^2+2a^2g/z+2gz=\text{const.}$$

Initially, $v=(ag)^{\frac{1}{2}}$, since $m(ag)^{\frac{1}{2}}$ is the applied impulse, and $z=a$. Hence

$$v^2+2a^2g/z+2gz=5ag\ .$$

Then $v=0$ when

$$2a^2+2z^2-5az=0\ ,$$

i.e.

$$z=a/2 \text{ or } 2a\ .$$

4.5 THE CYCLOID AND ITS DYNAMICAL PROPERTIES

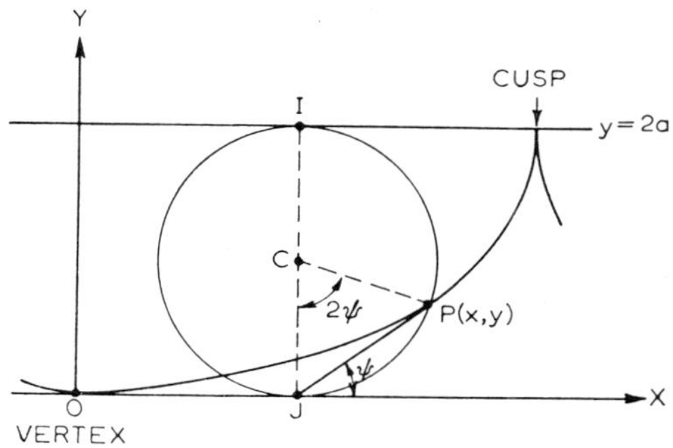

Fig. 4.17

Fig. 4.17 shows a circle of radius a rolling without slipping on the line $y=2a$. P is a point fixed on the rim of the circle and initially in contact with the origin of co-ordinates O. I is the instantaneous centre of rotation, C the centre of the circle. IC meets the x-axis in J. Let $\angle PJX=\psi$: then $\angle PCJ=2\psi$. Hence the co-ordinates (x, y) of P are given by

$$\left.\begin{aligned} x &= OJ+CP\sin 2\psi = a(2\psi+\sin 2\psi), \\ y &= JC-CP\cos 2\psi = a(1-\cos 2\psi). \end{aligned}\right\}$$

The locus traced out by P is a *cycloid* and is given parametrically by the above equations. The gradient at P is

$$dy/dx = dy/d\psi \div dx/d\psi = \sin 2\psi/(1+\cos 2\psi) = \tan\psi \ .$$

Hence PJ is tangential to the cycloid at P and PI is normal to it. It is readily shown that the curve has cusps on the line $y=2a$ and vertices on $y=0$.

To derive the *intrinsic equation* (connecting arc length s with ψ), we have

$$\begin{aligned} (ds/d\psi)^2 &= (dx/d\psi)^2+(dy/d\psi)^2 \\ &= 4a^2[(1+\cos 2\psi)^2+\sin^2 2\psi] = 16a^2\cos^2\psi \ . \\ &\therefore\ ds/d\psi = 4a\cos\psi \ . \\ \therefore\ \underline{s=4a\sin\psi}&, \text{ taking } s=0 \text{ when } \psi=0 \ . \end{aligned}$$

We now investigate the motion of a particle on a cycloid. Fig. 4.18 shows the position P at time t of a particle of mass m constrained to move on a smooth cycloidal arc $s=4a\sin\psi$ lying in a vertical plane, having vertex downwards.

Let R be the normal reaction of the curve on the particle at P and v the velocity along the tangent in the sense of s increasing. For the tangential motion we have

$$-mg\sin\psi = m\ddot{s} \ , \tag{1}$$

and for the normal motion, since the curve is concave upwards at P,

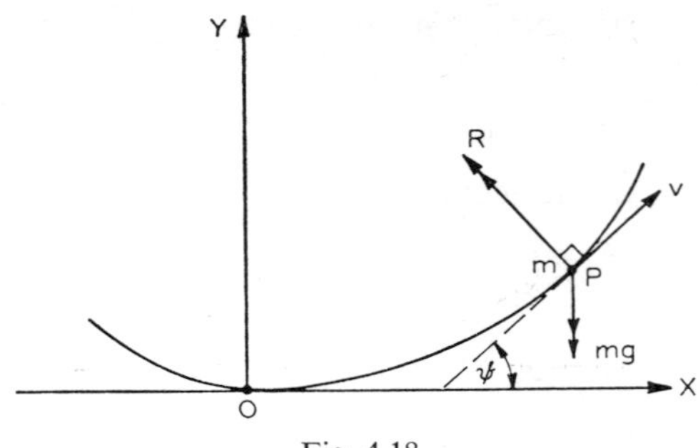

Fig. 4.18

$$R - mg\cos\psi = m\dot{s}^2/\rho\ , \tag{2}$$

where $\rho =$ radius of curvature $= ds/d\psi = 4a\cos\psi$.

Since $s = 4a\sin\psi$, (1) may be written

$$\ddot{s} = -(g/4a)s\ . \tag{1'}$$

(1′) is a S.H.M. and is true for any value of s which does not take the particle beyond the first cusp.

We have seen that the simple pendulum only furnishes an approximate periodic motion: the angular amplitude must be sufficiently small. To obviate this difficulty, the *isochronous pendulum* was designed by Huyghens. Essentially, it consists of a small bob P of mass m attached to a fixed point O by a string which is constrained to wind on and off a pair of fixed congruent cycloidal arches OA, OB having O as cusp and A and B as vertices (Fig. 4.19).

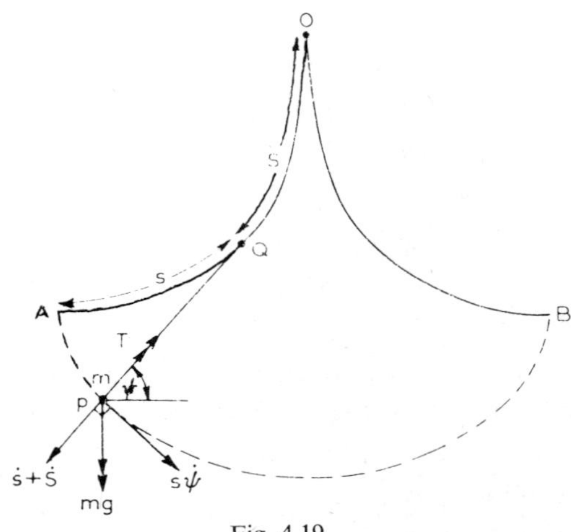

Fig. 4.19

The length of each arch and of the string is taken to be $4a$ and in the diagram, the string wraps round the section OQ of the arch OA, the remaining portion QP being straight. Let arc $AQ = s = 4a \sin \psi$, where ψ is the inclination of PQ to the horizontal. Let arc $QO = S = 4a(1 - \sin \psi)$. Then $s + S = 4a$ and $PQ = \text{arc } AQ = s$. The velocity of Q is $\dot{S}$ along $\overline{OP}$. The velocity of P relative to Q has components $\dot{s}$ along QP and $s\dot{\psi}$ in the transverse direction, so that the total velocity of P is simply $s\dot{\psi}$ along this transverse direction. Let σ be the length of the arc AP of the locus of P. Then

$$\dot{\sigma} = s\dot{\psi} = 4a \sin \psi \, \dot{\psi} \ ;$$

$$\therefore \ \sigma = 4a(1 - \cos \psi), \text{ since } \sigma = 0 \text{ when } \psi = 0 \ . \tag{3}$$

This shows that P also traces out a cycloid. The full length of its path is $8a$.

The equation of motion of the bob along the tangent to its locus is

$$mg \cos \psi = m \ddot{\sigma} \ . \tag{4}$$

From (3), (4), we have

$$\ddot{\sigma} = g[1 - (\sigma/4a)] \ . \tag{5}$$

(5) is seen to represent a S.H.M. about $\sigma = 4a$ (i.e. about the mid-point of the locus) with period $2\pi(4a/g)^{\frac{1}{2}}$. Thus in the absence of air resistance, the period of the motion is isochronous, whatever the amplitude.

EXERCISE 4

1. A particle is projected with initial velocity V from a point O at angle of elevation α. Show that the range on a plane through O inclined at β to the horizontal ($\beta < \alpha < \frac{1}{2}\pi$) is $V^2[\sin(2\alpha - \beta) - \sin \beta]/(g \cos^2 \beta)$ and hence deduce that for a given velocity of propagation V, the maximum range is $V^2/g(1 + \sin \beta) = R$, say. If we now vary β, the locus of the point having polar co-ordinates (R, β) is seen to be a parabola having O as focus, semi-latus rectum V^2/g, and axis $\beta = \pi/2$. This parabola is the parabola of safety.

2. A particle is projected with velocity u in a direction inclined at an angle α to the horizontal. Determine the horizontal and vertical displacements after time t on the assumption that gravity is the only force acting.

Any number of particles are projected from the same point at the same instant. Prove that (i) at any subsequent instant the directions of their velocities are concurrent; (ii) the point of concurrence rises with constant acceleration. (L.U.)

3. An aeroplane is flying with uniform speed v in an arc of a vertical circle of radius a, whose centre is at a height h vertically above a point O of the ground. If a bomb is dropped from the aeroplane when at a height Y and strikes the ground at O, show that Y satisfies the equation

$$KY^2 + Y(a^2 - 2hK) + K(h^2 - a^2) = 0 \ ,$$

where $K = h + (a^2 g/2v^2)$. (L.U.)

4. Drops of water are thrown tangentially off the rim of a wheel of radius a which is rotating in a vertical plane about its axis, which is fixed and horizontal, with angular velocity $\omega > (g/a)^{\frac{1}{2}}$. Show that if a horizontal ceiling, at height $h > a$ above the axis of the wheel, is not to be spattered with water,

$$a\omega < [gh + g(h^2 - a^2)^{\frac{1}{2}}]^{\frac{1}{2}} ,$$

the effects of air resistance being neglected. (L.U.)

5. A particle is projected under gravity with velocity V in a direction making an angle θ with an upward line of greatest slope of a plane of inclination α, the point of projection being on the plane. The motion takes place in the vertical plane through the line of greatest slope. Show that if the particle strikes the plane at an angle α with the horizontal then θ is given by the equation $(1 + 3 \tan^2 \alpha) \tan \theta = 2 \tan \alpha$. In this case show that the velocity with which the particle strikes the plane is $V(1 + 8 \sin^2 \alpha)^{-\frac{1}{2}}$. (L.U.)

6. A particle is projected at very high velocity, whose horizontal component is u, in a medium whose resistance per unit mass is $k \times (\text{velocity})^2$. If the angle of elevation α be small, show that the approximate equation of the trajectory is

$$y = x \tan \alpha - g(e^{2kx} - 2kx - 1)/(4k^2u^2) .$$

7. A particle whose mass increases through condensation of moisture at a constant time rate m_o/τ, where m_o is the initial mass of the particle and τ is a constant, moves freely under gravity. The particle is projected from the origin of co-ordinates with a velocity whose horizontal and vertical resolutes are u and v respectively. Prove that the co-ordinates of the particle when its mass is m are given by

$$x = u\tau \log (m/m_o) ,$$
$$y = \tfrac{1}{4}g\tau^2(1 - m^2/m_o^2) + (v\tau + \tfrac{1}{2}g\tau^2) \log (m/m_o) ,$$

the y-axis being vertically upwards. Show also that, if $v > 0$, the greatest height attained by the particle is

$$\tfrac{1}{2}(v\tau + \tfrac{1}{2}g\tau^2) \log (1 + 2v/g\tau) - \tfrac{1}{2}v\tau .$$

(L.U.)

8. The propulsive force on a heavy rocket moving in a vertical plane is so adjusted that its speed v is constant. Rectangular axes Ox, Oy are taken in the plane of the motion with origin O at the highest point of the path. Ox is horizontal and Oy is vertically downwards. If any change of mass is neglected, prove that the equation of the path is $y = (v^2/g) \log (gx/v^2)$.

Hence deduce that the horizontal range cannot exceed $\pi v^2/g$, and that to approach this range, the direction of projection must be nearly vertical. (L.U.)

9. A particle P is projected under gravity in a medium whose resistance per unit mass is kv^n where v is the speed. If ψ is the angle between the

direction of motion of P and the horizontal, show (i) that the horizontal velocity u of P is determined by

$$u^{-n} = u_o^{-n} - \frac{nk}{g}\int_{\psi_o}^{\psi} \sec^{n+1}\psi \, d\psi \,,$$

where u_o, ψ_o are the values of u, ψ at a point P_o of the curve, and (ii) that $d^2y/dx^2 = -g/u^2$, where x, y are the horizontal and vertical co-ordinates of P, positive y being taken upwards. If $n=4$, show that

$$d^3y/dx^3 = -2gk\{1+(dy/dx)^2\}^{3/2} \,.$$

(L.U.)

10. A particle moves in a plane so that its velocity is the vector sum of a velocity of magnitude V perpendicular to its radius vector from a fixed point and of a velocity of magnitude U in a fixed direction, where U and V are constants. Show that the acceleration of the particle is always directed towards the fixed point and varies as the inverse square of the distance from the point. Express the eccentricity of the orbit in terms of U, V. (L.U.)

11. A smooth wire in the shape of a cardioid $r = a(1+\cos\theta)$ is fixed in a horizontal plane. One end of a light elastic string, of natural length a and modulus mg, is fastened to the pole and the other end is fastened to a bead of mass m which slides on the wire. The bead is slightly displaced from rest at the vertex of the wire. Prove that the angular velocity of the string when it is inclined at an angle θ to the axis of the cardioid is $(g/a)^{\frac{1}{2}}\sin(\theta/2)$, and that the reaction of the wire on the bead is instantaneously zero when $\cos\theta = 3/5$ (L.U.)

12. A particle A of mass m moves on a smooth horizontal table and is connected by a light inextensible string passing through a smooth hole O in the table to a particle B of the same mass m, which moves along the vertical through O. Initially B is at rest and A is distant a from O, moving with velocity $\surd(ga/3)$ perpendicular to OA. Show that in the subsequent motion the distance $r = OA$ lies between a and $\frac{1}{2}a$. Show also that the tension in the string is $\frac{1}{6}mg(3+a^3/r^3)$. (L.U.)

13. A particle P is free to move on a smooth horizontal table and is attached to one end of an elastic string of natural length a the other end O of which is fixed to a point of the table. The particle is moving in a circle, centre O and radius $6a/5$, with uniform angular velocity, when it is subjected to a *small* impulse in the direction OP. Show that P will proceed to oscillate about its original path and that the period of oscillation is $\frac{1}{3}$ of the original period of revolution. (L.U.)

14. A particle P is moving in a plane under an attraction of magnitude $2/r^3$ per unit mass towards a fixed point O in the plane, and at $t=0$, $r=2$ and the radial and transverse components of velocity are $(3/2)^{\frac{1}{2}}$ and 1 respectively. Show that $\ddot{r} = 2/r^3$, and find r as a function of t. (L.U.)

15. A thin smooth straight tube OA is constrained to rotate with constant angular velocity ω about a fixed vertical axis through O, and a particle is free to move in the tube. The angle between OA and the upward vertical is

a fixed acute angle α. Show that the particle can describe a fixed horizontal circle of radius a, where $a\omega^2 = g \cot \alpha$.

While the particle is in a state of steady motion the angular velocity of the tube is suddenly reduced to $\frac{1}{2}\omega$ and is then maintained constant at the new value. Find the time the particle takes to reach O, and show that its speed relative to the tube at O will be $a\omega(7/4)^{\frac{1}{2}}$. (L.U.)

16. A circular wire of radius a is kept rotating with constant angular velocity ω about a fixed vertical diameter, and a particle of mass m, moving without friction on the wire is at an angular distance θ from the lowest point O of the wire at time t. If T is the K.E. of the particle and V the P.E. due to gravity, prove that $T+V-ma^2\omega^2 \sin^2 \theta$ is constant throughout the entire motion, and explain why $T+V$ is not constant in this case. The particle when at rest relative to the wire at O is slightly displaced. Prove that, if $a\omega^2 > g$, it next comes to rest relative to the wire at an angular distance from O of $\cos^{-1}(2g/a\omega^2 - 1)$. (L.U.)

17. (i) The bob P of a simple pendulum moves in a vertical circle of radius l, the speed at the lowest point A being $(2gd)^{\frac{1}{2}}$. Taking the height of P above A to be $2l \sin^2 \psi$, show that $\dot{\psi}^2 = (g/lk^2)(1-k^2 \sin^2 \psi)$, where $k^2 = 2l/d$.

(ii) The bob P of a spherical pendulum moves on a sphere of radius a between planes cutting the sphere at depths $b \pm c$ below the point of suspension O. If the depth of P below O is z, show that

$$a^2\dot{z}^2 = 2g(z-b-c)(z-b+c)(f-z), \text{ where}$$
$$2bf = c^2 - a^2 - b^2 \;.$$

Putting $z = b + c \cos 2\psi$, show that ψ varies synchronously with the angular motion of a spherical pendulum of length a^2/c, the speed of the bob at the lowest point being

$$(a/c)(g/b)^{\frac{1}{2}}\{a^2 + (b+c)(3b-c)\}^{\frac{1}{2}} \;.$$

(L.U.)

18. The interior surface of a bowl is given by $a^3z = a^2r^2 + r^4$, where a is a positive constant and r denotes the distance from Oz which is fixed vertically upwards. A particle of unit mass moves on this surface under gravity and surface reaction. Show that in steady motion in a horizontal circle of radius a, the moment of momentum about Oz is $(6ga^3)^{\frac{1}{2}}$. If the particle suffers a small disturbance while describing this path, determine the period of small oscillations produced. (L.U.)

19. A particle of mass m can slide freely on a light wire bent into the shape of the curve $a^3y = x^4$, which can rotate about the y-axis, which is vertically upwards. If a steady motion takes place with the particle at $x = a$, find the angular velocity Ω of the wire. The particle is slightly disturbed from its steady motion at $x = a$. Prove that the period of oscillation about this position is $2\pi(17a/8g)^{\frac{1}{2}}$ if the wire is constrained to rotate with angular velocity Ω. Find the period of the oscillation if the wire is not constrained but moves freely about the y-axis. (L.U.)

20. A heavy bead slides on a smooth cycloidal wire

$$x = a(2\psi + \sin 2\psi)\ , \quad y = a(1 - \cos 2\psi)\ ,$$

which is fixed with the (x, y)-plane vertical and the axes of x, y inclined at angles α, $(\frac{1}{2}\pi - \alpha)$ with the upward vertical. The bead moves from rest at the vertex. Prove that it does not reach a cusp if $2 \tan \alpha > \pi$, but moves to rest when

$$\sin \alpha + 2\psi \cos \alpha + \sin (2\psi - \alpha) = 0\ .$$

(L.U.)

21. A particle can slide on a rough cycloidal arch fixed with its axis vertical and vertex downwards. If the resistance is proportional to the speed, show that the oscillations of the particle about the lowest point are isochronous.

22. A smooth narrow tube, with open ends, in the form of one complete arch of a cycloid $s = 4a \sin \psi$ is fixed with the tangent at the midpoint $A(\psi = 0)$ of the arch vertical. A particle of mass m is placed in the tube at A and is connected to a light inextensible string. The string, whose length exceeds that of the arch, lies in the tube and after passing out through the upper end of the tube is attached to a particle of mass m' which hangs freely. If the system is released from rest and $m' > m$ show that

$$(m + m')\dot{s}^2 = 8m'ga \sin \psi - 2mga \sin 2\psi - 4mga\psi\ ,$$

and that if the reaction between m and the tube vanishes when $\psi = \frac{1}{4}\pi$, then $m'/m = (2\sqrt{2} + 1)(\pi + 4)/14$.

If $m' < m$, show that m will emerge from the lower end of the tube provided $4m' < \pi m$. (L.U.)

Chapter 5

Orbital Motion

'The Sun, whose rays
Are all ablaze
With ever-living glory.'
—W. S. Gilbert: The Mikado

5.1 MOTION OF A PARTICLE UNDER A CENTRAL FORCE

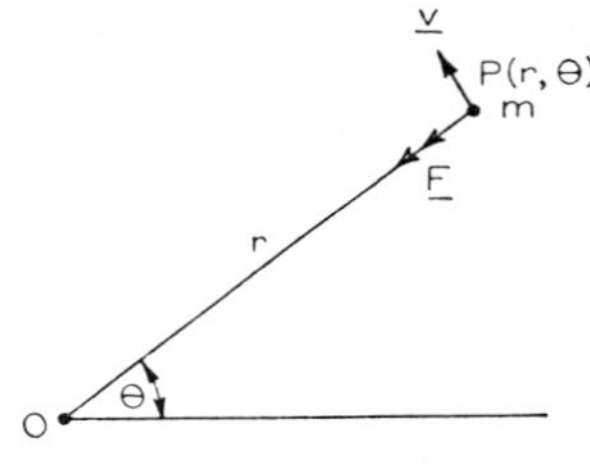

Fig. 5.1

In the last chapter, the topic of central forces and orbits was introduced. We here include some further remarks on the nature of orbits.

Fig. 5.1 shows the motion of a particle P of constant mass m under the action of a central force $\mathbf{F}$ along $\overline{PO}$, O being fixed. Let $\mathbf{v}$ be the velocity of the particle P. Then $\mathbf{F} = m\, d\mathbf{v}/dt$.

Now let $\hat{\mathbf{a}}$ be a unit vector normal to the plane of $\mathbf{F}$ and $\mathbf{v}$. Then since $\hat{\mathbf{a}}.\mathbf{F} = 0$, $\hat{\mathbf{a}}.d\mathbf{v}/dt = 0$, i.e. the acceleration component in the direction of $\hat{\mathbf{a}}$ is zero. Further, the velocity component $\hat{\mathbf{a}}.\mathbf{v}$ in this direction is zero. Hence the particle must continue to move in the plane determined by $\mathbf{F}$ and $\mathbf{v}$, i.e. *the orbit produced by a central force is a plane curve.*

Specifying the polar co-ordinates of the point P by (r, θ) at time t, we have seen that $r^2\dot{\theta} = h =$ const. Since the area of an elemental sector OPP', where P' is the point $(r + \delta r, \theta + \delta\theta)$, is $\frac{1}{2}r^2\delta\theta$ to first order, the rate of description of area is $\frac{1}{2}r^2\dot{\theta}$. Thus the angular momentum constant h is twice the rate of areal description. We have thus *established that the rate of description of area by the radius vector OP is constant.*

Positions of P for which OP is normal to the orbit are termed *apses* and the corresponding radius vector OP is called an *apse line*. The angle between two consecutive apse lines is called an *apsidal angle*. The point of nearest approach to O is called the *perihelion*: the point of farthest approach, if finite, is called the *aphelion*.

5.2 USE OF RECIPROCAL POLAR CO-ORDINATES

In the above case (Fig. 5.1), let the central force **F** be specified by $\mathbf{F}=mP(-\hat{\mathbf{r}})$, so that P is the magnitude of the force per unit mass and is directed towards O. The vector equation of motion is then

$$-mP\hat{\mathbf{r}}=m\{(\ddot{r}-r\dot{\theta}^2)\hat{\mathbf{r}}+\hat{\boldsymbol{\theta}}(1/r)d(r^2\dot{\theta})/dt\}\ ,$$

whence

$$\ddot{r}-r\dot{\theta}^2=-P\ , \qquad (1)$$

$$r^2\dot{\theta}=h\ . \qquad (2)$$

Now write $u=1/r$: the co-ordinates (u, θ) of P are termed *reciprocal polar co-ordinates*. Then $\dot{\theta}=hu^2$ and

$$\dot{r}=(dr/d\theta)(d\theta/dt)=hu^2d(1/u)/d\theta=-h\ du/d\theta\ ;$$

$$\ddot{r}=-h(d^2u/d\theta^2)(d\theta/dt)=-h^2u^2d^2u/d\theta^2\ .$$

Substitution in (1) now gives

$$-h^2u^2\ d^2u/d\theta^2-h^2u^3=-P\ .$$

Thus the equations of motion for m in reciprocal polar co-ordinates are

$$h^2u^2(u+d^2u/d\theta^2)=P\ , \qquad (3)$$

$$\dot{\theta}=hu^2\ . \qquad (4)$$

When the form of P is known as a function of u, substitution into (3) yields a second order differential equation which can often be solved to give $u=u(\theta)$ and hence the form for r deduced.

At an apse, r is stationary, and so $\dot{r}=0$ or $du/d\theta=0$.

Example 1—Description of a central conic under a central force towards a focus.

The polar equation of such a conic referred to a focus as pole and the axis as initial line is $l/r=1+e\cos\theta$ ($e=$eccentricity), or

$$u=(1+e\cos\theta)/l\ .$$

$$\therefore\ u''=-(e/l)\cos\theta\ .$$

$$\therefore\ P=h^2u^2(u+d^2u/d\theta^2)=(h^2/l)u^2=\mu/r^2\ ,$$

where $\mu=h^2/l=$constant. Thus in this case the central force is one which varies inversely as the square of the distance of the particle from the focus.

Example 2

Establish the formulae $d^2u/d\theta^2+u=(P/h^2u^2)$, $\dot{\theta}=hu^2(u=1/r)$ for the motion of a particle describing a central orbit under an attraction P per unit mass.

If $P=\mu u^5$, find the speed v with which the particle can describe the circle $r=a$.

If the particle moves under this attraction with the same areal constant as in the circular path, and $\dot{r}=-(3v/4\sqrt{2})$ when $r=2a$, $\theta=0$, find the equation of the spiral path of the particle and show that as $\theta\to\infty$ the path is asymptotic to the circle $r=a$. (L.U.)

(i) Circular motion of particle.

We have $u=1/a$, $P=v^2/a$, if v be the speed of the particle in its circular orbit. Hence

$$v^2/a=\mu/a^5, \text{ or } v=\sqrt{\mu}/a^2 .$$

$$\therefore\ \underline{h=av=\sqrt{\mu}/a} .$$

(ii) General motion of particle under $P=\mu/r^5$ law. Equation (3) writes

$$(u^2\mu/a^2)(d^2u/d\theta^2+u)=\mu u^5 ,$$

or

$$d^2u/d\theta^2+u=a^2u^3 .$$

Multiplying through by $2du/d\theta$ and integrating gives

$$(du/d\theta)^2+u^2=\tfrac{1}{2}a^2u^4+A .$$

Now

$$du/d\theta=[d(1/r)/dt][dt/d\theta]=-(\dot{r}/r^2\dot{\theta})=-\dot{r}/h=-a\dot{r}/\sqrt{\mu} .$$

$$\therefore \text{ When } \theta=0,\ u=(1/2a);\ du/d\theta=+(3av/4\sqrt{2})/\sqrt{\mu}=+(3/4\sqrt{2}a)$$

$$\therefore\ A=(9/32a^2)+(1/4a^2)-(1/32a^2)=(1/2a^2) .$$

$$\therefore\ (du/d\theta)^2+u^2=\tfrac{1}{2}a^2u^4+(1/2a^2) .$$

$$\therefore\ du/d\theta=\pm(a^2u^2-1)/(\sqrt{2}a)$$

To satisfy initial conditions, we must take the (—) sign.

$$\therefore\ d\theta=-\sqrt{2}a\,du/(a^2u^2-1)=(a/\sqrt{2})[(1+au)^{-1}+(1-au)^{-1}]du .$$

Integrating and using $u=(1/2a)$ when $\theta=0$,

$$\theta=(1/\sqrt{2})[\log(1+au)-\log(1-au)-\log 3] ,$$

i.e.

$$\theta=(1/\sqrt{2})\log[(r+a)/3(r-a)] .$$

$$\therefore\ \underline{\text{As } \theta\to\infty,\ r\to a} .$$

Example 3—Stability of a nearly circular orbit.

Suppose a particle is describing a central orbit according to the law of force $P=\mu u^s$, s being a constant. Then

$$d^2u/d\theta^2+u=P/h^2u^2=\mu u^{s-2}/h^2 .$$

If the particle describes a circular orbit of radius R, then $u=1/R$ and the above equation leads to $h^2=\mu/R^{s-3}$. Hence for perturbations from this circular orbit,

$$d^2u/d\theta^2+u=R^{s-3}u^{s-2} .$$

Now suppose the perturbations are small and that $u=\epsilon+1/R$, where ϵ is small. Then

$$d^2\epsilon/d\theta^2+\epsilon+1/R=(1/R)(1+\epsilon R)^{s-2}$$
$$=(1/R)[1+\epsilon(s-2)R+\ldots\ldots] .$$

Thus, retaining only first order terms, we have

$$d^2\epsilon/d\theta^2+(3-s)\epsilon=0\ .$$

The solution of this equation depends on the value of s.

(i) *For* $s<3$, the solution is

$$\epsilon=a\cos\,[(3-s)^{\frac{1}{2}}\theta+\alpha]\ ,$$

where a, α are integration constants. This gives a *stable orbit*, i.e. small oscillations about the circular orbit occur. The apses are found from $du/d\theta=0$ or

$$\sin\,[(3-s)^{\frac{1}{2}}\theta+\alpha]=0\ .$$

The roots of this equation differ by $\pi(3-s)^{-\frac{1}{2}}$, the apsidal angle.

(ii) *For* $s>3$, the solution is

$$\epsilon=a\,\mathrm{ch}[(s-3)^{\frac{1}{2}}\theta+\alpha]\ .$$

Thus $\epsilon\to\infty$ as $\theta\to\infty$ and the circular orbit is *unstable*.

(iii) *For* $s=3$, $\epsilon=A+B\theta$ and the orbit is again *unstable*.

5.3 USE OF PEDAL CO-ORDINATES AND EQUATIONS

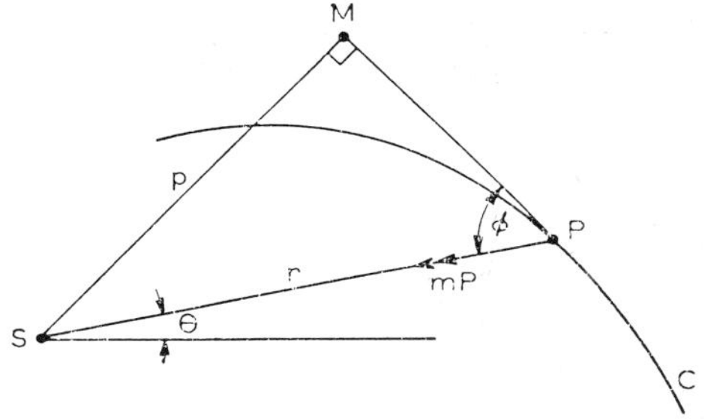

Fig. 5.2

Let P be a point on the orbit C described by a particle of mass m moving under a central force mP along $\overline{PS}$ towards a fixed point S. Further, suppose $(r,\,\theta)$ are the polar co-ordinates of P and that the length of the perpendicular SM from S on the tangent at P to C is p. Then $(p,\,r)$ are called the *pedal co-ordinates* of P and the functional relation between them is called the pedal equation of C.

If $\angle MPS=\varphi$, the equation for the motion of P along the normal to C is

$$mP\sin\varphi=mv^2/\rho\ ,$$

where ρ is the radius of curvature of C at P. Further,

$$h=pv\ .$$

Since $\sin \varphi = p/r$, $\rho = r\,dr/dp$, we obtain the important result

$$\underline{P = (h^2/p^3)(dp/dr)}\ .$$

For P known as a function of p, r, this leads to a first order differential equation in contrast to the second order equation which one obtains in (u, θ) co-ordinates. This suggests some types of orbital problem are more easily treated in (p, r) co-ordinates. Problems involving the inverse square law form for P leading, as we shall see, to conics having S as focus, are more tractable via this co-ordinate system. We therefore give the (p, r) equations for conic sections, referred to focus as origin, as follow:

Ellipse: $$b^2/p^2 = (2a/r) - 1\ , \tag{1}$$

Parabola: $$p^2 = \tfrac{1}{2}lr\ , \tag{2}$$

Hyperbola (near branch): $$b^2/p^2 = (2a/r) + 1\ , \tag{3}$$

Hyperbola (far branch): $$b^2/p^2 = 1 - (2a/r)\ . \tag{4}$$

Throughout, a, b denote the semi-major and minor axes, while l denotes the semi-latus rectum of the conic.

We shall here establish (1): the other results may be deduced by similar methods. The polar equation of the ellipse is

$$\begin{aligned}
&l/r = 1 + e\cos\theta \quad (0 < e < 1)\ .\\
\therefore\ & -(l/r^2)(dr/d\theta) = -e\sin\theta\ .\\
\therefore\ & \cot\varphi = (1/r)(dr/d\theta) = (er\sin\theta)/l\ .\\
& \sin\varphi = p/r\ .\\
\therefore\ & \cos\varphi = (ep\sin\theta)/l\ .\\
\therefore\ & (p/r)^2 + (ep\sin\theta)^2/l^2 = 1\ .\\
\therefore\ & 1/p^2 = (1/l^2)[(1+e\cos\theta)^2 + e^2\sin^2\theta]\\
& \qquad = (1/l^2)[1 + 2e\cos\theta + e^2]\\
& \qquad = (1/l^2)[1 + e^2 + 2l/r - 2]\\
& \qquad = (2/rl) - (1-e^2)/l^2\ .
\end{aligned}$$

Using $l = a(1-e^2)$, the result (1) now follows.

Suppose we have an orbit described under the inverse square law of force $P = \mu/r^2$ towards S, μ being a constant. Then

$$(h^2/p^3)(dp/dr) = \mu/r^2\ .$$

$$\therefore\ (h^2/2p^2) = (\mu/r) + A\ . \tag{5}$$

When $\mu > 0$, (5) has the form of either (1), (2) or (3), according to the nature of A. Thus if $A < 0$, the orbit is an ellipse; if $A = 0$, it is a parabola; and if $A > 0$, it is the near branch of a hyperbola. If, however, $\mu < 0$ and $A > 0$, then (5) is of the form (4), i.e. the orbit is the far branch of a hyperbola. Thus, the *inverse square law of force directed towards a fixed point always produces a conic type of orbit.*

If $\mu>0$, $A<0$, equation (5) is of the same form as (1), i.e. the orbit is elliptic. Identifying the two,

$$2b^2/h^2=2a/\mu=-1/A\ .$$
$$\therefore\ h^2=\mu b^2/a=\mu a(1-e^2)=\mu l\ ;$$
$$A=-(\mu/2a)\ .$$

$\therefore$ (5) may be written

$$(h^2/2p^2)=\mu/r-\mu/2a\ ,$$

i.e.

$$v^2=\mu(2/r-1/a)\ .$$

The reader can easily show that in all cases where the orbit is a con c, the law of attraction being $P=\mu/r^2$, the relation

$$h^2=|\mu|l \tag{6}$$

always holds and the velocity is given by the equations:

Ellipse: $$v^2=\mu(2/r-1/a)\ , \tag{7}$$
Hyperbola: $$v^2=\mu(2/r+1/a)\ , \tag{8}$$
Parabola: $$v^2=2\mu/r\ . \tag{9}$$

When the particle executes an ellipse of semi-axes a, $b(a>b)$ under the inverse square law of attraction $P=\mu/r^2(\mu>0)$, since the area of the ellipse is πab and $\frac{1}{2}h$ is the rate of description of area by the radius vector SP, the periodic time T is

$$T=(\pi ab)/(\tfrac{1}{2}h)=2\pi a^2(1-e^2)^{\frac{1}{2}}/(\mu l)^{\frac{1}{2}}=2\pi a^{3/2}/\mu^{\frac{1}{2}}\ ,$$

using $l=a(1-e^2)$.

Example 1

A particle is describing an ellipse of eccentricity e about a centre of force at a focus. Prove, with the usual notation,

$$v^2=\mu(2/r-1/a)\ ,\quad h^2=\mu a(1-e^2)\ .$$

When the particle is at one end of a minor axis, its velocity is doubled. Prove that the new path is a hyperbola of eccentricity $(9-8e^2)^{\frac{1}{2}}$. (L.U.)

The equation of the ellipse is

$$b^2/p^2=2a/r-1\ .$$
$$\therefore\ -(2b^2/p^3)(dp/dr)=-2a/r^2\ ,$$

so that

$$P=(h^2/p^3)(dp/dr)\ =(ah^2/b^2)(1/r^2)\ .$$
$$\therefore\ \mu=ah^2/b^2\ ,$$

or

$$\underline{h^2=\mu a(1-e^2)\ .}$$

Since $pv=h$, $p^2=h^2/v^2=\mu a(1-e^2)/v^2$.

$$\therefore\ b^2/p^2=a^2(1-e^2)/p^2=av^2/\mu\ .$$

Substituting in the (p, r)-equation of the ellipse,

$$av^2/\mu = 2a/r - 1$$

or

$$\underline{v^2 = \mu(2/r - 1/a)} \ .$$

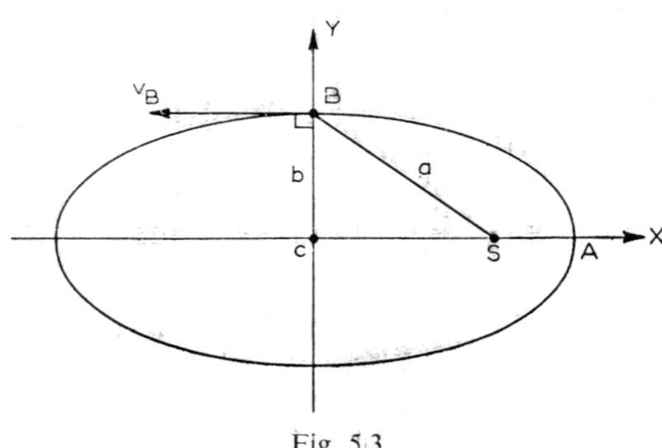

Fig. 5.3

When the particle is at B, the end of the semi-minor axis (Fig. 5.3), $r = a$, $p = b$, so that

$$v_B{}^2 = \mu(2/a - 1/a) = \mu/a \ .$$

$$\therefore \ v_B = (\mu/a)^{\frac{1}{2}} \ .$$

$\therefore$ New velocity becomes $v_B{}' = 2(\mu/a)^{\frac{1}{2}}$.

The angular momentum constant is

$$bv_B{}' = 2(\mu a)^{\frac{1}{2}}(1 - e^2)^{\frac{1}{2}} = h \ .$$

For the new orbit,

$$v^2 - 2\mu/r = v_B{}'^2 - 2\mu/a = 2\mu/a > 0 \ .$$

$\therefore$ The new orbit is a hyperbola.

Since $v = h/p$, we have

$$4\mu a(1 - e^2)/p^2 = 2\mu(1/r + 1/a) \ .$$

$$\therefore \ 2a^2(1 - e^2)/p^2 = a/r + 1 \ .$$

This is the (p, r) equation of the near branch of a hyperbola having semi-axes a', b', where

$$2a' = a, \quad b'^2 = 2a^2(1 - e^2) \ .$$

Let e' be its eccentricity so that

$$b'^2 = a'^2(e'^2 - 1) \ .$$

$$\therefore \ e'^2 = 1 + (b'^2/a'^2) = \underline{9 - 8e^2} \ .$$

Example 2—Nuclear repulsion. Use of energy principle.

A fixed nucleus S, having positive charge Ze, repels a particle P having mass m and positive charge e'. P is projected from a great distance at infinity with initial speed v_o in a direction whose perpendicular distance from S is d, the medium being a vacuum. Show that its ultimate direction of motion makes an angle φ with an initial direction where $\cot \frac{1}{2}\varphi = (dmv_o^2/Zee')$.

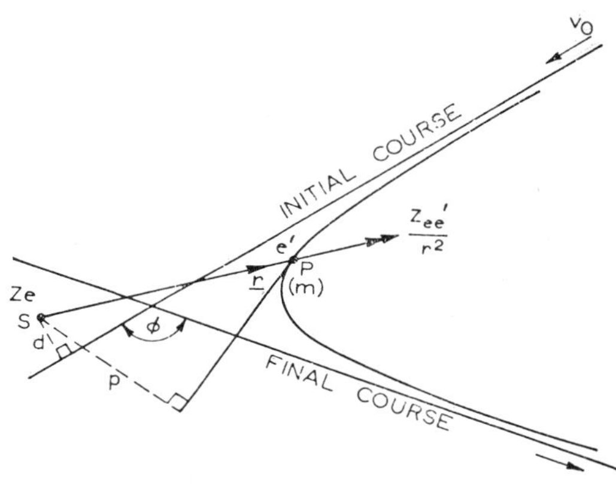

Fig. 5.4

Let (p, r) be the pedal co-ordinates of P at some stage, where $\overline{SP} \equiv \mathbf{r}$, its velocity then being $\mathbf{v}$ (Fig. 5.4). The force on P is then $(Zee'/r^2)\hat{\mathbf{r}}$, using Coulomb's law of electrostatic repulsion of like charges. Thus the work that is done on P by the repulsion in the motion from infinity to the point (p, r) is

$$\int_{\infty}^{r} (Zee'/r^2)dr = -Zee'/r \ .$$

The K.E. gained in the process is $\frac{1}{2}mv^2 - \frac{1}{2}mv_o^2$. Since the field of force is conservative, these two are equal and hence

$$v^2 = v_o^2 - (2Zee'/mr) \ .$$

Now $pv = dv_o = h$, so that

$$d^2/p^2 = 1 - (2Zee'/mv_o^2 r) \ .$$

This is the pedal equation of the far branch of a hyperbola having S as focus. The semi-axes a and b are given by $b = d$, $a = (Zee'/mv_o^2)$. Thus, taking φ to be the angle between the two asymptotes (Fig. 5.4),

$$\tan \tfrac{1}{2}(\pi - \varphi) = b/a = (dmv_o^2/Zee') \ .$$

5.4 KEPLER'S LAWS OF PLANETARY MOTION AND NEWTON'S LAW OF GRAVITATION

As the result of observations on the motions of planets of the solar system, Kepler (1571–1630) published the following laws governing their motions.

(1) Relative to the Sun the planets describe ellipses with the Sun as one focus.
(2) The radius vector drawn from the Sun to a planet sweeps out areas at a constant rate.
(3) The squares of the periodic times are proportional to the cubes of the semi-major axes of the elliptic orbits, and hence also to the cubes of the mean distances from the Sun.

Sir Isaac Newton later unified these three laws through his *Law of Gravitation* which states that *every particle of the universe attracts every other particle with a force proportional to the product of their masses and inversely proportional to the square of their distance apart.* Thus for two particles of masses m_1, m_2 distant r apart, the force of attraction between them is $\gamma m_1 m_2/r^2$ where γ is the *gravitational constant* (sometimes denoted by G).

If we assume the Sun to be at rest, the truth of Kepler's laws, based on Newton's law of gravitation, has already been established. Actually, (1) is not quite correct: we shall show that it is the centroid of the Sun and planet which is at rest, when there are no other bodies present. However, since the mass of the Sun greatly exceeds that of any planet in the solar system, the centroid very nearly coincides with the Sun.

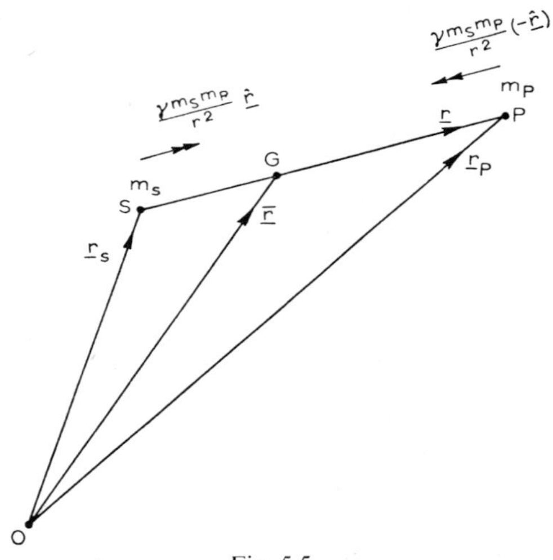

Fig. 5.5

Let S denote the Sun and P a planet of the solar system, no other bodies being present. Let m_S, m_P denote the masses of the Sun and planet and let $\overline{OS} \equiv \mathbf{r}_S$, $\overline{OP} \equiv \mathbf{r}_P$, $\overline{SP} \equiv \mathbf{r}$. G is the centroid of the system so that

$$\overline{OG} \equiv \bar{\mathbf{r}} = (m_S \mathbf{r}_S + m_P \mathbf{r}_P)/(m_S + m_P) .$$

By Newton's law of gravitation, S and P are subjected to forces $(\gamma m_S m_P/r^2)\hat{\mathbf{r}}$ and $(\gamma m_S m_P/r^2)\ (-\hat{\mathbf{r}})$, respectively. Hence their equations of motion are

$$(\gamma m_S m_P/r^2)\hat{\mathbf{r}} = m_S\ddot{\mathbf{r}}_S \ , \qquad (1)$$

$$(\gamma m_S m_P/r^2)(-\hat{\mathbf{r}}) = m_P\ddot{\mathbf{r}}_P \ . \qquad (2)$$

Adding (1) and (2), we find

$$m_S\ddot{\mathbf{r}}_S + m_P\ddot{\mathbf{r}}_P = \mathbf{0} \ , \qquad (3)$$

or $\ddot{\bar{\mathbf{r}}} = \mathbf{0}$, i.e. *the centroid of the two-body system has no acceleration* (We shall see later that this is a particular aspect of a more general principle governing the motion of a system of particles under no external forces.)

From (1), (2),

$$\ddot{\mathbf{r}} = \ddot{\mathbf{r}}_P - \ddot{\mathbf{r}}_S = -\gamma(m_S + m_P)\hat{\mathbf{r}}/r^2 \ .$$

Thus the planet is subjected to a force per unit mass of $-\mu\hat{\mathbf{r}}/r^2$, where $\mu = \gamma(m_S + m_P)$ and so relative to S it describes a conic with S as focus. Were the Sun at rest, we should have $\ddot{\mathbf{r}}_P = \ddot{\mathbf{r}}$ and (2) would then give $\ddot{\mathbf{r}} = -\gamma m_S\hat{\mathbf{r}}/r^2$, i.e. the planet would be subjected to a force of attraction per unit mass of $-\mu'\hat{\mathbf{r}}/r^2$, where $\mu' = \gamma m_S$. Again a conic orbit is described with S as focus. The ratio $\mu'/\mu = m_S/(m_S + m_P)$ is near to unity since $m_S \gg m_P$, and so the error involved in treating S as fixed is often negligible.

Example

Two gravitating particles of masses m and M move under the force of their mutual attraction. Show that the centre of mass of the two particles moves with constant velocity, and that if $\mathbf{r}$ is the position vector of m relative to M, $\ddot{\mathbf{r}} = -\gamma(M+m)\hat{\mathbf{r}}/r^3$, where γ is the gravitational constant.

If the orbit of m relative to M is a circle of radius a described with velocity v, show that $v = [\gamma(M+m)/a]^{\frac{1}{2}}$.

If the relative velocity of the particle m is suddenly changed to $v+u$, without change of direction, show that, if $(u/v)^2$ is negligible, the particle will describe relative to M an ellipse of eccentricity $2u/v$. (L.U.)

(i) This part of the question is already established.

(ii) The orbit of m relative to M is a circle of radius a and the speed is v.

$$\therefore \ \ddot{\mathbf{r}} = (-v^2/a)\hat{\mathbf{r}} + \dot{v}\hat{\mathbf{t}} \ ,$$

where $\hat{\mathbf{t}}$ is the unit vector in the tangent to the circle. Since

$$\ddot{\mathbf{r}} = -\gamma(M+m)\mathbf{r}/a^3 \ , \quad \mathbf{r} = a\hat{\mathbf{r}} \ ,$$

we have $v^2/a = \gamma(M+m)/a^2 \ , \quad \dot{v} = 0 \ .$

$$\text{i.e. } \underline{v = [\gamma(M+m)/a]^{\frac{1}{2}}} \ .$$

(iii) For the motion of m relative to M,

$$(h^2/p^3)(dp/dr) = \mu/r^2 \ , \quad \mu = \gamma(M+m) = av^2 \ .$$

$$\therefore \ h^2/p^2 = 2\mu/r + C \ .$$

Further, $h = a(v+u)$ and when $r = a$, $p = a$, and so

$$C = (v+u)^2 - 2\gamma(M+m)/a = (v+u)^2 - 2v^2 = u^2 + 2uv - v^2 \ .$$

Since $C<0$, the orbit of m relative to M is an ellipse. The (p, r) equation of the ellipse may be written

$$a^2(v+u)^2/p^2 = 2av^2/r - v^2 + 2vu + u^2 \ ,$$

or

$$\{a^2(v+u)^2/(v^2-2vu-u^2)\}/p^2 = 2\{av^2/(v^2-2vu-u^2)\}/r - 1 \ .$$

This is an ellipse with semi-major axis α and semi-minor axis β where $\alpha = av^2/(v^2-2vu-u^2)$, $\beta^2 = a^2(v+u)^2/(v^2-2vu-u^2)$. The eccentricity e is given by $\beta^2 = \alpha^2(1-e^2)$, so that

$$e^2 = 1-\beta^2/\alpha^2 = 1-(1+u/v)^2(1-2u/v-u^2/v^2)$$
$$= 1-(1-4u^2/v^2+\ldots.) \simeq 4u^2/v^2 \ .$$
$$\therefore \ \underline{e \simeq 2u/v} \ .$$

5.5 DISTURBED ORBITS

When a particle describing an orbit is subjected to a blow, in general the magnitude and direction of its velocity will change with consequent change of orbit. Here we shall treat the case of a particle moving in an elliptic orbit under the inverse square law of attraction and subjected to a small blow (i) in the tangential direction, (ii) in the normal direction.

(i) *Tangential blow. Motion in medium of small resistance*

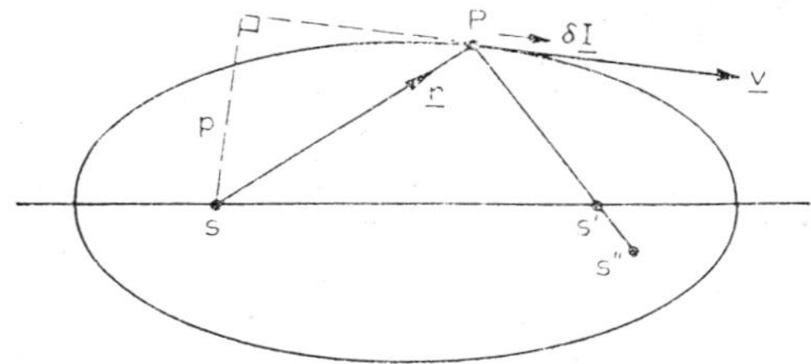

Fig. 5.6

Fig. 5.6 shows the particle P of mass m describing such an orbit, S and S' being the foci, the force per unit mass at P being μ/r^2 towards S. The velocity of P is $\mathbf{v}$ and the particle receives a small impulse $\delta\mathbf{I}$ in the direction of $\mathbf{v}$. With the usual notation the velocity at P before the impact is administered is given by

$$v^2 = \mu(2/r-1/a) \ .$$
$$\therefore \ 2v\delta v = \mu[-(2/r^2)\delta r + (1/a^2)\delta a] = (\mu/a^2)\delta a \ ,$$

since r will not change when the blow is administered.

$$\therefore \ \delta a = (2a^2v/\mu)\delta v, \text{ where } \delta v = |\delta\mathbf{I}|/m \ .$$
$$\therefore \ \text{Length of new semi-major axis} \simeq \underline{a + 2a^2v\delta v/\mu} \ .$$

The tangential direction of motion of P will not alter the moment the blow is administered. Since SP, $S'P$ make equal angles with $\mathbf{v}$, they must also be inclined at these same angles immediately after $\delta\mathbf{I}$ is applied. Thus the focus S' will move along PS' to some new position S''. Now $SP+PS'=2a$; $SP+PS''=2(a+2a^2v\delta v/\mu)$.

$$\therefore \; S'S''=4a^2v\delta v/\mu = \text{focal displacement along } \overline{PS'} \; .$$

The new value of h is $p(v+\delta v)$, where p is the perpendicular from S on $\mathbf{v}$. Hence $\delta h = p\delta v = h\delta v/v$. Further, since $h^2=\mu l=\mu b^2/a$,

$$\therefore \; 2\log h = \log \mu + 2\log b - \log a \; .$$
$$\therefore \; 2\delta h/h = 2\delta b/b - \delta a/a \; .$$
$$\therefore \; \delta b/b = \delta v/v + av\delta v/\mu \; .$$

This gives the increment in the minor axis.

Since the periodic time in the undisturbed orbit is

$$T=(\pi ab)/(h/2) \; ,$$
$$\log T = \log(2\pi)+\log a+\log b-\log h \; ,$$
$$\therefore \; \delta T/T = \delta a/a+\delta b/b-\delta h/h$$
$$\therefore \; \delta T/T = [2av/\mu+1/v+av/\mu-1/v]\delta v = (3av/\mu)\delta v \; .$$

This gives the change in periodic time.

Further, since $\angle S''SS'$ is small,

$$\begin{aligned}\angle S''SS' &= (S'S''/SS')\sin\angle PS'S \\ &= (S'S''/SS')(PS/PS')\sin\angle PSS' \\ &= (2av\delta v/\mu e)[r/(2a-r)]\sin\angle PSS' \\ &= [2arv\delta v/\mu e(2a-r)]\sin\angle PSS' \; . \\ &= \underline{(2\delta v/e)[ar/\mu(2a-r)]^{\frac{1}{2}}\sin\angle PSS'} \; .\end{aligned}$$

This gives the rotation of the apse line.

The above results may be adapted to discuss the change produced in orbital elements due to motion in a medium of *small resistance*. Let f be the tangential resistance per unit mass. Then in time δt, the velocity increment is $\delta v = -f\delta t$, correct to the first order. Thus the change δa in the semi-major axis is given by

$$\delta a=(2a^2v/\mu)\delta v = -2a^2vf\delta t/\mu \; .$$

Hence $da/dt = -2a^2vf/\mu$. This gives the rate of change of the semi-major axis. Similarly the rates of change of the other elements may be computed.

(ii) *Normal blow*

Now suppose that blow $\delta\mathbf{I}$ is applied normally to the ellipse at P. Then P is given a normal velocity component $\delta v=|\delta\mathbf{I}|/m$. This has the effect of changing the direction of the velocity $\mathbf{v}$ to some new direction $\mathbf{v}'$ (Fig. 5.7).

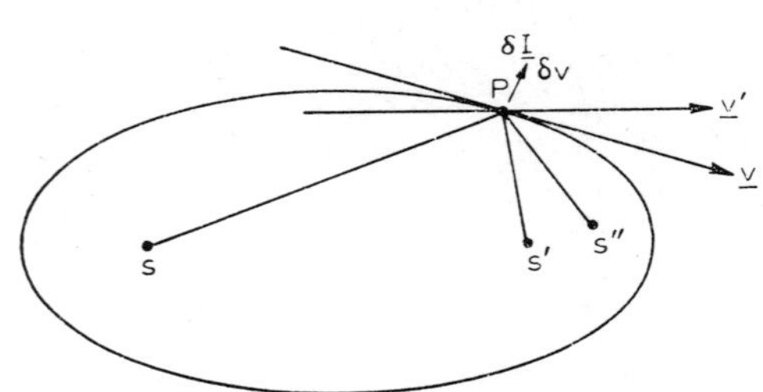

Fig. 5.7

Rotation of **v** takes place through an angle $\tan^{-1}(\delta v/v)$ or $\delta v/v$, to first order. Since SP is unaltered, PS' must change to a new orientation PS'' so that PS, PS'' both make equal angles with $\mathbf{v}'$. Thus $\angle S'PS''=2\delta v/v$, to first order.

Also we have

$$v^2=2\mu/r-\mu/a \ ,$$
$$v^2+(\delta v)^2=2\mu/r-\mu/a+(\mu/a^2)\delta a \ ,$$

so that δa is of second order smallness. Thus, to first order,

$$S''P+SP=2a=S'P+SP \ .$$

Thus S'' is located and the remaining orbital elements may be evaluated.

5.6 ELLIPTIC HARMONIC MOTION

Suppose a particle P moves under a central force μr per unit mass along $\overline{PO}$, where O is the centre of the force and $\overline{OP}\equiv\mathbf{r}$. The equation of motion is

$$-\mu\mathbf{r}=\ddot{\mathbf{r}} \ . \tag{1}$$

Introducing co-ordinate axes Ox, Oy through O in the plane of the motion and letting $\mathbf{r}=x\mathbf{i}+y\mathbf{j}$, where $\mathbf{i}$ and $\mathbf{j}$ are unit vectors in $\overline{Ox}$, $\overline{Oy}$, we obtain from (1) the equations

$$\left.\begin{aligned}\ddot{x}&=-\omega^2 x,\\ \ddot{y}&=-\omega^2 y,\end{aligned}\right\} \tag{2}$$

where $\omega^2=\mu$. These equations (2) are a pair of superposed simple harmonic motions. The general solutions of (2) are

$$x=A_1\cos\omega t+A_2\sin\omega t \ ,$$
$$y=A_3\cos\omega t+A_4\sin\omega t \ .$$

Solving these for $\cos\omega t$, $\sin\omega t$ we find

$$\cos\omega t=(A_4x-A_2y)/\triangle,\ \sin\omega t=(A_1y-A_3x)/\triangle,\ \triangle=A_1A_4-A_2A_3 \ .$$

Since $\cos^2\omega t+\sin^2\omega t=1$, we have for the locus of P

$$(A_4x-A_2y)^2+(A_1y-A_3x)^2=(A_1A_4-A_2A_3)^2 \ .$$

This locus, being of the second degree in x, y and having no first order terms in x and y, represents a conic having O as centre. The discriminant of the second degree terms is

$$\begin{aligned}&4(A_2A_4+A_1A_3)^2-4(A_3{}^2+A_4{}^2)(A_1{}^2+A_2{}^2)\\&=4\{2A_1A_2A_3A_4-A_3{}^2A_2{}^2-A_1{}^2A_4{}^2\}\\&=-4(A_1A_4-A_2A_3)^2<0, \text{ in general.}\end{aligned}$$

Thus the conic is an ellipse, centre O. The type of motion is called *elliptic harmonic*. The period of description is $2\pi/\omega$.

5.7 SATELLITE ORBITS

When two particles of masses m, M move under the force of their mutual gravitational attraction (assumed to be of inverse square law type), the worked example of Section 5.4 shows that if $\mathbf{r}$ is the position vector of m relative to M, then

$$\ddot{\mathbf{r}}=-\gamma(M+m)\mathbf{r}/r^3 \ , \tag{1}$$

where γ is the gravitational constant. For a satellite of mass m orbiting the Earth, of mass M_E, this gives

$$\ddot{\mathbf{r}}=-\frac{\mu}{r^3}\mathbf{r}=-\frac{\mu}{r^2}\hat{\mathbf{r}} \ , \tag{2}$$

where $\mu=\gamma(M_E+m)$ and, in the usual notation, $\mathbf{r}=\mathbf{r}/r$.

As $M_E\gg m$, we may take $\mu=\gamma M_E$. Scalar multiplying (2) through by $2\dot{\mathbf{r}}$ gives

$$2\dot{\mathbf{r}}.\ddot{\mathbf{r}}=-\frac{2\mu}{r^3}\mathbf{r}.\dot{\mathbf{r}} \ ,$$

or

$$\frac{d}{dt}(\dot{\mathbf{r}}^2)=-\frac{2\mu}{r^3}\mathbf{r}.\dot{\mathbf{r}} \ .$$

Now

$$2\mathbf{r}.\dot{\mathbf{r}}=\frac{d}{dt}(\mathbf{r}^2)=\frac{d}{dt}(r^2)=2r\dot{r} \ ,$$

and so

$$\frac{d}{dt}(\dot{\mathbf{r}}^2)=-\frac{2\mu}{r^2}\dot{r}$$

which integrates to give at once

$$\dot{r}^2=A+\frac{2\mu}{r} \ . \tag{3}$$

The nature of the integration constant A governs the form of orbit obtained. In accordance with equations (7), (8), (9) of Section 5.3, we see that, if $a>0$, then

$A = -\mu/a$ gives an elliptic orbit with semi-major axis a;

$A = 0$ gives a parabolic orbit ;

$A = +\mu/a$ gives a hyperbolic orbit .

The last case will not feature in the work that follows. With $v = |\dot{\mathbf{r}}|$, $mA = 2E$, the equation (3) may be written

$$\tfrac{1}{2}mv^2 - \frac{\mu m}{r} = E \ . \tag{4}$$

(4) will be referred to as the *energy equation*, since $\tfrac{1}{2}mv^2$ is the kinetic energy of the satellite (treated as a particle of mass m) and $-\mu m/r$ is its potential energy in the gravitational field.

At an apse on the satellite's orbit, the velocity of the satellite is perpendicular to its position vector so that

$$\mathbf{r}.\dot{\mathbf{r}} = 0 \ .$$

As above, $\mathbf{r}.\dot{\mathbf{r}} = r\dot{r}$ and as $r \neq 0$, we obtain

$$\dot{r} = 0$$

as apsidal condition. (The reader can establish that the condition is both necessary and sufficient.) Therefore points at maximum and minimum distances of the satellite from the Earth are apsidal points. The point on the orbit nearest to the Earth is called a *perigee* and that at greatest distance is called an *apogee*. For an elliptic orbit the apsidal positions are as illustrated in Fig. 5.8 in which r_1 and r_2 are the distances of the perigee and apogee.

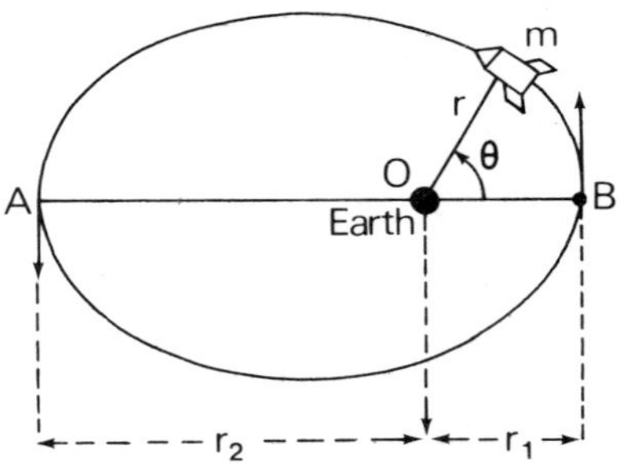

Fig. 5.8

Of necessity the angular momentum h per unit mass of the satellite about the normal to the orbital plane through O, the Earth's centre, stays constant. With reference to plane polar coordinates (r, θ) as indicated, $h = r^2\dot{\theta}$. We prove that r_1 and r_2 are the roots of the quadratic equation in r:

$$r^2 + \frac{\mu m}{E} r - \frac{mh^2}{2E} = 0 \ . \tag{5}$$

We have

$$E = \tfrac{1}{2}mv^2 - \frac{\mu m}{r}$$

$$= \tfrac{1}{2}m(\dot{r}^2 + r^2\dot{\theta}^2) - \frac{\mu m}{r}$$

$$= \tfrac{1}{2}m\left(\dot{r}^2 + \frac{h^2}{r^2}\right) - \frac{\mu m}{r} ,$$

and on taking $\dot{r} = 0$ in this we obtain equation (5). This confirms there are but two apses in the case of the elliptic satellite orbit.

Example

A satellite describes a circular orbit of radius NR_E with uniform speed round the centre of the Earth which has radius R_E and $N > 1$. At a given instant of time, the direction of motion of the satellite is changed through an angle α towards the Earth, without change of speed. Find α so that the orbit just touches the Earth's surface.

If v is the speed of description of the circular orbit of radius NR_E, then v^2/NR_E is the acceleration of the satellite towards the centre of the Earth. The force on the satellite due to the Earth's attraction is $\gamma m M_E/(NR_E)^2$. So the equation of motion of the satellite is

$$\frac{\gamma m M_E}{(NR_E)^2} = \frac{mv^2}{NR_E}$$

or

$$v^2 = \mu/NR_E .$$

The energy constant E is

$$E = \tfrac{1}{2}mv^2 - m\mu/r$$

$$= \frac{m\mu}{2NR_E} - \frac{m\mu}{NR_E} = -\frac{m\mu}{2NR_E} .$$

In the subsequent motion, E remains constant, but h will change. The new value of h is found from Fig. 5.9, showing an angular deflection α of the satellite towards the Earth at some instant.

The velocity at change is along a tangent distant $NR_E \cos \alpha$ from O the Earth's centre. Hence after change

$$h = (NR_E \cos \alpha)v = (\mu NR_E)^{1/2} \cos \alpha .$$

Substituting the values of E, h into the quadratic equation (5), the apsidal distances are found to be the roots of the quadratic equation

$$r^2 - 2NR_E r + N^2 R_E^2 \cos^2 \alpha = 0 .$$

For critical tangential contact of the new orbit with the Earth, we require that $r = R_E$ shall satisfy this equation,

i.e. $$\cos \alpha = (2N - 1)^{1/2}/N .$$

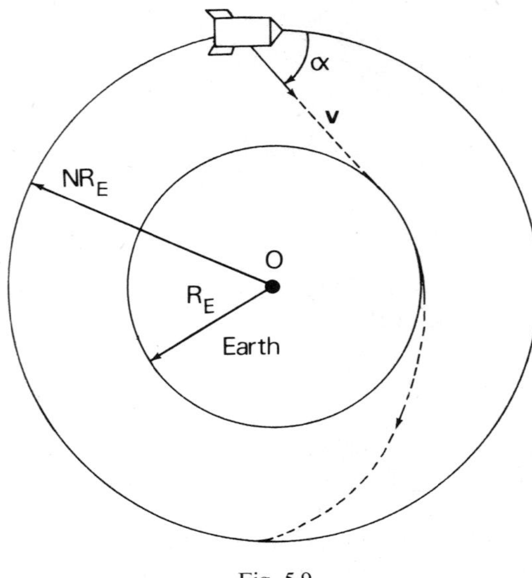

Fig. 5.9

The reader will easily confirm that the problem is solvable for all $N \geqslant 0$.

We note that in this example, since the energy equation

$$\tfrac{1}{2}mv^2 - \frac{\mu m}{r} = E$$

or

$$v^2 = \mu\left(\frac{2}{r} - \frac{2E}{m\mu}\right)$$

remains unchanged, the semi-major axis a of the elliptic orbit is

$$a = [\mu R_E(2N-1)/N]^{1/2} .$$

The length b of the semi-minor axis of the ellipse is given by

$$h = bv_B ,$$

where

$$v_B{}^2 = \mu\left(\frac{2}{a} - \frac{2E}{m\mu}\right) ,$$

as is evident from Fig. 5.3.

(i) *Communication satellites*

Two distinct types of communication satellite are used for transmitting signals from one part of the Earth's surface to another. Fig. 5.10(i) illustrates the *passive system* in which a signal transmitted from the ground station T is simply reflected from the satellite in orbit round the Earth to the receiving

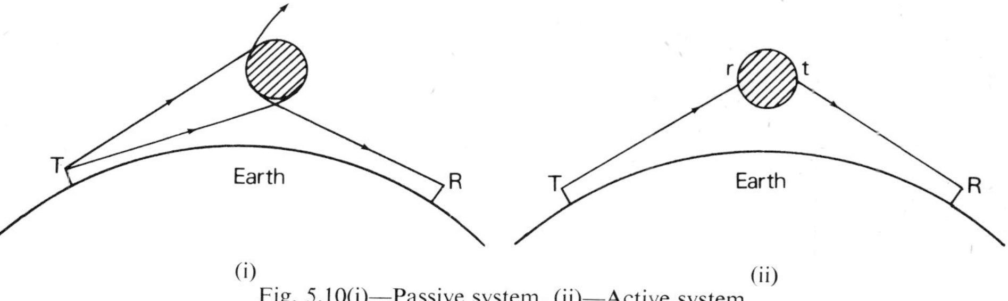

Fig. 5.10(i)—Passive system. (ii)—Active system.

ground station R. In the *active system* of Fig. 5.10(ii), however, the signal from T is received at r on the surface of the satellite and then undergoes amplification by means of the equipment which the satellite is designed to carry. Subsequently transmission of the amplified signal takes effect at t on the satellite's surface to R on the ground. In either case the satellites may be in motion or at rest relative to the Earth. The active system is of greater practical import.

Suppose the satellite is at rest relative to the Earth. Then it must revolve round the Earth in a circular orbit. From Section 5.3 the periodic time for description of a circular orbit of radius R_E+x is

$$T=2\pi\,(R_E+x)^{3/2}\mu^{-1/2}\ ,$$

confirming the existence of a unique x, the height of the satellite above the Earth, for circular orbital stability. Thus

$$x=-R_E+(\gamma M_E)^{1/3}(T/2\pi)^{2/3}\ .$$

The period of the satellite is 1 day or $T=8\cdot 64\times 10^4$ s. Further, $M_E=5\cdot 976\times 10^{24}$ kg, $\gamma=6\cdot 67\times 10^{-11}$ m^3 kg^{-1} s^2, $R_E=6368$ km. Calculation gives

$$x=3\cdot 58\times 10^4 \text{ km.}$$

Now suppose such a satellite is orbiting in a circle in the same plane as the equator and that the satellite covers an angle of 2α along the equator (Fig. 5.11). Then

$$\cos\alpha=R_E/(x+R_E)=6368/42\,140$$

and so $$\alpha\simeq 81^\circ\ .$$

Thus one satellite covers an angle of 162°. Two satellites, stationary relative to the Earth, would be insufficient to cover all points on the equator, but coverage could certainly be obtained using three of them.

(ii) *Evaluation of satellite orbit from initial conditions*

Satellites are usually launched into orbit by means of multi-staged rockets. A satellite will be the payload of such a rocket (Section 3.8). If we take as zero time the instant when the payload just enters the required orbit, then the nature of the orbit may be evaluated when certain of the initial conditions

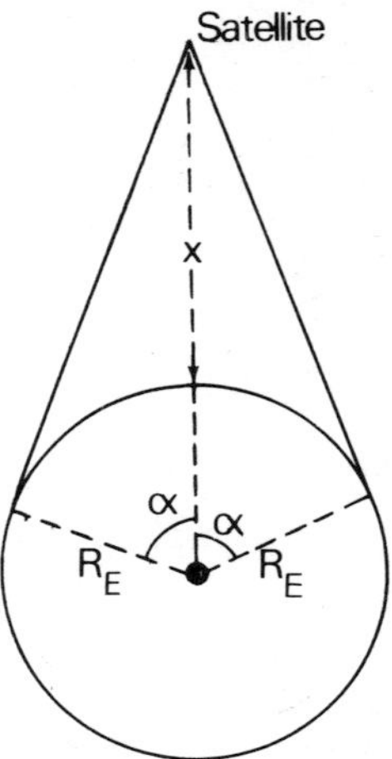

Fig. 5.11

are prescribed. Let r be the distance of the satellite from the Earth's centre at time t and v its speed. If $\mathbf{r}$ and $\mathbf{v}$ are the vectors corresponding to r and v, we measure β, the angle between the normal to $\mathbf{r}$ and the velocity $\mathbf{v}$; β is termed the *heading angle*. At $t=0$ we suppose the quantities r, v, β have the initial values

$$r=r_0 \ , \quad v=v_0 \ , \quad \beta=\beta_0 \ .$$

Complete orbital evaluation necessitates finding the eccentricity e, the semi-latus rectum l and the angle θ_0 between the orbital perigee and the satellite's initial position (Fig. 5.12). We recall that the polar equation of the orbit is

$$l/r=1+e\cos\theta \ .$$

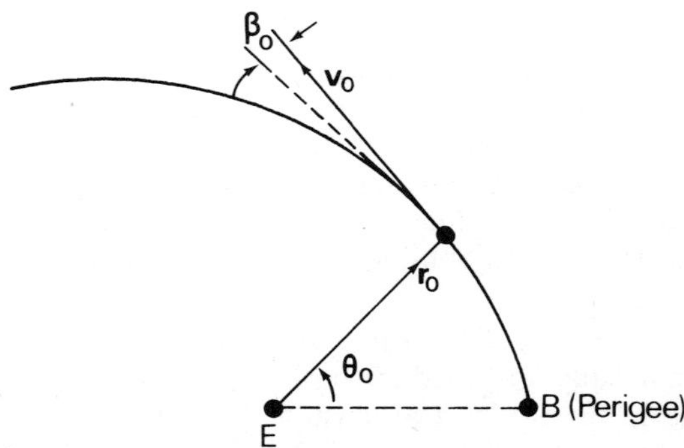

Fig. 5.12—Initial conditions.

Let us first find the eccentricity e of the orbit in terms of the initial conditions. The worked Example 1 of Section 5.3 shows that

$$h^2 = \mu a(1-e^2)$$

in the usual notation. This together with the relations

$$v_0{}^2 = \mu\left(\frac{2}{r_0} - \frac{1}{a}\right)$$

and

$$h = r_0 v_0 \cos\beta \qquad \text{(from Fig. 5.12)}$$

gives

$$e^2 = 1 + \frac{r_0{}^3 v_0{}^2 \cos^2\beta(r_0 v_0{}^2 - 2\mu)}{\mu^2} \ .$$

For the semi-latus rectum, we use

$$h^2 = \mu l$$

to give

$$l = (r_0{}^2 v_0{}^2 \cos^2\beta)/\mu \ .$$

From

$$l/r_0 = 1 + e\cos\theta_0 \ ,$$

we readily obtain

$$\underline{\cos\theta_0 = (r_0 v_0{}^2 \cos^2\beta - 1)/e} \ .$$

EXERCISE 5

(In the following examples, the central force per unit mass is denoted by either P or $f(r)$. Bookwork is often omitted from examination questions.)

1. A particle P, of mass m, is moving in a circle, radius a and centre O, under a force $\mu m(r+2a^3/r^2)$ directed towards O. If P is acted on by an impulse tangential to the path and of magnitude $(3\mu m^2 a^2)^{\frac{1}{2}}$, show that the velocity of P is immediately doubled, and that the greatest and least distances from O in the ensuing motion are a and $3a$. (L.U.)

2. Obtain the polar equation of the orbit for the case $f(r) = \mu/r^2$, and show that in this case, if a particle is projected from a great distance with speed v_o in a line whose perpendicular distance from O is p, then the closest approach of the particle to O is $[\mu + (\mu^2 + p^2 v_o{}^4)^{\frac{1}{2}}]/v_o{}^2$. (L.U.)

3. If $P = \mu(u^2 - au^3)$, where $a > 0$, and a particle is projected from an apse at a distance a from the centre of force with a velocity $(\mu c/a^2)^{\frac{1}{2}}$, where $a > c$, prove that the other apsidal distance of the orbit is $a(a+c)/(a-c)$ and find the apsidal angle. (L.U.)

4. A particle of mass m describes the curve $r = a/(1+2\cos^2\theta)$ under a force to the origin O. If the speed of the particle at the further apse is v, find the law of force.

When passing A, one of the near apses, the particle receives an impulse

mkv along the outward normal. Show that if $k < 2\sqrt{3}$, the particle continues to describe a closed curve about the origin, but that if $k > 2\sqrt{3}$ the new orbit has one asymptote making an angle θ with OA given by $\cos(\alpha + 2\theta) + 2\cos\alpha = 0$, where $\tan\alpha = \frac{1}{2}k$. (L.U.)

5. Establish a criterion for the stability of a circular orbit of radius a, with centre at the centre of force, described under the influence of the force $f(r)$.

If $f(r) = (\mu/r^2)e^{-\lambda r}$, where μ and λ are positive constants, prove that a circular orbit of radius a is stable provided that $a\lambda < 1$. Show also that if $a\lambda$ is small, the apse-line in a nearly circular orbit of mean radius a will advance in each revolution through an angle $\pi a\lambda$ approximately. (L.U.)

6. A particle P moves with acceleration μ/SP^2 directed towards a fixed point S. Prove that its orbit is an ellipse or hyperbola according as $V^2 <$ or $> 2\mu/R$, where V is the speed of the particle when $SP = R$.

A particle which is describing a circle under this law of force, collides and coalesces with an equal particle which is at rest. Show that, if the composite particle moves under the same law of force, it will describe an ellipse of eccentricity 3/4. Show also that the periods of description of the circle and the ellipse are in the ratio $7\sqrt{7} : 8$. (L.U.)

7. A particle P, of unit mass, moves under the action of a force of magnitude μ/SP^2 directed to a fixed point S. If the velocity of P is V when SP is r, show that the path of P is a conic having S as focus, and that the conic is an ellipse, parabola or hyperbola according as V^2 is less than, equal to, or greater than $2\mu/r$.

If the path of P is an ellipse, and ω_1, ω_2 are the greatest and least angular velocities of SP, show that the mean angular velocity of SP is $2(\omega_1\omega_2)^{\frac{3}{4}}/(\omega_1^{\frac{1}{2}} + \omega_2^{\frac{1}{2}})$. (L.U.)

8. A particle describes an ellipse of major axis $2a$ and eccentricity e under a force directed to a focus S. Find an expression for the velocity of the particle at any point.

When the particle is at the end of the latus rectum through S its velocity is suddenly turned through a right angle without change of magnitude. Prove that the particle will now describe an ellipse whose major axis is still of length $2a$ but is inclined at an angle $\tan^{-1}\{(1-e^2)/e\}$ to the major axis of the original orbit. Find the eccentricity of the new orbit. (L.U.)

9. A particle is describing a circle of radius a under an inverse square law of force μ/r^2 per unit mass directed towards the centre of the circle. An outward radial velocity $(\mu/5a)^{\frac{1}{2}}$ is suddenly impressed on it in such a way as to leave the transverse velocity unaltered; prove that it will describe an ellipse and find the new periodic time. (L.U.)

10. A particle of mass m moving in a circle of radius c under an attractive force μ/r^2 per unit mass towards the centre, collides and coalesces with a particle of mass λm which is at rest. Show that the orbit of the combined mass is an ellipse with major axis $c \operatorname{cosec}^2\alpha$, latus rectum $4c\cos^2\alpha$, and eccentricity $-\cos 2\alpha$, where $\sec^2\alpha = 2(1+\lambda)^2$. (L.U.)

11. A particle P of unit mass is moving under a central force μ/r^2 directed towards a fixed point S. Establish the relations

$$V^2 = \mu(2/r - 1/a) \ ,$$
$$h^2 = \mu a(1 - e^2) \ , \quad (e < 1)$$

where the initial conditions are such that $V^2 < 2\mu/r$ and the symbols have their usual meanings.

If when the particle is at P its radial component of velocity is suddenly destroyed, show that the focal chord PSP' of the original elliptical orbit becomes the major axis of the new elliptical orbit. (L.U.)

12. Two particles of equal mass are moving in coplanar ellipses under a central attraction μ/r^2 per unit mass. The major axis of each orbit is $2a$ and the latera recta are $2l_1$, $2l_2$ respectively. The particles collide at a point at a distance c from the centre of force and coalesce. Prove that the latus rectum of the orbit of the combined particles is $\frac{1}{4}(\sqrt{l_1} + \sqrt{l_2})^2$ and that the major axis of the new ellipse is $2[(2/c) \sin^2 (\alpha/2) + (1/a) \cos^2 (\alpha/2)]^{-1}$, α being the angle between the velocities of the masses before the collision. (L.U.)

13. Show that the speed v of a particle moving under a central force μ/r^2 per unit mass towards the pole (origin) is given by $v^2 = C + 2\mu/r$, and show that the path is one branch of a hyperbola when $C > 0$. When at a great distance from the sun, a comet moves towards it with speed V in the direction of the straight line whose perpendicular distance from the sun's centre is p. Show that the angle through which the comet's path is ultimately deflected is $2 \cot^{-1}(pV^2/M\gamma)$, where M is the mass of the sun, and γ is the constant of gravitation. (L.U.)

14. A star Q of mass M, initially at rest, is approached by a star P of mass m whose velocity $\mathbf{V}$ at a great distance from Q is along a line at a perpendicular distance p from Q. Show that the orbit of P relative to Q is hyperbolic with semi-major axis of length $a = \gamma(M + m)/V^2$ and that the velocity of P relative to Q is deflected through an angle 2α, where $\tan \alpha = a/p$, γ being the constant of gravitation.

Show further that Q acquires a velocity of magnitude $2\gamma m V[\gamma^2(M + m)^2 + p^2V^4]^{-\frac{1}{2}}$ and determine its direction. (L.U.)

15. A missile P is projected from the Earth's surface at A with speed $(2gd)^{\frac{1}{2}}$, the Earth being supposed spherical, of radius c and centre O, and $d < c$. The missile is assumed to be attracted by a force gc^2/r^2 per unit mass towards O, where $OP = r$. Show that the path of P is an ellipse having O as focus, and that the speed v of P is given by $v^2 = 2g(d - c + c^2/r)$.

Show that the locus of the second focus H, for all paths leaving A in a vertical plane with the same speed, is a circle of centre A and radius $cd/(c - d)$.

Hence, or otherwise, deduce that for the greatest range on a given line through A in the vertical plane, the direction of propagation bisects the angle between the vertical and the line. (L.U.)

16. A particle P, of mass m, is attracted towards the origin O of rectangular co-ordinates x, y by a force mn^2OP. Initially, the particle is projected from the point $(a, 0)$ with velocity u parallel to the axis of y. Prove that the

particle describes the ellipse

$$(x/a)^2+(ny/u)^2=1\ ,$$

and the eccentric angle of P on the ellipse increases at a constant rate.

Show that the hodograph of the motion is an ellipse which is similar to the orbit. (L.U.)

17. A particle P moves under a force $\omega^2\overline{PO}$ per unit mass towards the fixed point O. If its initial position is A and its initial velocity is represented by $\omega\overline{OC}$, show that its position at time t is given by

$$\overline{OP}=\overline{OA}\cos\omega t+\overline{OC}\sin\omega t\ ,$$

and that the orbit is an ellipse with centre O.

If the particle passes through a given point B, where OB is perpendicular to OA, show that the point C must lie on a hyperbola with semi axes OB, OA. (L.U.)

18. A particle of unit mass is describing an elliptic orbit of axes $2a$ and $2b$ under an attraction μr to the centre C when it is at a distance r from C; prove that its speed v is given by the equation $v^2=\mu(a^2+b^2-r^2)$, and that the moment of its velocity about C is $\sqrt{\mu}ab$.

If, in the above orbit, $a=2b$ and the intensity of attraction is suddenly increased in the ratio $1:4$ when the particle is at a distance $3b/2$ from C, prove that the axes of the new orbit are $(\sqrt{79}\pm\sqrt{15})b/4$. (L.U.)

19. Two particles P and Q, each of unit mass, attract one another with a force n^2PQ. The two particles move under their mutual attraction and also under the influence of a fixed centre of force O which attracts with a force n^2 times the distance from O. Show that G, the centre of mass of P and Q, describes an ellipse about O as centre with period $2\pi/n$, and that the orbit of P or Q relative to G is an ellipse with G as centre described in the period $(2\pi/n\sqrt{3})$.

If initially OPQ is a straight line with $OP=PQ=a$, and Q is at rest while P is projected at right angles to OPQ with speed v, show that the axes of the ellipse described by G are of lengths $3a$ and v/n. Find the lengths of the axes described by P relative to G. (L.U.)

20. A comet travelling in an elliptic orbit round the Sun under an attraction μ/r^2 per unit mass has its tangential velocity increased a small amount δv. Taking $2a$ to be the major axis and e the eccentricity of the former orbit, show that the comet's least distance from the Sun is increased by $4\delta v[a^3(1-e)/\mu(1+e)]^{\frac{1}{2}}$.

21. If the Sun's mass decreases slightly so that μ changes to $\mu-\delta\mu$, prove, with usual notation, the following results:

$$\delta a=a(2a/r-1)\delta\mu/\mu\ ;$$
$$\angle S''SS'=(\delta\mu/e\mu)\sin\angle PSS'\ ;$$
$$\delta e=(1/e-e)(1-a/r)\delta\mu/\mu\ .$$

If μ decreases at a continuous rate $n\mu$ per unit time, derive corresponding expressions for the rates of change of these elements.

22. *Escape velocity of a particle under a central force.*

Let v_o be the speed at $r=a$ of a particle subject to a central force $f(r)$ per unit mass and let v be the velocity at r from the centre of attraction. Show by energy methods or otherwise that

$$v^2=v_o^2-V^2+2\int_r^\infty f(r)dr \ ,$$

where $V^2=2\int_a^\infty f(r)dr$. This shows that when $v_o<V$, v is imaginary for $r=\infty$, but that v is real for $r=\infty$ when $v_o\geqslant V$. V is called the *escape velocity*: for $v_o\geqslant V$ the particle can escape to infinity. Taking $f(r)=\mu/r^2$, we find the escape velocity is $\sqrt{(2\mu/a)}$.

Chapter 6

Motion of a System of Particles

6.1 LINEAR MOMENTUM OF A SYSTEM OF PARTICLES

Let m be the constant mass of a typical particle P of a system of n such particles moving generally in space, such that at time t, $\overline{OP} \equiv \mathbf{r}$, O being a fixed origin. Then if G denote the position of the centroid of the system and if $\overline{OG} \equiv \bar{\mathbf{r}}$, we have, writing $M = \Sigma m$,

$$\Sigma m\mathbf{r} = M\bar{\mathbf{r}} \ ,$$

the summations here and in the following being taken over the entire system. Differentiating the last result with respect to t gives

$$\Sigma m\dot{\mathbf{r}} = M\dot{\bar{\mathbf{r}}} \ .$$

Now $m\dot{\mathbf{r}}$ is the linear momentum of the single particle P, and so $\Sigma m\dot{\mathbf{r}}$ is that of the entire system of particles. Further, $M\dot{\bar{\mathbf{r}}}$ represents the linear momentum of a single particle of mass M concentrated at G. We have thus shown *that at any instant the total linear momentum of a system of particles moving generally in space is the same as that of a single particle concentrated at the centroid of the system and having a velocity equal to that of the centroid and a mass equal to the total mass of the system.*

Denoting the total linear momentum of the system by $\mathbf{p}$, we have

$$\dot{\mathbf{p}} = \Sigma m\ddot{\mathbf{r}} = M\ddot{\bar{\mathbf{r}}} \ ,$$

showing that at any instant the rate of change of linear momentum of the system is equal to the mass of it multiplied by the acceleration of the centroid.

Further suppose the particle P to be subjected to an external force $\mathbf{F}$ and an internal force $\mathbf{F}'$. Then, since the total force on P is $\mathbf{F}+\mathbf{F}'$, its equation of motion is

$$\mathbf{F}+\mathbf{F}' = m\ddot{\mathbf{r}} \ .$$

Summing over the entire system and remembering that the sum of the internal forces $\Sigma\mathbf{F}'$ is zero, since by Newton's third law action and reaction are equal and opposite, we have

$$\Sigma\mathbf{F} = \Sigma m\ddot{\mathbf{r}} = \dot{\mathbf{p}} \ ,$$

showing that at any instant the vector sum of the external forces acting on a system of particles is equal to the rate of change of linear momentum of the system. In particular, when $\Sigma\mathbf{F}=\mathbf{0}$, $\dot{\mathbf{p}}=\mathbf{0}$, so that *the linear momentum is constant for a system having no resultant external force acting.* This last result is the *principle of conservation of linear momentum* established for a system of particles. In the previous chapter this result was established for two-particle systems moving solely under their mutual gravitational attraction.

Example

The cross-section of a right cylinder of mass M is bounded by a parabolic arc and the latus rectum (length $2l$) of the parabola. The surface of the cylinder is smooth and it is placed with its rectangular face on a horizontal plane. A particle of mass m is placed at the midpoint of the highest generator of the cylinder and slightly disturbed.

Using the principles of conservation of energy and linear momentum, find the speed of the cylinder and the speed of the particle relative to the cylinder when the particle, still in contact with the surface, has descended a vertical distance x. (L.U.)

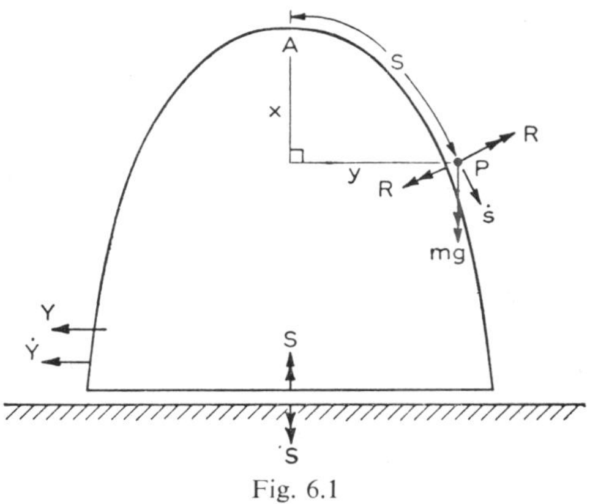

Fig. 6.1

In time t, suppose the particle has descended a distance x and become displaced horizontally a distance y relative to its original position A, and suppose Y denotes the horizontal movement of the wedge, where y, Y are measured in opposite senses (Fig. 6.1). Then, from the geometry of the parabola,

$$y^2=2lx\ . \tag{1}$$

Since the net displacement of m in the direction of motion of the cylinder is $Y-y$, the total linear momentum of the system in this direction is $m(\dot{Y}-\dot{y})+M\dot{Y}$. This quantity is constant and therefore zero, since the resultant external force in the horizontal direction is zero.

$$\therefore\ m(\dot{Y}-\dot{y})+M\dot{Y}=0\ . \tag{2}$$

The P.E. lost by the particle is mgx. The K.E. gained by the particle is $\frac{1}{2}m[\dot{x}^2+(\dot{Y}-\dot{y})^2]$ and that of the cylinder is $\frac{1}{2}M\dot{Y}^2$. Since the forces are conservative,

$$mgx=\tfrac{1}{2}M\dot{Y}^2+\tfrac{1}{2}m[\dot{x}^2+(\dot{Y}-\dot{y})^2] \ . \tag{3}$$

From (1), we find $\dot{y}=\dot{x}(l/2x)^{\frac{1}{2}}$ and from (2), $\dot{Y}=[m\dot{x}/(M+m)](l/2x)^{\frac{1}{2}}$, and substitution in (3) gives

$$\dot{x}=x\{4g(M+m)/[Ml+2(M+m)x]\}^{\frac{1}{2}} \ .$$

Further,

$$\dot{y}=\dot{x}(l/2x)^{\frac{1}{2}}=\{2lgx(M+m)/[Ml+2(M+m)x]\}^{\frac{1}{2}} \ ,$$

$$\dot{y}-\dot{Y}=[M/(M+m)]\{2lgx(M+m)/[Ml+2(M+m)x]\}^{\frac{1}{2}} \ .$$

Thus, denoting by v the speed of the particle,

$$\begin{aligned} v^2 &= \dot{x}^2+(\dot{y}-\dot{Y})^2 \\ &= \underline{2(M+m)gx\{2x+M^2l/(M+m)^2\}\{Ml+2(M+m)x\}^{-1}} \ . \end{aligned}$$

Further, $$\dot{Y}=\underline{[m/(M+m)]\{2lgx(M+m)/[Ml+2(M+m)x]\}^{-1}} \ .$$

6.2 ANGULAR MOMENTUM AND RATE OF CHANGE OF ANGULAR MOMENTUM OF A SYSTEM

Continuing in the notation of the last section, the angular momentum or moment of momentum of the particle P about the fixed point O is $\mathbf{r}_\wedge m\dot{\mathbf{r}}$. Hence the total (vector) angular momentum about O is

$$\mathbf{H}=\Sigma(\mathbf{r}_\wedge m\dot{\mathbf{r}}) \ , \tag{1}$$

the summation, as always, being over the entire particle system. The physical angular momentum about a line through O specified by the unit vector $\hat{\mathbf{a}}$ is then $\hat{\mathbf{a}}.\mathbf{H}$.

Differentiating (1) with respect to t,

$$\dot{\mathbf{H}}=\Sigma(\dot{\mathbf{r}}_\wedge m\dot{\mathbf{r}})+\Sigma(\mathbf{r}_\wedge m\ddot{\mathbf{r}}) \ ,$$

i.e.

$$\dot{\mathbf{H}}=\Sigma(\mathbf{r}_\wedge m\ddot{\mathbf{r}}) \ . \tag{2}$$

(2) *shows that the rate of change of angular momentum about the fixed point O is equal to the total moment of the rate of change of linear momentum about O.*

Since $\mathbf{F}+\mathbf{F}'=m\ddot{\mathbf{r}}$, in the notation of the last section,

$$\therefore \ \dot{\mathbf{H}}=\Sigma(\mathbf{r}_\wedge\mathbf{F})+\Sigma(\mathbf{r}_\wedge\mathbf{F}') \ .$$

We now show that the last summation on the R.H.S. is zero. To this end, let suffixes i, j refer to different particles of the system and let $\mathbf{F}'_{ij}$ denote the force which particle P_i exerts on particle P_j. Then $\mathbf{F}'_{ij}=-\mathbf{F}'_{ji}$, in virtue of Newton's Third Law of Motion ($i\neq j$). Further,

$$\Sigma(\mathbf{r}_\wedge\mathbf{F}') = \sum_i \sum_j \mathbf{r}_{i\wedge}\mathbf{F}'_{ji} \qquad (i \neq j)$$
$$= \sum_i \sum_j (\mathbf{r}_i - \mathbf{r}_j)_\wedge \mathbf{F}'_{ji} + \sum_i \sum_j (\mathbf{r}_{j\wedge}\mathbf{F}'_{ji})$$
$$= \sum_i \sum_j (\mathbf{r}_{j\wedge}\mathbf{F}'_{ji}) \ .$$

This last result follows since the vectors $(\mathbf{r}_i - \mathbf{r}_j)$ and $\mathbf{F}'_{ji}$ are parallel—the interaction $\mathbf{F}'_{ji}$ being along the line joining P_i, P_j. Since $\mathbf{F}'_{ji} = -\mathbf{F}'_{ij}$,

$$\sum_i \sum_j (\mathbf{r}_{j\wedge}\mathbf{F}'_{ji}) = -\sum_i \sum_j (\mathbf{r}_{j\wedge}\mathbf{F}'_{ij}) = -\sum_j \sum_i (\mathbf{r}_{j\wedge}\mathbf{F}'_{ij}) \ .$$

This last sum is essentially

$$-\Sigma(\mathbf{r}_\wedge\mathbf{F}').$$

Thus

$$\Sigma(\mathbf{r}_\wedge\mathbf{F}') = \mathbf{0} \ ,$$

as required.

We have thus established that

$$\dot{\mathbf{H}} = \Sigma(\mathbf{r}_\wedge\mathbf{F}) \ , \qquad (2)$$

i.e. *the rate of change of vector angular momentum about a fixed point for a system of particles moving generally in space is equal to the sum of the moments of the external forces acting on that system about the point.*

One can further establish that *the rate of change of angular momentum about any axis through a fixed point for such a system is equal to the sum of the physical moments of the external forces acting on that system about the line.* (Scalar multiply equation (2) by the unit vector specifying the axial direction.)

6.3 USE OF CENTROID

Let G be the centroid, so that

$$\overline{OP} \equiv \mathbf{r} \ , \quad \overline{OG} \equiv \bar{\mathbf{r}} \ , \quad \overline{GP} \equiv \mathbf{r}' = \mathbf{r} - \bar{\mathbf{r}} \ .$$

The total kinetic energy of the system is

$$T = \tfrac{1}{2}\,\Sigma m\dot{\mathbf{r}}^2$$
$$= \tfrac{1}{2}\,\Sigma m(\dot{\mathbf{r}}' + \dot{\bar{\mathbf{r}}})^2$$
$$= \tfrac{1}{2}\,\Sigma m\dot{\mathbf{r}}'^2 + \Sigma m\dot{\mathbf{r}}'.\dot{\bar{\mathbf{r}}} + \tfrac{1}{2}\,\Sigma m\dot{\bar{\mathbf{r}}}^2$$
$$= \tfrac{1}{2}\,\Sigma m\dot{\mathbf{r}}'^2 + \dot{\bar{\mathbf{r}}}.\Sigma m\dot{\mathbf{r}}' + \tfrac{1}{2}\,\dot{\bar{\mathbf{r}}}^2\,\Sigma m \ .$$

Now $\Sigma m\mathbf{r}' = \mathbf{0}$, and so $\Sigma m\dot{\mathbf{r}}' = d(\Sigma m\mathbf{r}')/dt = \mathbf{0}$.
Further, $\Sigma m = M$. Hence we have shown

$$\underline{T = \tfrac{1}{2}M\dot{\bar{\mathbf{r}}}^2 + \tfrac{1}{2}\,\Sigma m\dot{\mathbf{r}}'^2} \ , \qquad (1)$$

i.e. *the K.E. of a system of particles moving generally in space is equal to the sum of the K.E. of a single particle of total mass equal to that of the system, concentrated at its centroid and moving with the centroid's velocity, together with the K.E. of the system in its motion relative to the centroid.*

The angular momentum about O is

$$\begin{aligned}\mathbf{H} &= \Sigma(\mathbf{r}_\wedge m\dot{\mathbf{r}}) \\ &= \Sigma\{(\bar{\mathbf{r}}+\mathbf{r}')_\wedge m(\dot{\bar{\mathbf{r}}}+\dot{\mathbf{r}}')\} \\ &= \bar{\mathbf{r}}_\wedge(M\dot{\bar{\mathbf{r}}})+\bar{\mathbf{r}}_\wedge(\Sigma m\dot{\mathbf{r}}')+(\Sigma m\mathbf{r}')_\wedge\dot{\bar{\mathbf{r}}}+\Sigma(\mathbf{r}'_\wedge m\dot{\mathbf{r}}') \ .\end{aligned}$$

Now $\Sigma m\mathbf{r}'=\mathbf{0}$ and so $\Sigma m\dot{\mathbf{r}}'=d(\Sigma m\mathbf{r}')/dt=\mathbf{0}$.

$$\therefore \ \underline{\mathbf{H}=\bar{\mathbf{r}}_\wedge(M\dot{\bar{\mathbf{r}}})+\Sigma(\mathbf{r}'_\wedge m\dot{\mathbf{r}}')} \ , \tag{2}$$

i.e. *the angular momentum of the system about a fixed point O is equal to that about O of a single particle of total mass equal to that of the entire system, concentrated at its centroid and moving with the centroid's velocity, together with the angular momentum about the centroid of the system in its motion relative to the centroid.*

On differentiating (2) with respect to t, we find

$$\dot{\mathbf{H}}=\bar{\mathbf{r}}_\wedge(M\ddot{\bar{\mathbf{r}}})+\Sigma(\mathbf{r}'_\wedge m\ddot{\mathbf{r}}') \ , \tag{3}$$

showing that the rate of change of angular momentum of the system about O is equal to the moment of the rate of change of linear momentum about O of a single particle of total mass equal to that of the entire system, concentrated at its centroid and having the centroid's motion, together with the moment of the rate of change of momentum about the centroid of the system in its motion relative to the centroid. Further, this is equal to the total moment about O of the external forces acting on the system.

The reader will easily show that

$$\Sigma(\mathbf{r}'_\wedge m\ddot{\mathbf{r}}')=d\{\Sigma(\mathbf{r}'_\wedge m\dot{\mathbf{r}}')\}/dt \ ,$$

so that in the above enunciation, the phrase "moment of the rate of change of momentum about the centroid of the system in its motion relative to the centroid" may be replaced by "rate of change of the moment of momentum of the system in its motion relative to the centroid". We shall subsequently show that such reciprocity applies only for fixed points and centroids of systems, but not for moving origins.

6.4 MOVING ORIGINS

Now suppose O is an origin moving with velocity $\mathbf{v}_o$. Let P be a particle of mass m of the system, G the centroid of the system and let $\overline{OP}\equiv\mathbf{r}$, $\overline{OG}\equiv\bar{\mathbf{r}}$, $\overline{GP}\equiv\mathbf{r}'$, so that $\mathbf{r}=\bar{\mathbf{r}}+\mathbf{r}'$ Let $\mathbf{v}$ be the velocity of P and M the total mass of the system (Fig. 6.2). The angular momentum of the system about O is

$$\begin{aligned}\mathbf{H} &= \Sigma(\mathbf{r}_\wedge m\mathbf{v})=\Sigma\{(\bar{\mathbf{r}}+\mathbf{r}')_\wedge m(\mathbf{v}_o+\dot{\mathbf{r}})\} \\ &= \bar{\mathbf{r}}_\wedge M\mathbf{v}_o+\bar{\mathbf{r}}_\wedge d(\Sigma m\mathbf{r})/dt+(\Sigma m\mathbf{r}')_\wedge\mathbf{v}_o+\Sigma(\mathbf{r}'_\wedge m\dot{\mathbf{r}}) \ .\end{aligned}$$

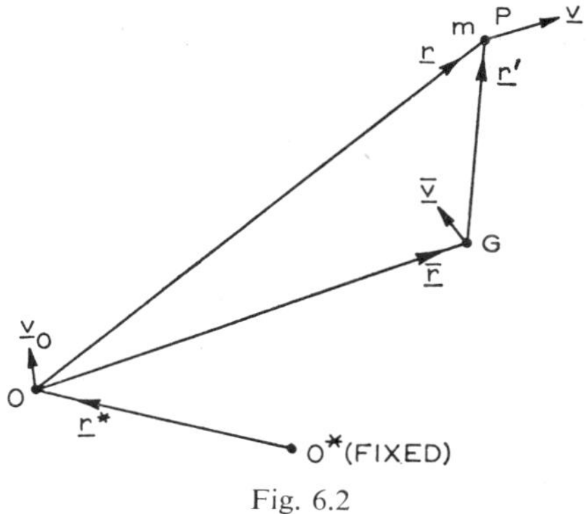

Fig. 6.2

Now $\Sigma m\mathbf{r} = M\bar{\mathbf{r}}$, $\Sigma m\mathbf{r}' = \mathbf{0}$ and

$$\begin{aligned}\Sigma(\mathbf{r}'{}_\wedge m\dot{\mathbf{r}}) &= \Sigma\{\mathbf{r}'{}_\wedge m(\dot{\bar{\mathbf{r}}} + \dot{\mathbf{r}}')\} = (\Sigma m\mathbf{r}')_\wedge \dot{\bar{\mathbf{r}}} + \Sigma(\mathbf{r}'{}_\wedge m\dot{\mathbf{r}}') \\ &= \Sigma(\mathbf{r}'{}_\wedge m\dot{\mathbf{r}}') \ . \\ \therefore\ \mathbf{H} &= \bar{\mathbf{r}}_\wedge M\mathbf{v}_o + \bar{\mathbf{r}}_\wedge M\dot{\bar{\mathbf{r}}} + \Sigma(\mathbf{r}'{}_\wedge m\dot{\mathbf{r}}') \\ &= \bar{\mathbf{r}}_\wedge M\bar{\mathbf{v}} + \Sigma(\mathbf{r}'{}_\wedge m\dot{\mathbf{r}}') \ , \end{aligned} \tag{1}$$

i.e. *the angular momentum of the particle system about O is equal to the angular momentum about O of a single particle of total mass equal to that of the entire system, concentrated at its centroid and moving with its centroid's velocity, together with the angular momentum about the centroid of the system in its motion relative to the centroid.*

The principle has thus been established for both fixed and moving points O.

In the equation (1) above, write $\mathbf{H}' = \Sigma(\mathbf{r}'{}_\wedge m\dot{\mathbf{r}}')$. Then

$$\begin{aligned}\dot{\mathbf{H}} &= \dot{\bar{\mathbf{r}}}_\wedge M\bar{\mathbf{v}} + \bar{\mathbf{r}}_\wedge M\dot{\bar{\mathbf{v}}} + \dot{\mathbf{H}}' \\ &= (\bar{\mathbf{v}} - \bar{\mathbf{v}}_o)_\wedge M\bar{\mathbf{v}} + \bar{\mathbf{r}}_\wedge M\dot{\bar{\mathbf{v}}} + \dot{\mathbf{H}}' \\ &= \dot{\mathbf{H}}' + \bar{\mathbf{r}}_\wedge M\dot{\bar{\mathbf{v}}} - \mathbf{v}_{o\wedge} M\bar{\mathbf{v}} \ .\end{aligned}$$

Now let $O \equiv G$. Then $\bar{\mathbf{r}} = \mathbf{0}$, $\mathbf{v}_o = \bar{\mathbf{v}}$ and so

$$\dot{\mathbf{H}} = \dot{\mathbf{H}}' = d\Sigma(\mathbf{r}'{}_\wedge m\dot{\mathbf{r}}')/dt = \Sigma(\mathbf{r}'{}_\wedge m\ddot{\mathbf{r}}'') \ ,$$

i.e. *the rate of change of angular momentum of the system of particles about its centroid G is equal to the total moment of the rate of change of momentum about G of the system in its motion relative to G.*

Taking $\bar{\mathbf{r}} = \mathbf{0}$ in equation (3) of the previous section, we see that the value of $\dot{\mathbf{H}}$ obtained above for the case $O \equiv G$ is the same as if G were at rest. Thus *when calculating the rate of change of the angular momentum of the particle system about its centroid, we may treat the centroid as if it were at rest.*

Now introduce a *fixed* origin O^* in Fig. 6.2. Let $\mathbf{F}$ denote the total external force on the particle at P and $\mathbf{L}^*$ the total moment about O^*, $\mathbf{L}$ that about O.

Then if $\overline{O^*P} \equiv \mathbf{r}^* + \mathbf{r}$,

$$\mathbf{L}^* = \Sigma(\mathbf{r}^* + \mathbf{r})_\wedge \mathbf{F} = \mathbf{r}^*_\wedge \Sigma \mathbf{F} + \mathbf{L} \ .$$

If $\mathbf{H}^*$ be the angular momentum about O^*, then from the previous section,

$$\dot{\mathbf{H}}^* = \Sigma\{(\mathbf{r}^* + \mathbf{r})_\wedge \mathbf{F}\} = \mathbf{r}^*_\wedge \Sigma \mathbf{F} + \mathbf{L} \ .$$

But

$$\begin{aligned} \mathbf{H}^* &= \Sigma\{(\mathbf{r}^* + \mathbf{r})_\wedge m\mathbf{v}\} = \mathbf{r}^*_\wedge \Sigma m\mathbf{v} + \mathbf{H} \\ &= \mathbf{H} + \mathbf{r}^*_\wedge M\mathbf{v} \ . \\ \therefore \ \dot{\mathbf{H}} &= \mathbf{r}^*_\wedge \Sigma \mathbf{F} + \mathbf{L} - \dot{\mathbf{r}}^*_\wedge M\mathbf{v} - \mathbf{r}^*_\wedge M\dot{\bar{\mathbf{v}}} \\ &= \mathbf{L} - \mathbf{v}_{o\wedge} M\bar{\mathbf{v}} \ , \end{aligned}$$

since $\Sigma \mathbf{F} = M\dot{\bar{\mathbf{v}}}$. Now let $O \equiv G$. Then $\dot{\mathbf{H}} = \mathbf{L}$.

i.e. *The rate of change of the angular momentum of the particle system about its centroid is always equal to the vector sum of the moments about the centroid of all the external forces, irrespective of whether G be moving or at rest.*

Letting $\mathbf{F}'$ be the total internal force on the particle P, we have for the motion of P, since $\mathbf{F} + \mathbf{F}'$ is the total force on it,

$$m\dot{\mathbf{v}} = \mathbf{F} + \mathbf{F}' \ .$$

$$\therefore \ \Sigma(\mathbf{r}_\wedge m\dot{\mathbf{v}}) = \Sigma(\mathbf{r}_\wedge \mathbf{F}) + \Sigma(\mathbf{r}_\wedge \mathbf{F}') \ .$$

We have seen, however, that $\Sigma(\mathbf{r}_\wedge \mathbf{F}') = \mathbf{0}$ and so

$$\Sigma(\mathbf{r}_\wedge m\dot{\mathbf{v}}) = \Sigma(\mathbf{r}_\wedge \mathbf{F}) \ , \qquad (2)$$

i.e. *the total moment of the rate of change of momentum of the system about any point, moving or fixed, is always equal to the total moment of the external forces about that point.*

When O is fixed or coincident with the centroid G, the R.H.S. of (2) has been shown to equal the rate of change of moment of momentum about G. *Thus when we take moments about a point which is either at rest or coincident with the centroid of the system of particles, the rate of change of moment of momentum (or rate of change of angular momentum) about the point is always equal to the total moment of the external forces about the point.*

Example (Notation as above)

Prove that the angular momentum of the system about O is equal to the sum of the angular momenta about O of the motion relative to O and that of a particle of mass $M = \Sigma m$ at G moving with the velocity of O.

$$\begin{aligned} \mathbf{H} = \Sigma(\mathbf{r}_\wedge m\mathbf{v}) &= \Sigma\{(\mathbf{r}_\wedge m(\mathbf{v} - \mathbf{v}_o)\} + \Sigma(\mathbf{r}_\wedge m\mathbf{v}_o) \\ &= \Sigma\{\mathbf{r}_\wedge m(\mathbf{v} - \mathbf{v}_o)\} + (\Sigma m\mathbf{r})_\wedge \mathbf{v}_o \\ &= \Sigma\{\mathbf{r}_\wedge m(\mathbf{v} - \mathbf{v}_o)\} + (M\bar{\mathbf{r}})_\wedge \mathbf{v}_o \\ &= \underline{\Sigma\{\mathbf{r}_\wedge m(\mathbf{v} - \mathbf{v}_o)\} + \bar{\mathbf{r}}_\wedge (M\mathbf{v}_o)} \ . \end{aligned}$$

6.5 IMPULSIVE FORCES

Now suppose the particles of the considered system be subjected to impulsive forces. Let $\mathbf{I}$, $\mathbf{I}'$ be the resultant external and internal impulses acting on the particle P of mass m. Let the instantaneous velocity of P change from $\mathbf{v}_o$ to $\mathbf{v}$. Then since $\mathbf{I}+\mathbf{I}'$ is the resultant impulse applied to P and its linear momentum gain is $m(\mathbf{v}-\mathbf{v}_o)$,

$$\mathbf{I}+\mathbf{I}'=m(\mathbf{v}-\mathbf{v}_o) \ .$$

$$\therefore \ \Sigma\mathbf{I}+\Sigma\mathbf{I}'=\mathbf{p}-\mathbf{p}_o \ ,$$

where $\mathbf{p}=\Sigma m\mathbf{v}$, $\mathbf{p}_o=\Sigma m\mathbf{v}_o$. From Newton's Third Law of Motion and the nature of impulsive force, $\Sigma\mathbf{I}'=\mathbf{0}$.

$$\therefore \ \Sigma\mathbf{I}=\mathbf{p}-\mathbf{p}_o \ .$$

Thus the total external impulse applied to the system of particles is equal to the total change of linear momentum produced.

Letting $\overline{OP}\equiv\mathbf{r}$, we have

$$\Sigma\{\mathbf{r}_\wedge(\mathbf{I}+\mathbf{I}')\}=\Sigma\{\mathbf{r}_\wedge m(\mathbf{v}-\mathbf{v}_o)\} \ .$$

By arguments similar to those previously invoked, $\Sigma\mathbf{r}_\wedge\mathbf{I}'=\mathbf{0}$.

$$\therefore \ \Sigma(\mathbf{r}_\wedge\mathbf{I})=\mathbf{H}-\mathbf{H}_o \ ,$$

where $\mathbf{H}=\Sigma(\mathbf{r}_\wedge m\mathbf{v})$, $\mathbf{H}_o=\Sigma(\mathbf{r}_\wedge m\mathbf{v}_o)$.

i.e. *Irrespective of whether the origin is stationary or moving, the total vector sum of the moments of the external impulses about it is equal to the instantaneous increase in angular momentum produced about the origin.*

Example 1

Two particles, of masses m_1 and m_2 at A and B are connected by a rigid massless rod AB. Their velocities $\mathbf{v}_1$ and $\mathbf{v}_2$ are suddenly changed by the application of external impulses $\mathbf{J}_1$ and $\mathbf{J}_2$. Prove that the magnitude J of of the impulsive reaction of the rod on m_1 is

$$\{m_1m_2\mathbf{e}/(m_1+m_2)\}.\ (\mathbf{J}_2/m_2-\mathbf{J}_1/m_1)$$

where $\mathbf{e}$ is a unit vector in $\overline{AB}$. Also prove that the energy of the system is increased by an amount

$$\mathbf{J}_1.\mathbf{v}_1+\mathbf{J}_2.\mathbf{v}_2+\tfrac{1}{2}(\mathbf{J}_1{}^2/m_1+\mathbf{J}_2{}^2/m_2)-\tfrac{1}{2}(m_1+m_2)J^2/(m_1m_2) \ .$$

(Imperial College)

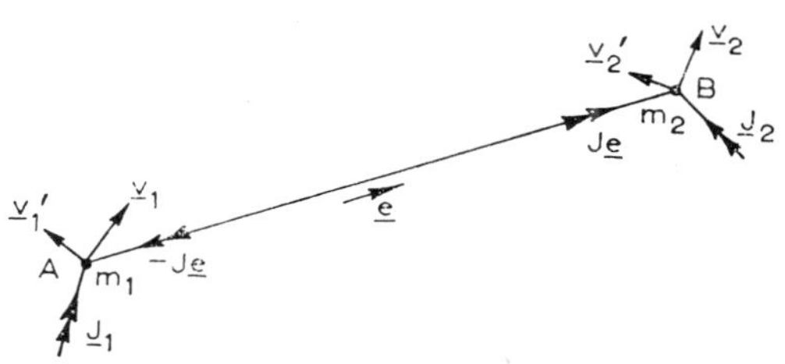

Fig. 6.3

Let $\mathbf{v}_1'$, $\mathbf{v}_2'$ be the velocities of A and B immediately after the action. The rod AB exerts an impulse $J\mathbf{e}$ on m_2 and $-J\mathbf{e}$ on m_1.

Since $\mathbf{J}_1 - J\mathbf{e}$ is the resultant impulse on m_1,

$$\mathbf{J}_1 - J\mathbf{e} = m_1(\mathbf{v}_1' - \mathbf{v}_1) \ . \tag{1}$$

Since $\mathbf{J}_2 + J\mathbf{e}$ is the resultant impulse on m_2,

$$\mathbf{J}_2 + J\mathbf{e} = m_2(\mathbf{v}_2' - \mathbf{v}_2) \ . \tag{2}$$

From (1) and (2),

$$\begin{aligned}&\mathbf{e}.(\mathbf{v}_1' - \mathbf{v}_1) - \mathbf{e}.(\mathbf{v}_2' - \mathbf{v}_2)\\ &= \mathbf{e}.\{(\mathbf{J}_1 - J\mathbf{e})/m_1 - (\mathbf{J}_2 + J\mathbf{e})/m_2\} \ .\end{aligned}$$

But since AB is a rigid connection,

$$\mathbf{e}.(\mathbf{v}_1' - \mathbf{v}_1) = \mathbf{e}.(\mathbf{v}_2' - \mathbf{v}_2) \ .$$

$$\therefore \ 0 = \mathbf{e}.(\mathbf{J}_1/m_1 - \mathbf{J}_2/m_2) - J(1/m_1 + 1/m_2) \ ,$$

or

$$\underline{J = -\{m_1 m_2 \mathbf{e}/(m_1 + m_2)\}. \{\mathbf{J}_2/m_2 - \mathbf{J}_1/m_1\}} \ .$$

The K.E. is originally

$$T = \tfrac{1}{2} m_1 \mathbf{v}_1^2 + \tfrac{1}{2} m_2 \mathbf{v}_2^2 \ ,$$

and after the application of the impulses it is

$$\begin{aligned}T' &= \tfrac{1}{2} m_1 \mathbf{v}_1'^2 + \tfrac{1}{2} m_2 \mathbf{v}_2'^2\\ &= \tfrac{1}{2} m_1 \{\mathbf{v}_1 + (\mathbf{J}_1 - J\mathbf{e})/m_1\}^2 + \tfrac{1}{2} m_2 \{\mathbf{v}_2 + (\mathbf{J}_2 + J\mathbf{e})/m_2\}^2\\ &= T + \mathbf{v}_1.(\mathbf{J}_1 - J\mathbf{e}) + \mathbf{v}_2.(\mathbf{J}_2 + J\mathbf{e})\\ &\quad + (\mathbf{J}_1 - J\mathbf{e})^2/(2m_1) + (\mathbf{J}_2 + J\mathbf{e})^2/(2m_2) \ .\end{aligned}$$

$$\begin{aligned}\therefore \ T' - T &= \mathbf{v}_1.\mathbf{J}_1 + \mathbf{v}_2.\mathbf{J}_2 + J(\mathbf{e}.\mathbf{v}_2 - \mathbf{e}.\mathbf{v}_1) + (\mathbf{J}_1^2 - 2J\mathbf{e}.\mathbf{J}_1 + J^2)/(2m_1)\\ &\quad + (\mathbf{J}_2^2 + 2J\mathbf{e}.\mathbf{J}_2 + J^2)/(2m_2)\\ &= \mathbf{v}_1.\mathbf{J}_1 + \mathbf{v}_2.\mathbf{J}_2 + \tfrac{1}{2}(\mathbf{J}_1^2/m_1 + \mathbf{J}_2^2/m_2) + (m_1 + m_2)J^2/(2m_1 m_2)\\ &\quad + J[(\mathbf{e}.\mathbf{J}_2)/m_2 - (\mathbf{e}.\mathbf{J}_1)/m_1] \ .\end{aligned}$$

From the derived expression for J, $(\mathbf{e}.\mathbf{J}_2)/m_2 - (\mathbf{e}.\mathbf{J}_1)/m_1 = -(m_1 + m_2) J/m_1 m_2$. Hence the required result.

Example 2

Two particles, of masses m_1 and m_2 at A and B, are connected by a rigid rod AB lying on a smooth horizontal table. If an impulse I is applied at A in the plane of the table and perpendicular to AB, find the initial velocities of A and B.

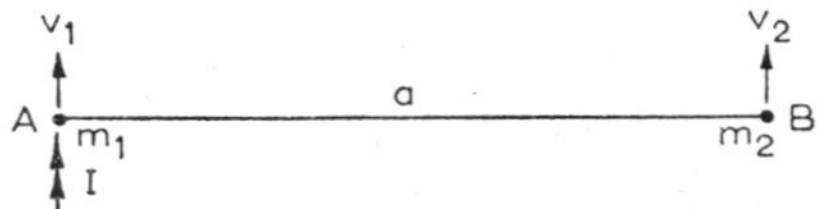

Fig. 6.4

Let a be the length of the rod. The centroid moves in the direction of the impulse and so A and B must also move in the same direction (Fig. 6.4).

The gain of linear momentum is $m_1v_1 + m_2v_2$.

$$\therefore \; m_1v_1 + m_2v_2 = I \; . \tag{1}$$

The moment of the impulse about B is equal to the gain of angular momentum of the system about B.

$$\therefore \; Ia = m_1v_1a \; ,$$

or

$$v_1 = I/m_1 \tag{2}$$

Thus,

$$v_2 = 0 \; .$$

Example 3

Particles A, B, C of masses m_1, m_2, m_3 lie on a smooth horizontal table and are connected together by taut inextensible strings AB, BC. The angle ABC is $\pi - \alpha$, α being acute. If an impulse I is applied to C along $\overline{BC}$, show that B starts to move in a direction making with $\overline{AB}$ an angle $\tan^{-1}\{(1 + m_1/m_2) \tan \alpha\}$. Find the initial velocity of A.

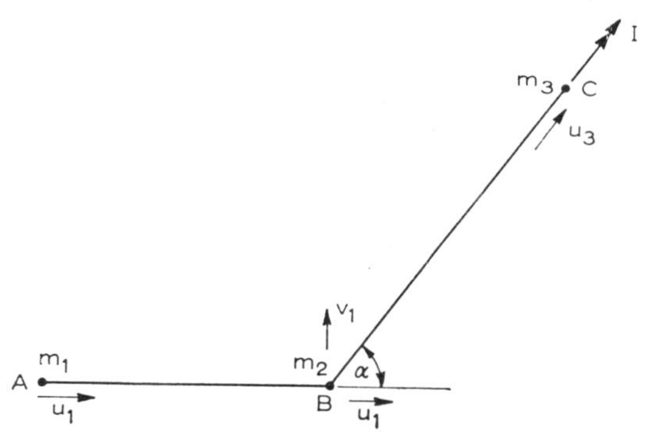

Fig. 6.5

Since an impulsive tension is produced in AB, the motion of A is along $\overline{AB}$. Let u_1 be its velocity in this direction. C is subjected to I along $\overline{BC}$ and an impulsive tension along $\overline{CB}$ and so its velocity is u_3, say, along $\overline{BC}$. Since A and B are connected by a taut string, B's velocity along $\overline{AB}$ will be u_1 and its velocity perpendicular to AB will be v_1, say (Fig. 6.5).

Since BC is taut, the component of B's velocity along $\overline{BC}$ is u_3.

$$\therefore \; u_1 \cos \alpha + v_1 \sin \alpha = u_3 \; . \tag{1}$$

Considering linear momentum changes along and perpendicular to BC,

$$m_3u_3 + m_2(u_1 \cos \alpha + v_1 \sin \alpha) + m_1u_1 \cos \alpha = I \; . \tag{2}$$

$$m_2v_1 \cos \alpha - m_2u_1 \sin \alpha - m_1u_1 \sin \alpha = 0 \; . \tag{3}$$

From (3), $v_1/u_1 = (1 + m_1/m_2) \tan \alpha$, as required.

From the three equations, we readily find

$$u_1 = Im_2 \cos \alpha / \{(m_1 + m_2 + m_3)m_2 + m_1 m_3 \sin^2 \alpha\} \ .$$

The impulsive tension in AB is $m_1 u_1$.

6.6 ELASTIC IMPACT

Newton's law of restitution states that *when two particles impinge, their relative velocity component along the common normal after impact, bears a constant ratio* $-e$ *to that before impact, where* e *is a constant for given materials and* $0 \leqslant e \leqslant 1$. This constant is called the *coefficient of restitution. For a perfectly elastic impact,* $e = 1$. For an inelastic impact, $e = 0$. In most cases, e lies between these two values. The law is experimental in nature.

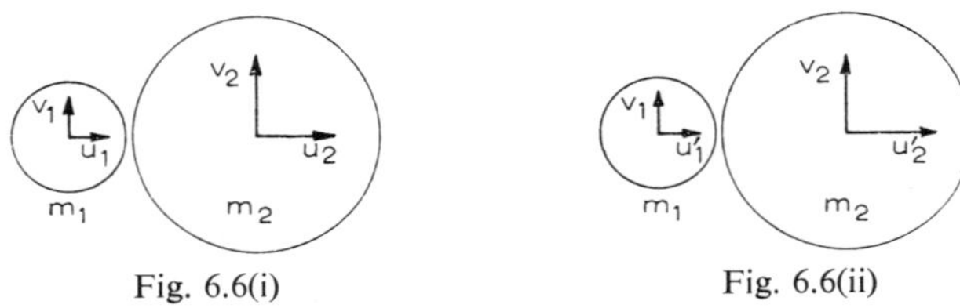

Fig. 6.6(i) Fig. 6.6(ii)

Fig. 6.6(i) shows two small spherical particles of masses m_1, m_2 just before impact occurs: m_1 has velocity components $[u_1, v_1]$ along and at right angles to the line of centres, while m_2 has components $[u_2, v_2]$. At the moment of impact, an impulsive action is set up at the point of contact of the spheres, thereby changing the line-of-centre components to values u_1', u_2' immediately after impact, the components v_1, v_2 at right angles being unaltered (Fig. 6.6(ii)).

Applying the principle of conservation of linear momentum along the line of centres gives

$$m_1 u_1 + m_2 u_2 = m_1 u_1' + m_2 u_2' \ . \tag{1}$$

Applying Newton's law of restitution,

$$u_1' - u_2' = -e(u_1 - u_2) \ . \tag{2}$$

The K.E. lost ΔT is given by

$$\Delta T = \tfrac{1}{2} m_1 (u_1^2 + v_1^2) + \tfrac{1}{2} m_2 (u_2^2 + v_2^2) - \tfrac{1}{2} m_1 (u_1'^2 + v_1^2) - \tfrac{1}{2} m_2 (u_2'^2 + v_2^2) \ ,$$

i.e.

$$\Delta T = \tfrac{1}{2} m_1 (u_1^2 - u_1'^2) + \tfrac{1}{2} m_2 (u_2^2 - u_2'^2) \ . \tag{3}$$

From equations (1) and (2) we find

$$u_1 - u_1' = m_2(1 + e)(u_1 - u_2)/(m_1 + m_2) \ ,$$
$$u_2 - u_2' = -m_1(1 + e)(u_1 - u_2)/(m_1 + m_2) \ ,$$
$$u_1 + u_1' = \{[2m_1 + (1 - e)m_2]u_1 + m_2(1 + e)u_2\}/(m_1 + m_2) \ ,$$
$$u_2' + u_2 = \{m_1(1 + e)u_1 + [2m_2 + (1 - e)m_1]u_2\}/(m_1 + m_2) \ .$$

Hence substitution into (3) gives

$$\Delta T = \tfrac{1}{2} m_1 m_2 (1 - e^2)(u_1 - u_2)^2/(m_1 + m_2) \ ,$$

showing *there is no loss of energy for a perfectly elastic impact.*

Example—Repeated impacts.

A ball is projected from a point O of a plane inclined at an angle α to the horizontal. The direction of projection makes an angle β with the plane, and the path lies in a vertical plane through a line of greatest slope. If the ball returns to O at the n-th impact, show that

$$(1 - e) \cot \alpha \cot \beta = 1 - e^n \ ,$$

where e is the coefficient of restitution between the ball and the plane.

If, further, the r-th impact is normal to the plane, show that

$$e^n - 2e^r + 1 = 0 \ .$$

(L.U.)

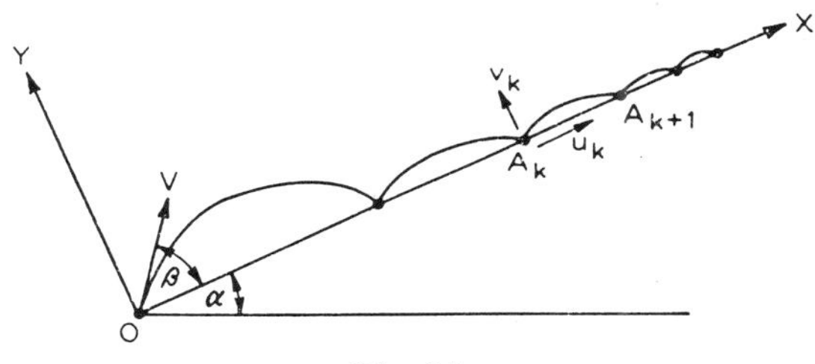

Fig. 6.7

Take axes OX along the plane, OY at right-angles to it (Fig. 6.7). Let V be the velocity of projection at O. Let A_k be the k-th point of impact and let $[u_k, v_k]$ be the components of velocity just after the k-th impact ($k = 0, 1, 2 \ldots .$).

Since the acceleration components in the x- and y-directions are $[-g \sin \alpha, -g \cos \alpha]$, the time t_{k+1} taken to traverse the distance $A_k A_{k+1}$ is given by

$$0 = v_k t_{k+1} - \tfrac{1}{2} g \cos \alpha \, t^2_{k+1} \ ,$$

and so $\quad t_{k+1} = 2v_k/(g \cos \alpha) \ .$

The total time of flight from O to A_k is

$$T_k = t_1 + t_2 + \ldots + t_k$$
$$= 2[v_o + v_1 + \ldots . + v_{k-1}]/(g \cos \alpha) \ .$$

Now $\quad v_o = V \sin \beta, v_1 = eV \sin \beta, v_2 = e^2 V \sin \beta, \ldots .$

$$\therefore \ T_k = 2V \sin \beta [1 + e + \ldots + e^{k-1}]/(g \cos \alpha)$$
$$= 2V \sin \beta (1 - e^k)/\{g \cos \alpha (1 - e)\} \ .$$
$$\therefore \ OA_k = V \cos \beta T_k - \tfrac{1}{2} g \sin \alpha \ T_k^2 \ .$$

Since $OA_n=0$, we require

$$2V\cos\beta=g\sin\alpha.T_n$$

i.e. $$2V\cos\beta=g\sin\alpha.2V\sin\beta(1-e^n)/\{g\cos\alpha(1-e)\}$$

i.e. $$\underline{(1-e)\cot\alpha\cot\beta=1-e^n\ .} \qquad (1)$$

Further,

$$u_k=V\cos\beta-(g\sin\alpha)T_k\ .$$

If the r-th impact is normal to the plane, $u_r=0$.

i.e. $$V\cos\beta=(g\sin\alpha)T_r=2V\sin\beta\tan\alpha(1-e^r)/(1-e)\ ,$$

or $$\underline{(1-e)\cot\beta\cot\alpha=2(1-e^r)\ .} \qquad (2)$$

From (1) and (2),

$$1-e^n=2(1-e^r)$$

or $$\underline{e^n-2e^r+1=0\ .}$$

EXERCISE 6

1. A gun is mounted on a railway truck which is free to run without friction on a straight horizontal railway track. The gun and truck, together of mass M, are moving along the track with velocity u, when a shell of mass m (not included in M) is fired from the gun with muzzle velocity v relative to the gun. If the gun barrel and track lie in the same vertical plane, and the former is inclined at an angle α to the direction in which the truck is moving, show that the shell has a horizontal range $2v\sin\alpha\ \{u+Mv\cos\alpha/(M+m)\}/g$, and that the distance between the shell and the truck when the former lands is independent of M and m. (L.U.)

2. A light rod of length a can turn freely about the end A, which is fixed, and to the other end B is attached a particle of mass m. The end B is joined by a string of length a to a small ring C of mass m, which can slide on a smooth horizontal wire through A. The system is released from rest when $AC=2a$. Show that when $\angle CAB=60°$, the angular acceleration of the rod is $-g/16a$, and find the tension in the string at this instant. (L.U.)

3. A wedge of mass M and angle α stands on a smooth horizontal plane. To the inclined face, which is smooth, is attached the upper end of a light elastic string of natural length a and modulus λ, at the lower end of which is attached a particle of mass m. The system is released from rest with the string of length a and lying along a line of greatest slope of the face of the wedge. Assuming that the ensuing motion is parallel to the vertical plane through the string and that the particle does not reach the horizontal plane, show that the wedge performs simple harmonic oscillations of period.

$$2\pi\surd[(M+m\sin^2\alpha)ma/(M+m)\lambda]\ ,$$

and find their amplitude.

Show also that the reaction between the wedge and the particle is never less than

$$(Mmg \cos \alpha)/(M + m \sin^2 \alpha) .$$

(L.U.)

4. A gun of weight W is mounted so that it is free to recoil along a horizontal railway track. When pointed at an elevation α and in a vertical plane making an angle ϕ with the track, it fires a shell of weight ϵW with muzzle velocity V. If ϵ is so small that its squares and higher powers may be neglected, prove that the shell leaves the gun with an absolute velocity $V(1 - \epsilon \cos^2 \phi \cos^2 \alpha)$ at an elevation $\alpha + \epsilon \sin \alpha \cos \alpha \cos^2 \phi$.

Prove that the maximum range is $R(1 - \epsilon \cos^2 \phi)$, where R is the range when there is no recoil. (L.U.)

5. The cross-section of a wedge, with three smooth plane faces and of mass $2m$, is a triangle whose angles are 30°, 60° and 90° respectively. A small smooth pulley is fixed at the middle point A of the edge at which the angles of all cross-sections are 60°; a fine string is passed over the pulley and has particles P and Q, of masses $3m$ and m respectively, tied to its ends. The wedge is placed on a smooth horizontal table with the face opposite to A in contact with the table and the particle Q held at rest vertically below A in contact with the vertical face of the wedge, the particle P lying on the inclined face. If Q is now set free, prove that the acceleration of the wedge is $\sqrt{3}g/23$, and find the tension in the string and the reaction of the wedge on Q. (L.U.)

6. A wedge of mass M and angle $\pi/6$ rests on a smooth horizontal table. One end of a light elastic string of natural length l and modulus $2mg$ is fastened to a point O on a smooth inclined face of the wedge. A particle of mass m is tied to the other end of the string and is held at O. The system is released from rest.

Write down the energy and horizontal momentum equations for the system and show that when it is next instantaneously at rest the particle is at a distance $2l$ from O. Find the velocity of the particle relative to the wedge when the string is just stretched, and show that the velocity of the wedge is greatest when the particle is at a distance $5l/4$ from O. (L.U.)

7. A system of particles is set in motion by given impulses. Prove that the initial motion of their centre of mass G is the same as that of a single particle, of mass equal to the sum of the masses of all the particles, placed at G and acted on by an impulse which is the resultant of a set of impulses equal and parallel to the given impulses and acting at G.

Two particles A and B, each of mass m, are connected by a light rigid rod, of length $2a$, and the system rests on a smooth horizontal table. A blow in the plane of the table acts on the particle A and gives it a velocity u perpendicular to AB. Describe the resulting motion and find the displacements of the particles from their original positions after a time t. Prove that the tension in the rod is equal to $mu^2/(4a)$. (L.U.)

8. Two particles A and B, each of mass m, are attached to the ends of a light inextensible string of length l, and B moves in a smooth straight

groove in a smooth horizontal table. Initially A is at rest on the table and B is at rest in the groove, AB being perpendicular to the groove and of length $\frac{1}{2}l$. If A is projected with velocity v along the table and parallel to the groove, show that when B begins to move its velocity is $3v/7$.

Find the impulsive tension in the string and the impulsive reaction of the groove. (L.U.)

9. Four particles of equal mass are attached to a light inextensible string at points A, B, C, D, where $AB = BC = CD$, and placed with the three parts of the string taut and forming three sides of a regular hexagon. An impulse is given to the particle at A so that it moves in the direction BA with speed u. Prove that the particle at D begins to move with speed $u/13$. (L.U.)

10. Two particles A, B, each of mass m, are connected to the ends of a light inelastic string of length l. The particles are initially at rest on a smooth horizontal table, with the particle B at the edge of the table, and the particle A at a distance l from B and such that AB is at right angles to the edge of the table at B. If B is pushed gently over the edge of the table, show that the impulsive tension in the string at the instant when A leaves the table is $\frac{1}{2}m\sqrt{(gl)}$ (it is to be assumed that there is no *impulsive* reaction, between the particle at A and the edge of the table), and that A and B will be instantaneously at the same horizontal level when they have fallen through a vertical distance $l(4 + 2\pi + \pi^2)/8$. (L.U.)

11. Particles A, B, each of mass m, are joined by an inelastic string of length $2a$. Initially AB is vertical with B at a distance a below A. A is given a horizontal velocity V while B is simultaneously released from rest. Find the inclination of the string to the vertical when it becomes taut and the time which elapses before the string is horizontal. (L.U.)

12. Four equal particles A, B, C, D, each of mass m, are connected by equal light inextensible strings AB, BC, CD, DA and lie at rest on a smooth horizontal table with the strings taut, so that $ABCD$ is a rhombus and angle $BAD = 2\alpha (\alpha < \pi/4)$. The particle A is given a velocity V along the table in the direction CA (produced). Find the initial velocity of the particle C and prove that the initial kinetic energy of the system is

$$2mV^2/(1 + 2\sin^2\alpha) \ .$$

(L.U.)

13. Three particles A, B, C of equal mass m, lie on a smooth horizontal plane and are smoothly connected by light rigid rods AB, BC so that the angle $ABC = \pi - \alpha$, $(\alpha < \pi/2)$. The whole system is moving parallel to AB with speed V, when C collides with a smooth inelastic wall fixed at right angles to AB. Show that the impulsive blow on the wall is

$$mV(3 + \sin^2\alpha)/(1 + 3\sin^2\alpha) \ .$$

(L.U.)

14. Two particles, A, B each of mass m, are free to slide in narrow straight smooth grooves on a horizontal table and are connected by a light rod of length l. The grooves are perpendicular and intersect at O. If the

system is started from rest by an impulse I applied to the particle A in the direction OA when the angle OAB is α, show that A and B will both return to their initial positions after a time $2\pi ml/(I \sin \alpha)$. Show also that the tension in the rod remains constant throughout the motion and find its magnitude. (L.U.)

15. Two particles A, B of masses $2m$, m respectively, are connected to the ends of a light inelastic string of length $2l$. The particle A is held at rest at a point on a smooth flat table inclined at 30° to the horizontal. From A the string lies along the line of steepest descent until it reaches the edge of the table at C, which is distant l from A and near which the edge of the table is straight and horizontal. The particle B hangs freely. If A is now released, show that the impulsive tension in the string when A leaves the table is $m(4gl/27)^{\frac{1}{2}}$, assuming that there is no *impulsive* reaction between the particle A and the edge of the table.

Using the principle of conservation of momentum, show further that the particles will be instantaneously at the same horizontal level after a time $\pi(l/g)^{\frac{1}{2}}$ from the moment when A leaves the table, and that both particles will then be at a vertical distance below C of $(\frac{2}{3}+\frac{4}{9}\sqrt{3}\pi+\frac{1}{2}\pi^2)l$. (L.U.)

16. Two equal particles A, B lie on a smooth horizontal table, joined by a taut inextensible string. A third equal particle C moving on the table at right angles to the string hits it at its midpoint. Prove that in the subsequent motion C does not leave the string before A, B collide.

Show also that at the time of this collision C has only one-ninth of its original kinetic energy. (L.U.)

17. Three particles A, B, C of masses m, $2m$, $3m$, respectively, lie on a smooth horizontal table at the vertices of an equilateral triangle. A and B, B and C, are connected by light inextensible strings. C is given a velocity V parallel to AB and in the sense from A to B. Show that A eventually begins to move with velocity $2V/19$. (L.U.)

18. Two small smooth elastic spheres, moving in a given manner on a smooth horizontal plane, impinge on one another. Write down equations from which the motion of the spheres after impact may be determined.

A small smooth sphere falls freely from rest through a distance h and then strikes a fixed smooth plane which is inclined at an angle α to the horizontal. If the coefficient of restitution between the sphere and the plane is $\frac{1}{2}$, prove that the distance between the first and fourth points of impact is $(105/16)h \sin \alpha$ and find the impulsive reactions of the plane at these impacts. (L.U.)

19. A smooth sphere A impinges on a stationary and equal smooth sphere B, the velocity of A making an angle θ with the common normal at the moment of impact. Show that the angle through which A is deflected is a maximum when $\tan^2 \theta = \frac{1}{2}(1-e)$ and is then equal to $\frac{1}{2}\pi - 2\theta$, where e is the coefficient of restitution between the spheres.

Show further that in this case the fractional loss of kinetic energy of the spheres is $(1-e^2)/(3-e)$. (L.U.)

20. A, B, C, D are the corners of a smooth horizontal rectangular table, bounded along all four edges by a smooth vertical rim. From the midpoint

of AB, a particle is projected along the surface of the table in a direction making an angle α with AB, to strike in turn the rims BC, CD, DA. If the sides AB, BC are of length a, b, respectively, and if e is the coefficient of restitution between the particle and the rim, show that the particle will return to its starting point if

$$\tan \alpha = 2eb/[a(1+e)] \ ,$$

and that if this condition is satisfied the particle will always pursue the same path. (L.U.)

21. A particle P, of unit mass, moves under the action of a repulsive force $\omega^2 OP$. The particle is initially projected from a point A, with speed $a\omega\sqrt{6}$, in the direction AOB, where $OA = a$, $OB = 2a$. Show that the time from A to B is $(1/\omega)\log(1+\sqrt{6})$.

If there is a fixed elastic obstacle at B, the coefficient of restitution between the particle and the obstacle being e, show that the particle will not return to A if $e < 2/3$. (L.U.)

22. A uniform spherical shell of mass $2m$ and internal radius a rests on a horizontal *inelastic* table. An elastic string of modulus $5mg$ and natural length a connects the highest point of the inside of the shell to a particle of mass m. The coefficient of restitution for impact between the particle and the shell is $\frac{1}{2}$. If the particle is held at the lowest point on the inside of the shell, so that the string is stretched and vertical, and then released, prove: (i) that the particle will strike the shell; (ii) that the shell will be projected into the air for a time $(a/g)^{\frac{1}{2}}$; (iii) that during this time the string remains slack.

Find the greatest depth of the particle below the centre of the shell in the subsequent motion. (L.U.)

23. Two equal smooth spheres are joined by an inextensible string and lie on a smooth horizontal table with the string taut. A third equal sphere moving on the table at an angle α with the string hits one of the spheres directly, so that the string remains taut. Prove that the third sphere rebounds if the coefficient of restitution is greater than $\frac{1}{2}(1+\sin^2\alpha)$. (L.U.)

24. Two massive particles A, B each of mass m are joined by a thin rigid rod of length $2a$ and the system is held with A resting on a smooth horizontal table. If it is allowed to fall from rest, prove that

$$(dy/dt)^2 = 2g(b-y)(a^2-y^2)(2a^2-y^2)^{-1} \qquad (0 < y < b)$$

where y is the height of the mass-centre above the table at time t and $b (< a)$ is the initial value of y.

If P is the thrust in the rod, prove that

$$P = -mga(4a^2b - 6a^2y + y^3)(2a^2 - y^2)^{-2} \qquad (0 < y < b)$$

and hence that P changes sign during the motion. (L.U.)

Chapter 7

Introduction to Rigid Body Dynamics

7.1 MOMENTS AND PRODUCTS OF INERTIA

Let $P(x, y, z)$ be a single particle of mass m of a given system, its co-ordinates being with respect to a tri-rectangular set of axes through a point O. Write

$$A = \Sigma m(y^2 + z^2)\ , \quad B = \Sigma m(z^2 + x^2)\ , \quad C = \Sigma m(x^2 + y^2),$$
$$D = \Sigma mzy\ , \qquad E = \Sigma mzx\ , \qquad F = \Sigma mxy\ ,$$

where the summations are taken throughout the system of particles. In the case when the distribution is *continuous*, the above expressions become replaced by $A = \int (y^2 + z^2)dm$, etc., the integrations now being taken throughout the continuous system. In general discussion the summation signs will be used but for computing $A, B, \ldots .$ in particular cases, the precise form will, of course, have to be adopted. The quantities A, B, C are called the *moments of inertia or second moments of mass distribution* about the x-, y-, z-co-ordinate axes, respectively. D, E, F are termed the *products of inertia* with respect to the pairs of axes (Oy, Oz); (Oz, Ox); (Ox, Oy), respectively. A table giving the moments of inertia for specific cases is included in the Appendix.

Example

Determine the moment of inertia of the distribution about the axis through O having direction cosines $[\lambda, \mu, \nu]$ in terms of these D.C.s and $A, B, \ldots, F$.

Let $\hat{\mathbf{a}} = [\lambda, \mu, \nu]$, $\mathbf{r} = [x, y, z]$. Then the distance d of $P(x, y, z)$ from $\hat{\mathbf{a}}$ is given by

$$d = |\mathbf{r}_\wedge \hat{\mathbf{a}}|\ .$$

Now

$$\mathbf{r}_\wedge \hat{\mathbf{a}} = \begin{vmatrix} \mathbf{i} & \mathbf{j} & \mathbf{k} \\ x & y & z \\ \lambda & \mu & \nu \end{vmatrix} = [(\nu y - \mu z), (\lambda z - \nu x), (\mu x - \lambda y)]\ .$$

Thus the required moment of inertia I is

$$I = \Sigma md^2 = \Sigma m\{(\nu y - \mu z)^2 + (\lambda z - \nu x)^2 + (\mu x - \lambda y)^2\}$$
$$= A\lambda^2 + B\mu^2 + C\nu^2 - 2\mu\nu D - 2\nu\lambda E - 2\lambda\mu F\ .$$

7.2 THE THEOREMS OF PARALLEL AND PERPENDICULAR AXES

(i) Let $G(\bar{x}, \bar{y}, \bar{z})$ be the centroid of the above system and let axes be taken through G parallel to those through O. Further, let (x', y', z') be the new co-ordinates of P referred to these parallel axes through G so that $x=\bar{x}+x'$, $y=\bar{y}+y'$, $z=\bar{z}+z'$.
Then

$$\begin{aligned} A &= \Sigma m\{(\bar{y}+y')^2+(\bar{z}+z')^2\} \\ &= M(\bar{y}^2+\bar{z}^2)+A'+2\bar{y}\Sigma my'+2\bar{z}\Sigma mz' \ , \end{aligned}$$

where $M=\Sigma m$, the total mass of the system; $A'=\Sigma m(y'^2+z'^2)$, the moment of inertia about the parallel x'-axis through G. But from the centroid property, $\Sigma my'=0=\Sigma mz'$.
Hence

$$A=A'+Mh_1{}^2 \ , \tag{1}$$

where $h_1=(\bar{y}^2+\bar{z}^2)^{\frac{1}{2}}$, the distance of the centroid from the x-co-ordinate axis. Equation (1) is the *parallel axes theorem for moments of inertia.*

Further, we have

$$D=\Sigma m(\bar{y}+y')(\bar{z}+z')=M\bar{y}\bar{z}+D' \ , \tag{2}$$

where $D'=\Sigma my'z'$, the product of inertia with respect to the y'- and z'-axes through G. (2) is *the parallel axes theorem for products of inertia.*

(ii) Now suppose the distribution be such that all the particles lie in the plane $z=0$. Then we have

$$\begin{aligned} &A=\Sigma my^2 \ , \quad B=\Sigma mx^2 \ , \quad C=\Sigma m(x^2+y^2) \ , \\ &D=0 \ , \qquad E=0 \ , \qquad F=\Sigma mxy \ . \end{aligned}$$

Thus $$C=A+B \ . \tag{3}$$
(3) is the *perpendicular axes theorem for a laminar distribution.* It states that *the moment of inertia of a plane distribution with respect to any normal axis is equal to the sum of the moments of inertia about any two perpendicular axes in the plane of the distribution and passing through the intersection of the normal with the distribution.*

7.3 ANGULAR MOMENTUM OF A RIGID BODY ABOUT A FIXED POINT AND ABOUT FIXED AXES

Suppose the point P has a velocity $\mathbf{v}$ and that the distribution is a rigid body rotating about an axis through O with vector angular velocity $\boldsymbol{\omega}$. If $\overline{OP}\equiv\mathbf{r}$, then $\mathbf{v}=\boldsymbol{\omega}_\wedge\mathbf{r}$ and the vector angular momentum about O is

$$\begin{aligned} \mathbf{H} &= \Sigma(\mathbf{r}_\wedge m\mathbf{v}) \\ &= \Sigma m[\mathbf{r}_\wedge(\boldsymbol{\omega}_\wedge\mathbf{r})] \\ &= \Sigma mr^2\boldsymbol{\omega}-\Sigma m(\mathbf{r}.\boldsymbol{\omega})\mathbf{r} \ . \end{aligned} \tag{1}$$

(1) gives the vector angular momentum of the rigid body about O. Writing

$$\begin{aligned}\mathbf{r} &= [x, y, z],\ \boldsymbol{\omega} = [\omega_1, \omega_2, \omega_3],\ \mathbf{H} = [h_1, h_2, h_3]\ ,\\ h_1 &= \Sigma m r^2 \omega_1 - \Sigma m(\omega_1 x + \omega_2 y + \omega_3 z)x\\ &= \{\Sigma m(y^2 + z^2)\}\omega_1 - \{\Sigma mxy\}\omega_2 - \{\Sigma mzx\}\omega_3\\ &= A\omega_1 - F\omega_2 - E\omega_3\ .\end{aligned}$$

Thus the angular momenta of the rigid body about the x-, y-, z-axes are respectively

$$\left.\begin{aligned} h_1 &= A\omega_1 - F\omega_2 - E\omega_3,\\ h_2 &= B\omega_2 - D\omega_3 - F\omega_1,\\ h_3 &= C\omega_3 - E\omega_1 - D\omega_2.\end{aligned}\right\} \tag{2}$$

The equations (2) may be more easily remembered in matrix form:

$$\begin{bmatrix} h_1 \\ h_2 \\ h_3 \end{bmatrix} = \begin{bmatrix} A & -F & -E \\ -F & B & -D \\ -E & -D & C \end{bmatrix} \begin{bmatrix} \omega_1 \\ \omega_2 \\ \omega_3 \end{bmatrix}.$$

The symmetrical 3×3 matrix on the R.H.S. will be termed the *inertia matrix* in this treatise.

7.4 PRINCIPAL AXES

The last equation may be written in the form

$$\mathbf{H} = J\boldsymbol{\omega} \tag{1}$$

where $\mathbf{H}$ is the column vector with components h_1, h_2, h_3, $\boldsymbol{\omega}$ is one with components ω_1, ω_2, ω_3 and J is the inertia matrix. Generally speaking the vectors $\mathbf{H}$ and $\boldsymbol{\omega}$ will differ in direction as well as in magnitude and J may be regarded as an operator which transforms the vector $\boldsymbol{\omega}$ into an entirely different vector $\mathbf{H}$. We now investigate the possibility of finding one or more vectors $\boldsymbol{\omega}$ which, for a given J at O, produce corresponding $\mathbf{H}$ that are in the same direction as $\boldsymbol{\omega}$. In the event the two vectors will be related as

$$\mathbf{H} = n\boldsymbol{\omega} \tag{2}$$

where n is a scalar multiplier. In matrix notation, (2) may be written

$$\mathbf{H} = nI\boldsymbol{\omega} \tag{2'}$$

where I is the 3×3 unit matrix having 1s as principal diagonal elements and 0s elsewhere. Comparison of (1) and (2′) leads at once to the matrix equation

$$(J - nI)\boldsymbol{\omega} = \mathbf{0} \tag{3}$$

where $\mathbf{0}$ is the zero column vector. The equivalent set of homogeneous linear algebraic equations is thus

$$\left.\begin{aligned}(A-n)\omega_1-F\omega_2-E\omega_3&=0,\\-F\omega_1+(B-n)\omega_2-D\omega_3&=0,\\-E\omega_1-D\omega_2+(C-n)\omega_3&=0.\end{aligned}\right\}\tag{3'}$$

If we reject the trivial solution $\omega_1=\omega_2=\omega_3=0$, then the determinant of coefficients of the ωs on the L.H.SS. of (3′) vanishes, i.e.

$$\det(J-nI)=0\ .\tag{4}$$

Moreover, ω_1, ω_2, ω_3 cannot be determined absolutely from equations (3′), but their ratio $\omega_1:\omega_2:\omega_3$ may be found from any pair of the three equations. Thus the directions for which the vector angular velocity $\boldsymbol{\omega}$ and the vector angular momentum $\mathbf{H}$ are the same, in accordance with (2), are found by solving the cubic equation (4) for n and then finding $\omega_1:\omega_2:\omega_3$. An instantaneous axis of rotation through O specifying $\boldsymbol{\omega}$ and having the property of being parallel to the instantaneous vector angular momentum $\mathbf{H}$ of the body about O is called a *principal axis* for the point O of the body. The corresponding value of n, as given by the cubic equation (4), is called the *principal moment of inertia* of the body about that axis. In the parlance of matrix algebra the roots of (4) are the *eigenvalues* of the inertia matrix and the corresponding ratios $\omega_1:\omega_2:\omega_3$ determine the directions of the *eigenvectors* of the matrix.

The nature of the roots of the cubic equation (4) in n requires some discussion. As the coefficients of the cubic equation are all real, at least one of its roots is real, say $n=n_1$. Choose the associated direction as the x-axis. Then $\omega_1\neq 0$, $\omega_2=0=\omega_3$ and equations (3′) show that $F=0=E$, $A=n_1$. Equation (4) now gives

$$\begin{vmatrix}n_1-n & 0 & 0\\ 0 & B-n & -D\\ 0 & -D & C-n\end{vmatrix}=0\ ,$$

or, as $n\neq n_1$, the other two roots of the cubic equations are given by

$$(B-n)(C-n)-D^2=0\ .$$

The discriminant of this quadratic equation for n is

$$(B+C)^2-4(BC-D^2)=(B-C)^2+4D^2\geqslant 0\ .$$

Therefore the other two roots are real. They are equal only when $B=C$, $D=0$.

Now suppose that (3) is satisfied by the pairs of values

$$n=n_1\ ,\quad \boldsymbol{\omega}=\boldsymbol{\omega}^{(1)}\ ;\quad n=n_2\ ,\quad \boldsymbol{\omega}=\boldsymbol{\omega}^{(2)}$$

so that

$$J\boldsymbol{\omega}^{(1)}=n_1\boldsymbol{\omega}^{(1)}\ ,\tag{5}$$

$$J\boldsymbol{\omega}^{(2)}=n_2\boldsymbol{\omega}^{(2)}\tag{6}$$

where $\boldsymbol{\omega}^{(1)}$, $\boldsymbol{\omega}^{(2)}$ are column vectors and J is the inertia matrix. If $\tilde{\boldsymbol{\omega}}^{(1)}$, $\tilde{\boldsymbol{\omega}}^{(2)}$ denote the corresponding row vectors, multiplication of (5) through by $\tilde{\boldsymbol{\omega}}^{(2)}$ and (6) by $\tilde{\boldsymbol{\omega}}^{(1)}$ gives

$$\tilde{\boldsymbol{\omega}}^{(2)}J\boldsymbol{\omega}^{(1)}=n_1\tilde{\boldsymbol{\omega}}^{(2)}\boldsymbol{\omega}^{(1)}\ , \tag{5'}$$

$$\tilde{\boldsymbol{\omega}}^{(1)}J\boldsymbol{\omega}^{(2)}=n_2\tilde{\boldsymbol{\omega}}^{(1)}\boldsymbol{\omega}^{(2)}\ . \tag{6'}$$

Taking the transpose of both sides of (5′) and remembering that J is symmetric, we obtain

$$\tilde{\boldsymbol{\omega}}^{(1)}J\boldsymbol{\omega}^{(2)}=n_1\tilde{\boldsymbol{\omega}}^{(1)}\boldsymbol{\omega}^{(2)} \tag{5''}$$

and subtraction of (5″) from (6) gives

$$0=(n_2-n_1)\boldsymbol{\omega}^{(1)}\tilde{\boldsymbol{\omega}}^{(2)}\ . \tag{7}$$

If $n_1 \neq n_2$, then

$$\tilde{\boldsymbol{\omega}}^{(1)}\boldsymbol{\omega}^{(2)}=0\ , \tag{8}$$

i.e. the vectors $\boldsymbol{\omega}^{(1)}$, $\boldsymbol{\omega}^{(2)}$ are orthogonal, since (8) expresses the vanishing of their scalar product and neither vector is presumed to be a zero one. This establishes orthogonality of the principal axes if $n_1 \neq n_2$. If, however, $n_1=n_2=n$, say, on multiplying (5) and (6) through by the scalars λ, μ respectively and adding,

$$(J-nI)(\lambda\boldsymbol{\omega}^{(1)}+\mu\boldsymbol{\omega}^{(2)})=\mathbf{0} \tag{9}$$

which shows that if $n_1=n_2$, then any axis in the plane of $\boldsymbol{\omega}^{(1)}$ and $\boldsymbol{\omega}^{(2)}$ is a principal axis.

In general, since the roots of (4) are real and distinct there will be three mutually orthogonal principal axes at O and these will be determined from the roots. Suppose the solution is specified by

$$n=n_k\ , \quad \boldsymbol{\omega}=\boldsymbol{\omega}^{(k)}=\omega^{(k)}\hat{\mathbf{a}}_k \quad (k=1,\,2,\,3)\ ,$$

where the $\hat{\mathbf{a}}_k$ $(k=1, 2, 3)$ are the unit vectors in the three principal axes at O. Then the $\hat{\mathbf{a}}_k$ are mutually orthogonal and the subscripts may be so ordered that

$$\hat{\mathbf{a}}_1{\wedge}\hat{\mathbf{a}}_2=\hat{\mathbf{a}}_3\ , \quad \text{etc.}$$

Further we may write

$$\mathbf{r}=r_1\hat{\mathbf{a}}_1+r_2\hat{\mathbf{a}}_2+r_3\hat{\mathbf{a}}_3=x\mathbf{i}+y\mathbf{j}+z\mathbf{k}\ .$$

From the vector equation (2), for rotation about the principal axis, specified by $\hat{\mathbf{a}}$, we have

$$\Sigma m\mathbf{r}{\wedge}(\boldsymbol{\omega}^{(1)}\mathbf{r})=n_1\boldsymbol{\omega}^{(1)}\ . \tag{10}$$

Scalar multiplication of (10) through by $\hat{\mathbf{a}}_1$ and division by $\omega^{(1)}$ gives

$$n_1=\Sigma m(\hat{\mathbf{a}}_1{\wedge}\mathbf{r})^2\ .$$

Since $\hat{\mathbf{a}}_1{\wedge}\mathbf{r}=r_2\hat{\mathbf{a}}_3-r_3\hat{\mathbf{a}}_2$,

$$n_1=\Sigma m(r_2{}^2+r_3{}^2)=A^*\ , \quad \text{say.}$$

Similarly

$$n_2=\Sigma m(r_3{}^2+r_1{}^2)=B^*\ , \quad \text{say,}$$

$$n_3 = \Sigma m(r_1{}^2 + r_2{}^2) = C^* \ , \quad \text{say.}$$

These expressions determine the principal moments of inertia. Further, on expanding the vector triple product in (10) we obtain

$$(\Sigma mr^2 - n_1)\boldsymbol{\omega}^{(1)} = \Sigma m(\mathbf{r}.\boldsymbol{\omega}^{(1)})\mathbf{r}$$

or
$$(\Sigma mr^2 - n_1)\hat{\mathbf{a}}_1 \quad = \Sigma mr_1\mathbf{r}$$

Scalar multiplication through by $\hat{\mathbf{a}}_2$ now gives

$$0 = \Sigma mr_1r_2 \ .$$

Similarly,
$$\Sigma mr_2r_3 = 0 = \Sigma mr_3r_1 \ .$$

Thus we have shown that *the products of inertia with respect to the principal axes are zero.* Hence the inertia matrix for the principal axes through O is simply

$$\begin{bmatrix} A^* & 0 & 0 \\ 0 & B^* & 0 \\ 0 & 0 & C^* \end{bmatrix} .$$

The components of angular momentum about the principal axes through O are

$$h_1 = A^*\omega^{(1)} \ , \quad h_2 = B^*\omega^{(2)} \ , \quad h_3 = C^*\omega^{(3)} \ ,$$

and the vector angular momentum about O is

$$\mathbf{H} = A^*\boldsymbol{\omega}^{(1)} + B^*\boldsymbol{\omega}^{(2)} + C^*\boldsymbol{\omega}^{(3)} \ . \tag{11}$$

This is a more convenient form than the one given in Section 7.3, since no products of inertia are involved.

From the above treatment, the reader will appreciate that for the principal axes at any point O of a rigid body three cases may arise:

(1) There are exactly three mutually perpendicular axes with the ns all different. This is the case generally occurring.

(2) A principal axis and every line perpendicular to it a principal axis, i.e. two of the ns are equal. A right circular cylinder with O the mid-point of the axis of symmetry is such a case. This axis and every line through O perpendicular to it are principal axes.

(3) Three perpendicular axes for which the ns are all the same. This case is realized when O is the centre of a homogeneous solid sphere.

The reader will easily confirm that if two perpendicular axes through a point O of a rigid body are axes of symmetry, then they are principal axes for the body. It is not true, however, that principal axes are necessarily axes of symmetry.

Further, it is easily confirmed from the foregoing analysis that if, for three mutually perpendicular axes through some point O of a rigid body, the three products of inertia with respect to these axes are all zero, then the axes are principal axes.

7.5 KINETIC ENERGY OF A RIGID BODY ROTATING ABOUT A FIXED POINT

Let a rigid body rotate instantaneously with vector angular velocity $\boldsymbol{\omega}$ about an axis through a fixed point O of the body. Let m be the mass of a particle P of the body where $\overline{OP} \equiv \mathbf{r}$. Then the velocity of P is $\boldsymbol{\omega}_\wedge\mathbf{r}$, its K.E. is $\frac{1}{2}m(\boldsymbol{\omega}_\wedge\mathbf{r})^2$, and so the total K.E. of the body is

$$T = \tfrac{1}{2}\Sigma m(\boldsymbol{\omega}_\wedge\mathbf{r})^2 \ .$$

Now $$(\boldsymbol{\omega}_\wedge\mathbf{r})^2 = (\boldsymbol{\omega}_\wedge\mathbf{r}).(\boldsymbol{\omega}_\wedge\mathbf{r}) = \boldsymbol{\omega}.[\mathbf{r}_\wedge(\boldsymbol{\omega}_\wedge\mathbf{r})] \ .$$

$$\therefore \ T = \tfrac{1}{2}\boldsymbol{\omega}.\Sigma\{\mathbf{r}_\wedge m(\boldsymbol{\omega}_\wedge\mathbf{r})\} = \tfrac{1}{2}\boldsymbol{\omega}.\mathbf{H} \ ,$$

since the vector angular momentum about O is

$$\mathbf{H} = \Sigma\{\mathbf{r}_\wedge m(\boldsymbol{\omega}_\wedge\mathbf{r})\} \ .$$

Suppose $\hat{\mathbf{a}}_1$, $\hat{\mathbf{a}}_2$, $\hat{\mathbf{a}}_3$ denote the three principal axes through O and let A^*, B^*, C^* be the corresponding principal moments of inertia. Let

$$\boldsymbol{\omega} = \omega_1\hat{\mathbf{a}}_1 + \omega_2\hat{\mathbf{a}}_2 + \omega_3\hat{\mathbf{a}}_3 \ .$$

Since the axes are principal axes,

$$\mathbf{H} = A^*\omega_1\hat{\mathbf{a}}_1 + B^*\omega_2\hat{\mathbf{a}}_2 + C^*\omega_3\hat{\mathbf{a}}_3 \ .$$

$$\therefore \ \underline{T = \tfrac{1}{2}\mathbf{H}.\boldsymbol{\omega} = \tfrac{1}{2}(A^*\omega_1{}^2 + B^*\omega_2{}^2 + C^*\omega_3{}^2)} \ .$$

Now suppose **i**, **j**, **k** specify unit vectors along any other set of mutually perpendicular axes through O and suppose

$$\mathbf{r} = x\mathbf{i} + y\mathbf{j} + z\mathbf{k} \ ,$$

$$\boldsymbol{\omega} = \omega(\lambda\mathbf{i} + \mu\mathbf{j} + \nu\mathbf{k}) \ ,$$

$[\lambda, \mu, \nu]$ being the D.Cs of the instantaneous axis of rotation.

$$\therefore \ \boldsymbol{\omega}_\wedge\mathbf{r} = \omega \begin{vmatrix} \mathbf{i} & \mathbf{j} & \mathbf{k} \\ \lambda & \mu & \nu \\ x & y & z \end{vmatrix}$$

and so $(\boldsymbol{\omega}_\wedge\mathbf{r})^2 = \omega^2\{(\mu z - \nu y)^2 + (\nu x - \lambda z)^2 + (\lambda y - \mu x)^2\}$.
Hence we find

$$T = \tfrac{1}{2}\Sigma m(\boldsymbol{\omega}_\wedge\mathbf{r})^2 = \tfrac{1}{2}\omega^2 I \ ,$$

where $I = A\lambda^2 + B\mu^2 + C\nu^2 - 2D\mu\nu - 2E\nu\lambda - 2F\lambda\mu$, A, B, C are the moments of inertia about the axes specified by **i**, **j**, **k**, and D, E, F the products of inertia with respect to pairs of these axes. I is, of course, the moment of inertia about the instantaneous axis of rotation. Comparison of the two formulae for T shows that the one referred to principal axes is much simpler; there are no products of inertia involved.

7.6 MOMENTAL ELLIPSOID—EQUIMOMENTAL SYSTEMS

In the notation of the preceding sections, we have seen that the moment of inertia of the given mass distribution about the line through O having D.Cs. $[\lambda, \mu, \nu]$ is

$$I = A\lambda^2 + B\mu^2 + C\nu^2 - 2D\mu\nu - 2E\nu\lambda - 2F\lambda\mu \ .$$

Consider the quadric

$$Ax^2 + By^2 + Cz^2 - 2Dyz - 2Ezx - 2Fxy = M\epsilon^4 \ , \tag{1}$$

where M is the total mass and ϵ a constant length, so that both sides of (1) are dimensionally consistent. Let P be a point on the line $[\lambda, \mu, \nu]$ distant R from O so that the co-ordinates of P are $x=\lambda R$, $y=\mu R$, $z=\nu R$. Substituting these in (1) gives

$$I = M\epsilon^4/R^2 \ . \tag{2}$$

(2) shows that for all $[\lambda, \mu, \nu]$, R is real and finite, since $I>0$. Thus the quadric (1) is an ellipsoid having O as centre: it is called the *momental ellipsoid* of the distribution for the point O.

By means of a suitable transformation of axes (1) can be reduced to the form

$$A^*x^2 + B^*y^2 + C^*z^2 = M\epsilon^4 \ , \tag{3}$$

where A^*, B^*, C^* are the principal moments of inertia. Thus the *principal axes of the distribution coincide with the principal axes of the momental ellipsoid.*

Example

A uniform solid rectangular block is of mass M and dimensions $2a \times 2b \times 2c$. Find the equation of the momental ellipsoid for a corner O of the block, referred to the edges through O as co-ordinate axes and hence determine the moment of inertia about OO', where O' is the point diagonally opposite to O.

Take the x-, y-, z-axes along the edges of lengths $2a$, $2b$, $2c$. Then

$$A = (M/3)(b^2+c^2) + M(b^2+c^2) = (4M/3)(b^2+c^2) \ ,$$
$$B = (4M/3)(c^2+a^2) \ ,$$
$$C = (4M/3)(a^2+b^2) \ .$$

Further, if ρ denote the density of the material,

$$D = \int_{z=0}^{2c} \int_{y=0}^{2b} (2a\rho \, dy \, dz . yz)$$

$$= \tfrac{1}{2}a\rho[y^2]_0^{2b}\,[z^2]_0^{2c}$$

$$= 8a\rho \, b^2c^2 = Mbc \ .$$

[*N.B.* This result for D could be obtained more simply using the parallel

axes theorem for products of inertia. See Example 1, Section 7.7 of this chapter.]
Similarly, $E = Mca$, $F = Mab$.

Thus the equation of the momental ellipsoid at O is

$$(4M/3)\{(b^2+c^2)x^2+(c^2+a^2)y^2+(a^2+b^2)z^2\} \\ -2M(bcyz+cazx+abxy) = M\epsilon^4 \ .$$

The co-ordinates of O' are $(2a, 2b, 2c)$ and the distance OO' is $R = 2(a^2+b^2+c^2)^{\frac{1}{2}}$. Hence the moment of inertia about OO' is

$$\underline{M\epsilon^4/R^2 = \tfrac{2}{3}M(b^2c^2+c^2a^2+a^2b^2)/(a^2+b^2+c^2)} \ .$$

Definition

If two systems are such that they have equal moments of inertia about every line of space, then they are said to be *equimomental.*

Necessary and sufficient conditions for the two systems to be equimomental are expressed in the following theorem.

If two systems are such that they have (i) *the same total mass*; (ii) *the same centroid*; (iii) *the same principal axes and principal moments of inertia at the centroid, then they are equimomental.*

Proof

(*a*) First suppose the stated conditions hold. Denote by M the total mass of each system, G the centroid, A^*, B^*, C^* the principal moments of inertia at G.

In both cases, the moment of inertia about the line through G having D.Cs. $[\lambda, \mu, \nu]$ with respect to the principal axes is $A^*\lambda^2+B^*\mu^2+C^*\nu^2$. Hence, by the parallel axes theorem, the M.I. about a parallel line distant h from this line is

$$A^*\lambda^2+B^*\mu^2+C^*\nu^2+Mh^2 \ ,$$

in both cases. Thus the two systems have equal moments of inertia about any line of space and the systems are equimomental.

(*b*) Assuming the systems to be equimomental, let us derive conditions (i), (ii), (iii).

Let G_1, G_2, be the centroids of the two systems, and M_1, M_2 their total masses. Since the systems are equimomental, they have equal moments of inertia I, say, about G_1G_2. The M.Is. about any parallel line distant d from G_1G_2 are respectively $I+M_1d^2$, $I+M_2d^2$, in virtue of the parallel axes theorem. Since the systems are equimomental.

$$I+M_1d^2 = I+M_2d^2 \ ,$$

and so $M_1 = M_2 = M$, say.

Let J be the M.I. of either system about the line through G_1 perpendicular to G_1G_2. Then the M.I. of the first system about the parallel line through G_2 is $J+M.G_1G_2{}^2$, and that of the second system is $J-M.G_1G_2{}^2$.

$$\therefore\ J+M.G_1G_2{}^2=J-M.G_1G_2{}^2\ ,$$

i.e.
$$G_1\equiv G_2\ .$$

The systems have equal moments of inertia about any line through their common centroid G. Thus their momental ellipsoids at G are the same and so are the principal axes and principal moments of inertia at G.

Example

Find an equimomental system of particles for a uniform rod AB of mass M.

Let O be the centroid of the rod, $2a$ its length.

Particles m, m, $M-2m$ at A, B, O have the same centroid and total mass as the rod. They also have the same M.I. (zero) about AB.

To find m, equate the M.Is. about a perpendicular bisector through O getting

$$2ma^2=\tfrac{1}{3}Ma^2\ .$$

$$\therefore\ m=\tfrac{1}{6}M\ .$$

$\therefore$ The required equimomental system is

$$\tfrac{1}{6}M \text{ at } A\ ,\quad \tfrac{1}{6}M \text{ at } B\ ,\quad \tfrac{2}{3}M \text{ at } O\ .$$

The reader will easily confirm that an alternative equimomental system would be particles of mass $\tfrac{1}{2}M$ on the rod at distances $a/\sqrt{3}$ on either side of O.

7.7 COPLANAR DISTRIBUTIONS

Although the treatment of two-dimensional mass distributions can be derived from the three-dimensional case considered, it is instructive to treat some aspects of two-dimensional distributions on their own merits.

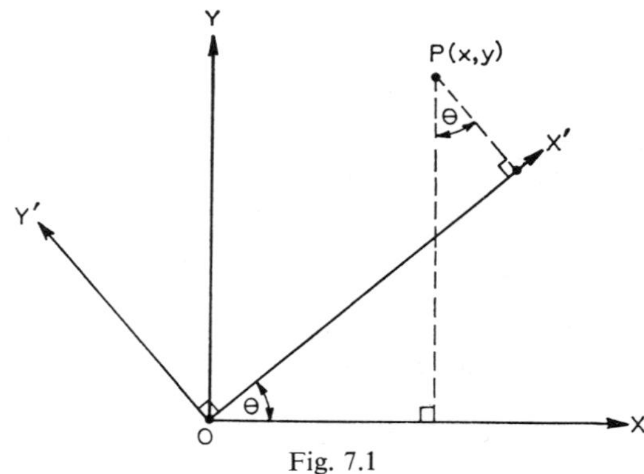

Fig. 7.1

Fig. 7.1 depicts a two-dimensional distribution in the plane of perpendicular co-ordinate axes OX, OY. The mass of the particle at $P(x, y)$ is m.

Then the moments of inertia about OX, OY are respectively

$$A=\Sigma my^2 \ , \quad B=\Sigma mx^2$$

and the product of inertia with respect to OX, OY is

$$F=\Sigma mxy \ ,$$

where all summations are taken over the entire distribution (and replaced by appropriate integrations, of course, if the distribution be continuous).

Through O, take axes OX', OY' inclined at θ, $\theta+\frac{1}{2}\pi$ to OX. The distance from P to OX' is $y\cos\theta-x\sin\theta$ and so the M.I. of the distribution about OX' is

$$I_{OX'}=\Sigma m(y\cos\theta-x\sin\theta)^2 \ ,$$

i.e.

$$I_{OX'}=A\cos^2\theta-2F\sin\theta\cos\theta+B\sin^2\theta \ . \tag{1}$$

To find the M.I. about OY', replace θ by $\theta+\frac{1}{2}\pi$ getting

$$I_{OY'}=A\sin^2\theta+2F\sin\theta\cos\theta+B\cos^2\theta \ . \tag{2}$$

Since the distance of P from OY' is $x\cos\theta+y\sin\theta$, the product of inertia with respect to OX', OY' is

$$P_{X'Y'}=\Sigma m(y\cos\theta-x\sin\theta)(x\cos\theta+y\sin\theta) \ ,$$

i.e.

$$P_{X'Y'}=F\cos 2\theta+\tfrac{1}{2}(A-B)\sin 2\theta \ . \tag{3}$$

If OX', OY' be principal axes of the distribution, then $P_{X'Y'}=0$. Let $\theta\equiv\alpha$ denote a solution for which $P_{X'Y'}=0$.

$$\therefore \ \tan 2\alpha=2F/(B-A) \ . \tag{4}$$

But $\tan 2\alpha=\tan 2(\alpha+\frac{1}{2}\pi)$, so that $\theta=\alpha+\frac{1}{2}\pi$ is appropriately also a solution.

Suppose that $B>A$. Then (4) shows 2α and hence also α are acute angles. Also from (1), (2)

$$I_{OX'}=\tfrac{1}{2}(A+B)-\tfrac{1}{2}[(B-A)\cos 2\theta+2F\sin 2\theta] \ ,$$
$$I_{OY'}=\tfrac{1}{2}(A+B)+\tfrac{1}{2}[(B-A)\cos 2\theta+2F\sin 2\theta] \ .$$

Thus

$$\text{min. } I_{OX'}=\tfrac{1}{2}\{A+B-\sqrt{[(B-A)^2+4F^2]}\} \ ,$$
$$\text{max. } I_{OY'}=\tfrac{1}{2}\{A+B+\sqrt{[(B-A)^2+4F^2]}\} \ ,$$

these extreme values being attained for $\theta=\alpha$, $\alpha+\frac{1}{2}\pi$, as defined by (4). Thus the greatest and least moments of inertia for lines through O in the plane of the distribution are attained along the principal axes.

Since for the coplanar distribution the z-axis through O is normal to the plane of the distribution, $C=A+B$, $D=0=E$, and the equation of the momental ellipsoid becomes

$$Ax^2+By^2+(A+B)z^2-2Fxy=M\epsilon^4.$$

The section with the plane $z=0$ gives the *momental ellipse*:

$$Ax^2 + By^2 - 2Fxy = M\epsilon^4 \ .$$

Example 1

A uniform rectangular lamina $ABCD$ is such that $AB = 2a$, $BC = 2b$. Find the directions of the principal axes at A.

Taking AB as x-axis, AD as y-axis, by the parallel axes theorem for moments of inertia, we have

$$A = \tfrac{1}{3}Mb^2 + Mb^2 = \tfrac{4}{3}Mb^2 \ ; \quad B = \tfrac{4}{3}Ma^2 \ ,$$

where M is the total mass of the lamina. From symmetry, the P.I. with respect to parallel axes through the centroid is zero, and so using the theorem of parallel axes for products of inertia, the P.I. with respect to the x- and y-axes is

$$F = Mab.$$

$$\therefore \ \tan 2\alpha = 2F/(B - A) = \underline{3ab/2(a^2 - b^2)} \ .$$

Example 2—Equimomental system for a uniform triangular lamina.

Let us first find the M.I. of a uniform triangular lamina ABC of mass M about a side BC. Let A be distant h from BC (Fig. 7.2).

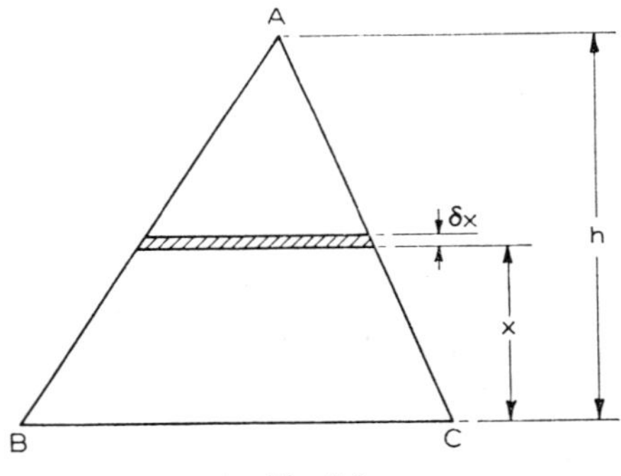

Fig. 7.2

Consider a strip of thickness δx, parallel to BC and distant x from it. Let σ be the surface density of the material and a the length of BC so that the length of the strip is $a(h - x)/h$ and its mass is $\sigma a(h - x)\delta x/h$. Since every particle of this element is distant x from BC, the moment of inertia of the element about BC is $\sigma ax^2(h - x)\delta x/h$ and so the total M.I. about BC is

$$\frac{\sigma a}{h}\int_0^h (hx^2 - x^3)dx = \tfrac{1}{12}\sigma ah^3 = \tfrac{1}{6}Mh^2 \ .$$

We now apply this result to the more general case of finding the M.I. about any line l in the plane of the triangle ABC. Let h_1, h_2, h_3 be the distances of A, B, C from l, where $h_1 < h_2 < h_3$ (Fig. 7.3).

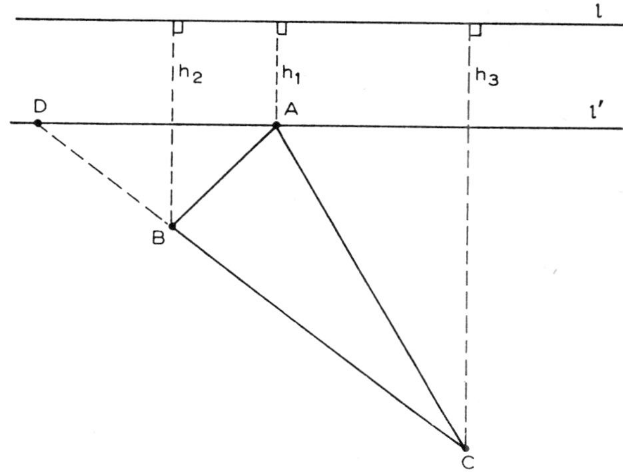

Fig. 7.3

Through A, draw l' parallel to l and let CB meet l' in D. C and B are distant h_3-h_1, h_2-h_1, from l'.

Suppose M_1, M_2 denote the masses of triangles ACD, ABD, when the surface density is uniform throughout. Then

$$M = M_1 - M_2 \ ,$$
$$M_1/M_2 = (h_3-h_1)/(h_2-h_1) \qquad \text{(common base } AD\text{)},$$

and so $M_1 = M(h_3-h_1)/(h_3-h_2)$, $M_2 = M(h_2-h_1)/(h_3-h_2)$. Thus the M.I. of triangle ABC about l' is

$$\begin{aligned} I_l' &= \tfrac{1}{6}M_1(h_3-h_1)^2 - \tfrac{1}{6}M_2(h_2-h_1)^2 \\ &= \tfrac{1}{6}M\{(h_3-h_1)^3-(h_2-h_1)^3\}/(h_3-h_2) \ . \\ &= \tfrac{1}{6}M\{3h_1^2+h_2^2+h_3^2+h_2h_3-3h_3h_1-3h_1h_2\} \ . \end{aligned}$$

Since the centroid is distant $\frac{1}{3}(h_1+h_2+h_3)$ from l and $\frac{1}{3}(h_2+h_3-2h_1)$ from l', by double application of the parallel axes theorem the M.I. about l is given by

$$\begin{aligned} I_l &= I_l' - \tfrac{1}{9}M(h_2+h_3-2h_1)^2 + \tfrac{1}{9}M(h_1+h_2+h_3)^2 \\ &= I_l' + \tfrac{1}{9}M(3h_1)(2h_2+2h_3-h_1) \\ &= \underline{\tfrac{1}{6}M\{h_1^2+h_2^2+h_3^2+h_2h_3+h_3h_1+h_1h_2\} \ .} \end{aligned}$$

We can write this last result in the alternative form

$$I_l = \tfrac{1}{3}M\{[(h_2+h_3)/2]^2+[(h_3+h_1)/2]^2+[(h_1+h_2)/2]^2\} \ ,$$

which *is the same as the* M.I. *about l of equal particles of mass* $\frac{1}{3}$M *at the mid-points of the sides of* ΔABC. But the total mass of these particles is the same as that of the triangle and the centroids of the two systems coincide. The systems have the same principal moments of inertia and principal axes in their planes. Also the moments of inertia about corresponding axes normal to their planes are equal, and so the two distributions are equimomental.

7.8 GENERAL MOTION OF A RIGID BODY

In the previous chapter, several results were obtained for the motion of a general system of particles. In many cases, the theorems that were established involved in some way the motion of the centroid, together with the motion of the system relative to the centroid. Since a rigid body is a collection of particles such that the distances between all pairs of them remain invariant, such results are also applicable to rigid body motion, though the contribution to motion relative to the centroid can be simplified to forms involving the angular velocity and the moments and products of inertia with respect to axes through the centroid.

Suppose O is a point fixed in space, P a point fixed in the rigid body, and G its centroid, so that $\overline{OP} \equiv \mathbf{r}$, $\overline{OG} \equiv \bar{\mathbf{r}}$, $\overline{GP} \equiv \mathbf{r}' = \mathbf{r} - \bar{\mathbf{r}}$. Then from Chapter 6, Section 6.3, the kinetic energy is

$$T = \tfrac{1}{2}M\dot{\bar{\mathbf{r}}}^2 + \tfrac{1}{2}\Sigma m\dot{\mathbf{r}}'^2 \ ,$$

where M is the total mass of the body, m the mass of the particle at P. Suppose that $\boldsymbol{\omega}$ denotes the vector angular velocity of the rigid body. Then $\dot{\mathbf{r}}' = (\boldsymbol{\omega}_\wedge\mathbf{r}')$. Now $(\boldsymbol{\omega}_\wedge\mathbf{r}')^2 = \boldsymbol{\omega}.\{\mathbf{r}'_\wedge(\boldsymbol{\omega}_\wedge\mathbf{r}')\}$,

$$\therefore \ T = \tfrac{1}{2}M\bar{\mathbf{v}}^2 + \tfrac{1}{2}\boldsymbol{\omega}.\Sigma m\{\mathbf{r}'_\wedge(\boldsymbol{\omega}_\wedge\mathbf{r}')\} \ , \quad \bar{\mathbf{v}} = \dot{\bar{\mathbf{r}}} \ .$$

But $\Sigma\{\mathbf{r}'_\wedge m(\boldsymbol{\omega}_\wedge\mathbf{r}')\} = \mathbf{H}'$, the angular momentum of the body in its motion relative to G.

$$\therefore \ T = \tfrac{1}{2}M\bar{\mathbf{v}}^2 + \tfrac{1}{2}\boldsymbol{\omega}.\mathbf{H}' \ .$$

If $\hat{\mathbf{a}}_1$, $\hat{\mathbf{a}}_2$, $\hat{\mathbf{a}}_3$ denote the unit vectors in the principal axes at G; A^*, B^*, C^*, the corresponding principal moments of inertia, and $\boldsymbol{\omega} = \omega_1\hat{\mathbf{a}}_1 + \omega_2\hat{\mathbf{a}}_2 + \omega_3\hat{\mathbf{a}}_3$,

then $$\mathbf{H}' = A^*\omega_1\hat{\mathbf{a}}_1 + B^*\omega_2\hat{\mathbf{a}}_2 + C^*\omega_3\hat{\mathbf{a}}_3 \ ,$$

and so $$\underline{T = \tfrac{1}{2}M\bar{\mathbf{v}}^2 + \tfrac{1}{2}(A^*\omega_1{}^2 + B^*\omega_2{}^2 + C^*\omega_3{}^2) \ .}$$

The term $\tfrac{1}{2}M\bar{\mathbf{v}}^2$ is the K.E. due to the translation of a single particle of mass M at G: the term $\tfrac{1}{2}(A^*\omega_1{}^2 + B^*\omega_2{}^2 + C^*\omega_3{}^2)$ is the K.E. of the rigid body rotating about the point G, now considered fixed, with vector angular velocity $\boldsymbol{\omega}$. Thus the total K.E. is the sum of these two, showing that the components of translation and rotation are independent of each other.

The angular momentum $\mathbf{H}$ about a moving point O is given by equation (1), Section 6.4, Chapter 6:

$$\mathbf{H} = \bar{\mathbf{r}}_\wedge M\bar{\mathbf{v}} + \mathbf{H}' \ ,$$

where $$\mathbf{H}' = \Sigma(\mathbf{r}'_\wedge m\dot{\mathbf{r}}') = \text{A.M. about } G.$$

But $$\mathbf{H}' = A^*\omega_1\hat{\mathbf{a}}_1 + B^*\omega_2\hat{\mathbf{a}}_2 + C^*\omega_3\hat{\mathbf{a}}_3 \ ,$$

and so $$\underline{\mathbf{H} = \bar{\mathbf{r}}_\wedge M\bar{\mathbf{v}} + A^*\omega_1\hat{\mathbf{a}}_1 + B^*\omega_2\hat{\mathbf{a}}_2 + C^*\omega_3\hat{\mathbf{a}}_3 \ .}$$

Again there is exhibited a component due to translation, viz. $\bar{\mathbf{r}}_\wedge M\mathbf{v}$, and one due to rotation, viz. $\mathbf{H}'$.

It was shown in the last chapter that for any system of particles the sum of the moments of the external forces about any point is equal to the total moment of the rate of change of momentum about that point, irrespective of whether the point be moving or fixed. It is important to derive the form of this 'moment of rate of change of momentum' for a rigid body. Let $\mathbf{v}_o$ be the velocity of a point O of the body and $\mathbf{v}$ the velocity of a particle P of mass m where $\overline{OP} \equiv \mathbf{r}$. Then $\mathbf{v} = \mathbf{v}_o + \dot{\mathbf{r}}$. The moment of the rate of change of momentum about O is $\Sigma(\mathbf{r}_\wedge m\dot{\mathbf{v}})$. Let G be the centroid of the body, $\overline{OG} \equiv \bar{\mathbf{r}}$, so that $\mathbf{r} = \bar{\mathbf{r}} + \mathbf{r}'$. Then

$$\Sigma(\mathbf{r}_\wedge m\dot{\mathbf{v}}) = \mathbf{r}_\wedge \Sigma m\dot{\mathbf{v}} + \Sigma(\mathbf{r}'_\wedge m\dot{\mathbf{v}}) \ .$$

Now $\mathbf{v} = \bar{\mathbf{v}} + \dot{\mathbf{r}}'$ and so $\Sigma(\mathbf{r}'_\wedge m\dot{\mathbf{v}}) = (\Sigma m\mathbf{r}')_\wedge \dot{\bar{\mathbf{v}}} + \Sigma(\mathbf{r}'_\wedge m\ddot{\mathbf{r}}')$

$$= \Sigma(\mathbf{r}'_\wedge m\ddot{\mathbf{r}}'). \quad \text{Also, } \Sigma m\mathbf{v} = M\bar{\mathbf{v}} \ .$$

$$\therefore \ \Sigma(\mathbf{r}_\wedge m\mathbf{v}) = \bar{\mathbf{r}}_\wedge M\dot{\bar{\mathbf{v}}} + \Sigma(\mathbf{r}'_\wedge m\ddot{\mathbf{r}}') \ .$$

But
$$\Sigma(\mathbf{r}'_\wedge m\ddot{\mathbf{r}}') = \dot{\mathbf{H}}', \text{ where } \mathbf{H}' = \Sigma(\mathbf{r}'_\wedge m\dot{\mathbf{r}}') \ .$$

$$\therefore \ \Sigma(\mathbf{r}_\wedge m\mathbf{v}) = \bar{\mathbf{r}}_\wedge M\dot{\bar{\mathbf{v}}} + \dot{\mathbf{H}}' \ ,$$

where $\mathbf{H}' = A^*\omega_1\hat{\mathbf{a}}_1 + B^*\omega_2\hat{\mathbf{a}}_2 + C^*\omega_3\hat{\mathbf{a}}_3$.

Example

A uniform rigid rod AB moves so that A and B have velocities $\mathbf{u}_A$, $\mathbf{u}_B$ at any instant. Show that the K.E. is then $T = \frac{1}{6}M(\mathbf{u}_A{}^2 + \mathbf{u}_A.\mathbf{u}_B + \mathbf{u}_B{}^2)$, M being the mass.

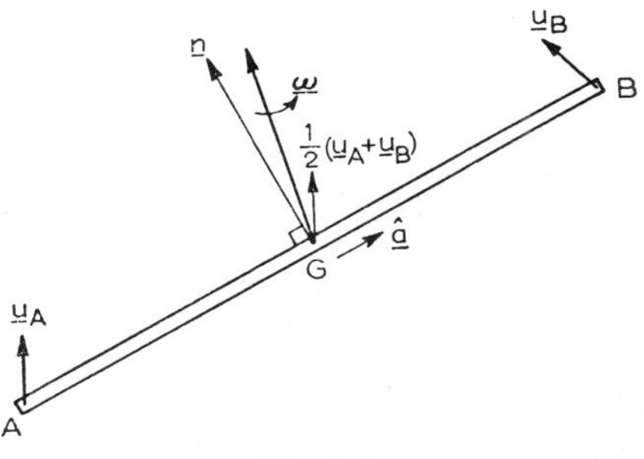

Fig. 7.4

Let $2a$ be the length of the rod, $\hat{\mathbf{a}}$ unit vector in AB. Let $\boldsymbol{\omega}$ be the vector angular velocity of the rod and let $\mathbf{n}$ be the unit normal to the rod through G in the plane of $\boldsymbol{\omega}$, $\hat{\mathbf{a}}$ (Fig. 7.4).

$$\text{Velocity of } G = \tfrac{1}{2}(\mathbf{u}_A + \mathbf{u}_B) = \bar{\mathbf{v}}, \text{ say.}$$

AB is a principal axis, $\mathbf{n}$ the direction of a second principal axis, a third being determined by $\hat{\mathbf{a}}_\wedge\mathbf{n}$. The principal moments of inertia about these three directions are respectively 0, $\frac{1}{3}Ma^2$, $\frac{1}{3}Ma^2$, where M is the mass of the rod. Since the angular velocity components are $\boldsymbol{\omega}.\hat{\mathbf{a}}$, $\boldsymbol{\omega}.\mathbf{n}$, 0,

$$\begin{aligned}\therefore\; T &= \tfrac{1}{2}M\bar{\mathbf{v}}^2 + \tfrac{1}{2}[(0)(\boldsymbol{\omega}.\hat{\mathbf{a}})^2 + (\tfrac{1}{3}Ma^2)(\omega.\mathbf{n})^2 + (\tfrac{1}{3}Ma^2)(0)] \\ &= \tfrac{1}{2}M\bar{\mathbf{v}}^2 + \tfrac{1}{6}Ma^2(\omega.\mathbf{n})^2 \;. \\ &= \tfrac{1}{8}M(\mathbf{u}_A + \mathbf{u}_B)^2 + \tfrac{1}{6}Ma^2|\boldsymbol{\omega}\wedge\hat{\mathbf{a}}|^2 \;.\end{aligned}$$

Now $\boldsymbol{\omega}\wedge(a\hat{\mathbf{a}})$ = velocity of B relative to $G = \frac{1}{2}(\mathbf{u}_B - \mathbf{u}_A)$.

$$\begin{aligned}\therefore\; T &= \tfrac{1}{8}M(\mathbf{u}_A + \mathbf{u}_B)^2 + \tfrac{1}{24}M(\mathbf{u}_B - \mathbf{u}_A)^2 \\ &= \underline{\tfrac{1}{6}M(\mathbf{u}_A{}^2 + \mathbf{u}_A.\mathbf{u}_B + \mathbf{u}_B{}^2)\;.}\end{aligned}$$

N.B. The problem can be done more simply by replacing the rod by the equimomental system of particles $\frac{1}{6}M$, $\frac{1}{6}M$, $\frac{2}{3}M$ at A, B, G moving with velocities $\mathbf{u}_A$, $\mathbf{u}_B$, $\frac{1}{2}(\mathbf{u}_A + \mathbf{u}_B)$.

EXERCISE 7

1. A rigid body is such that the moment of inertia about an axis of it is equal to the sum of the moments of inertia about two other axes, the three axes being concurrent and mutually perpendicular. Prove that the body is a lamina. (Converse of the perpendicular axes theorem.)

2. A square of side a has particles of masses m, $2m$, $3m$, $4m$ at its vertices. Show that the principal moments of inertia at the centre of the square are $2ma^2$, $3ma^2$, $5ma^2$, and find the directions of the principal axes.

3. $P_1P_2, \ldots, P_n$ is a regular n-gon enclosing a lamina of uniform surface density ($n \geqslant 3$). If O be the centre of the circumscribing circle, prove that the momental ellipsoid at O has an equation of the form $x^2 + y^2 + 2z^2 =$ const. (Observe the M.I.s about the n radii $OP_1, OP_2, \ldots OP_n$ are the same so that the section of the ellipsoid with the plane of the lamina is a circle.)

4. *Geometric construction of* M.I.*s and* P.I.*s for coplanar distributions.* Following the notation of the text (Section 7.7), we have, for the case $B > A$,

$$\begin{aligned}I_{OX'} &= \tfrac{1}{2}(A + B) - R\cos(2\alpha - 2\theta)\;, \\ I_{OY'} &= \tfrac{1}{2}(A + B) + R\cos(2\alpha - 2\theta)\;, \\ P_{X'Y'} &= R\sin(2\alpha - 2\theta)\;,\end{aligned}$$

where $\tan 2\alpha = 2F/(B - A)$, $R = \{F^2 + \frac{1}{4}(B - A)^2\}^{\frac{1}{2}}$.
Draw the line segment $OC = \frac{1}{2}(A + B)$. With centre C and radius R, draw a circle meeting OC in D, E, where $OD < OC < OE$. Draw the radius CP_1 to cut the circle in P_1, where $\angle P_1CE = 2\alpha$. Choose P on the minor arc EP_1 so that $\angle P_1CP = 2\theta$ and let P_1C meet the circle again in P_2. Then $\angle P_1P_2P = \theta$. Take rectangular axes through O, along OCE and normal to it. Then P has co-ordinates $(I_{OY'}, P_{X'Y'})$. This affords a construction for finding the M.I.s and P.I.s with respect to the axes OX', OY', when those with respect to OX, OY are known. Show further that $I_{OX'}$, $I_{OY'}$, assume maximum or minimum values when $\theta = \alpha$ or $\theta = \alpha + \frac{1}{2}\pi$, and that $P_{x'y'}$ is maximum or minimum when $\theta = \alpha \pm \frac{1}{4}\pi$, which establishes that *the axes for*

which the product of inertia is numerically greatest are inclined at $\pm\frac{1}{4}\pi$ *to the principal axes.*

5. For a coplanar distribution and the notation of No. 4, show that if $\Theta = A + B$, $\Phi = A - B + 2iF$ $(i^2 = -1)$, then the principal moments of inertia are $\frac{1}{2}(\Theta \pm |\Phi|)$ and the directions made by the principal axes with the positive x-axis are given by the distinct values of $\frac{1}{2}$ arg Φ.

6. The centre of mass of a plane lamina of mass M is G and GX, GY are principal axes of inertia at G. Ox, Oy are axes in the plane of the lamina parallel to GX, GY respectively. Prove that the product of inertia of the lamina with respect to Ox, Oy is $M\alpha\beta$, where α, β are the co-ordinates of G with respect to Ox, Oy.

If the principal moments of inertia of the lamina are $Mn\alpha^2$ about GX and $Mm\beta^2$ about GY, and one of the axes of inertia at O is inclined at an angle of 45° to Ox, prove that

$$(n-1)/(m-1) = \beta^2/\alpha^2 \ .$$

(L.U.)

7. Define the term *equimomental systems.*

The perimeter of a uniform lamina, of mass m, is a regular hexagon $ABCDEF$ of side $2a$. Find a system of particles equimomental with the lamina and hence, or otherwise, show that the moment of inertia of the lamina about AD is $5ma^2/6$. Show also that the moment of inertia of the lamina about *any* line in its plane passing through the intersection of AD and BE has this same constant value. (L.U.)

8. Show that a uniform solid cuboid of mass M is equimomental with (i) masses $\frac{1}{24}M$ at the midpoints of its edges and $\frac{1}{2}M$ at its centre; (ii) with masses $\frac{1}{24}M$ at its corners and $\frac{2}{3}M$ at its centre.

Chapter 8

Two-dimensional Rigid Body Dynamics

8.1 INTRODUCTION

A rigid body is said *to move in two dimensions when every particle of it moves parallel to a fixed plane*. In particular, a rigid lamina moving in its own plane is an important case of two-dimensional motion. We have seen in Chapter 2 that in general the motion of a rigid lamina consists of a translation and a rotation and that there is a point in its plane which is instantaneously at rest. Moreover, knowing the instantaneous angular velocity of the lamina and the linear velocity of its centroid, we can determine the velocity of any other particle of it at the considered instant.

For the motion of a rigid planar distribution, two co-ordinates are required to fix the position of the centroid and a further angular co-ordinate is required to fix the orientation of the lamina. Thus three independent variables are involved in the instantaneous motion of the lamina; two linear velocity components of its centroid and an angular velocity of the distribution. The analysis of rigid body motion in two dimensions consists of applying the principles developed in Chapters 6 and 7 to the determination of these three variables. We now apply these techniques to several classes of two-dimensional motions.

8.2 PROBLEMS ILLUSTRATING THE LAWS OF MOTION

Knowing the magnitudes and positions of forces acting on a rigid body, we find (i) the acceleration of its centroid; (ii) the angular acceleration. In (i), we apply the following result. The translation of the centroid of the rigid body under the system of prescribed forces is the same as that of a particle of equal mass concentrated at the centroid under the action of all forces transferred parallel to themselves to act at the centroid. In (ii), we may use the fact that the total moment of the external forces about any point of the distribution is equal to the moment of the total rate of change of momentum of the distribution about that point, irrespective of whether the point be fixed or moving. When the point is fixed or when it is the centroid, we have seen that this moment of rate of change of momentum is equal to the rate of change of moment of momentum about the considered point.

Example 1

Motion of a uniform solid circular cylinder down a rough inclined plane.

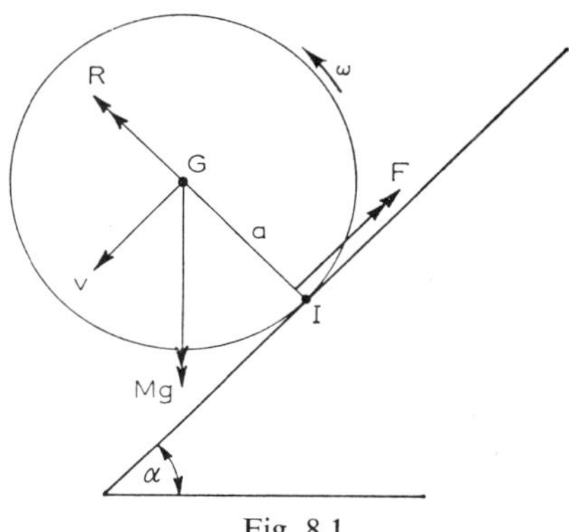

Fig. 8.1

Fig. 8.1 shows the cylinder of mass M and radius a rolling down a rough inclined plane at inclination α to the horizontal. At time t, v is the velocity of its centroid G down the plane and ω the instantaneous angular velocity. The point of contact I is the instantaneous centre of rotation and so we have

$$v = a\omega \ . \tag{1}$$

Let F be the friction force on the cylinder at I and R the normal component of reaction of the plane on the cylinder at this point. The resultant force component down the plane is $Mg \sin \alpha - F$. Hence transferring this force to G and replacing the distribution by a particle of mass M at G, we obtain for the motion in this direction

$$Mg \sin \alpha - F = M\dot{v} \ . \tag{2}$$

Further, since there is no acceleration perpendicular to the plane and since $R - Mg \cos \alpha$ is the resultant force in this direction, similar reasoning shows

$$R - Mg \cos \alpha = 0 \ . \tag{3}$$

Equations (2), (3) express the translatory motion in the two perpendicular directions.

The moment of inertia of the distribution about the axis through G is $\frac{1}{2}Ma^2$ and so its moment of momentum is $\frac{1}{2}Ma^2\omega$ and the rate of change of this is $\frac{1}{2}Ma^2\dot{\omega}$. The moment of the forces acting on the cylinder about this axis is Fa in the same sense. Since G is the centroid,

$$Fa = \tfrac{1}{2}Ma^2\dot{\omega} \ . \tag{4}$$

Equation (4) describes the rotatory motion of the body.

From (1), $\dot{v} = a\dot{\omega}$ and the reader will easily confirm the solution

$$\dot{v}=\tfrac{2}{3}g\sin\alpha\ ,\quad \dot{\omega}=2g\sin\alpha/(3a)\ ,$$

or

$$v=\tfrac{2}{3}gt\sin\alpha+v_o\ ,$$
$$\omega=2gt\sin\alpha/(3a)+\omega_o\ ,$$

where v_o, ω_o are initial values of v, ω at time $t=0$.

It is instructive to take moments about I to produce an equation of rotatory motion alternative to (4). In the anticlockwise sense, the moment of the rate of change of momentum about I is

$$aM\dot{v}+\tfrac{1}{2}Ma^2\dot{\omega}\ ,$$

the first term being due to the translation of the centroid, the second to the rotation of the cylinder about G. The moment of the external forces in the same sense is $Mga\sin\alpha$.

$$\therefore\ Mga\sin\alpha=Ma\dot{v}+\tfrac{1}{2}Ma^2\dot{\omega}\ . \tag{5}$$

It is seen that elimination of F between (2), (4) leads to (5).

Example 2

A circular hoop of radius a, rotating in a vertical plane with spin ω and with its centre at rest, is in contact with a rough plane inclined at angle α, the angle of friction for the surfaces in contact also being α. Show that, if the initial slip velocity is down the plane, the hoop remains stationary for a time $a\omega/(g\sin\alpha)$ and then the hoop rolls down the plane with acceleration $\tfrac{1}{2}g\sin\alpha$.

Show also that, if the initial slip velocity is up the plane, the hoop slides down the plane with acceleration $2g\sin\alpha$ for a time $a\omega/(3g\sin\alpha)$ and then rolls down the plane with acceleration $\tfrac{1}{2}g\sin\alpha$. (L.U.)

(i) *Slip velocity down the plane.*

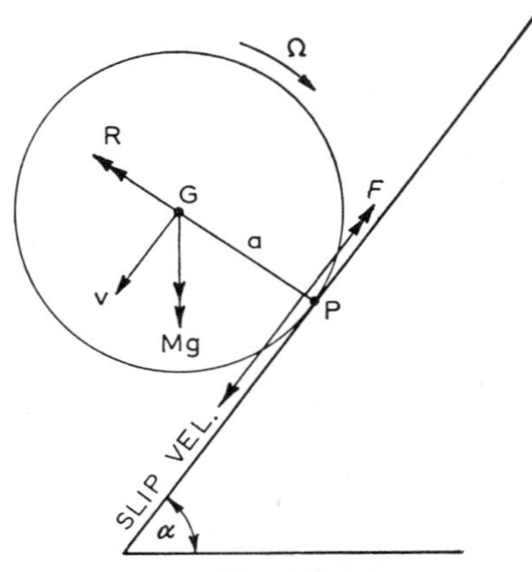

Fig. 8.2

Fig. 8.2 shows the hoop with angular velocity Ω in the sense appropriate to producing a slip velocity down the plane when G is at rest. R, F are the normal component and friction at the contact point P. F must oppose the slip velocity. During slipping, friction is limiting so that $F = R \tan \alpha$. Since the centroid has no motion perpendicular to the plane, $R = Mg \cos \alpha$ (M = mass of hoop) and so during slipping

$$F = Mg \sin \alpha \ . \tag{1}$$

Hence during slipping the resultant force down the plane is zero, *so that the hoop remains stationary while slipping occurs.*

Taking moments about G since $Ma^2\dot{\Omega}$ is the moment of rate of change of momentum about G,

$$-Fa = Ma^2\dot{\Omega} \ . \tag{2}$$

From (1), (2), $-g \sin \alpha = a\dot{\Omega}$ during slipping. Integrating this equation from $t = 0$ to $t = \tau$, the time during which slipping ensues,

$$-g\tau \sin \alpha = a \int_0^\tau \dot{\Omega}\, dt = a \int_\omega^0 d\Omega = -a\omega \ .$$

$$\therefore \ \underline{\tau = a\omega/(g \sin \alpha)} \ .$$

Now let Ω denote the angular velocity at any time t *after* slipping ceases. Fig. 8.1 is still applicable to this rolling phase if the slip velocity of P be taken as zero. This requires

$$v + a\Omega = 0 \ . \tag{3}$$

For the motion down the plane,

$$mg \sin \alpha - F = M\dot{v} \ . \tag{4}$$

Taking moments about G leads again to

$$F = -Ma\dot{\Omega} \ . \tag{5}$$

From (3), (4), (5) we derive

$$\underline{\dot{v} = \tfrac{1}{2} g \sin \alpha} \ .$$

(ii) *Slip velocity up the plane.*

Fig. 8.3 shows the slip velocity up the plane, F being the friction down the plane opposing it and Ω the angular velocity at time t. If v denote the velocity of G down the plane at time t, then the equations of translatory motion along and perpendicular to the plane are

$$F + Mg \sin \alpha = M\dot{v} \ , \tag{6}$$

$$R = Mg \cos \alpha \ . \tag{7}$$

During slipping, $F = R \tan \alpha$, and so

$$F = Mg \sin \alpha \ . \tag{8}$$

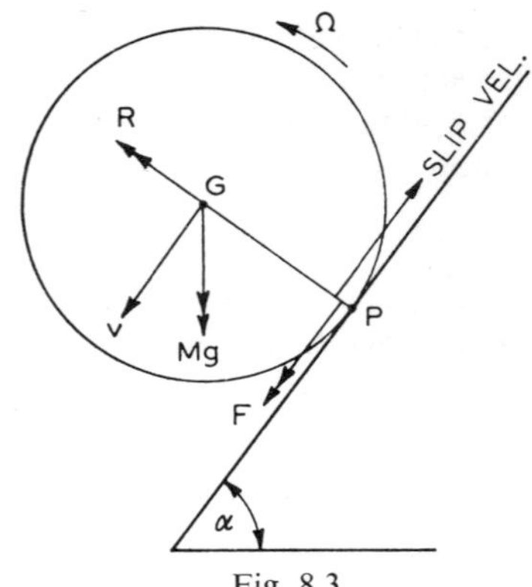

Fig. 8.3

From (6), (8),

$$\underline{\dot{v} = 2g \sin \alpha \ .}$$

Taking moments about G gives

$$-Fa = Ma^2\dot{\Omega} \ , \tag{9}$$

$$\text{or} \quad a\dot{\Omega} = -g \sin \alpha \ . \tag{10}$$

Hence at time t, the angular velocity during the slipping phase is Ω where

$$a(\Omega - \omega) = -gt \sin \alpha \ . \tag{11}$$

Thus if T be the time of slipping and Ω_T the angular velocity at this time,

$$a\Omega_T = a\omega - gT \sin \alpha \ .$$

Also, since $\dot{v}$ is constant and v initially zero, the linear velocity of G down the plane is then $2gT \sin \alpha$. Thus the velocity of the contact point P down the plane at time T is $v - a\Omega_T = 2gT \sin \alpha - (a\omega - gT \sin \alpha)$. Since P is instantaneously at rest at the end of the slipping phase,

$$\underline{T = a\omega/(3g \sin \alpha) \ .}$$

When rolling ensues, equations (6), (9) still apply together with the condition

$$v - a\Omega = 0$$

$$\text{or} \quad \dot{v} - a\dot{\Omega} = 0 \ . \tag{12}$$

From (6), (9), (12) we derive at once

$$\underline{\dot{v} = \tfrac{1}{2}g \sin \alpha \text{ during rolling.}}$$

8.3 PROBLEMS ILLUSTRATING THE LAW OF CONSERVATION OF ANGULAR MOMENTUM

When the total moment of the external forces about a point fixed in space is zero, the moment of momentum or angular momentum of the system about that point in constant (see Chapters 6, 7).

Example 1

A uniform circular disc of mass M and radius a is rotating in its plane with initial angular velocity ω, its centre being at rest. If a point on the rim be suddenly fixed, find the new angular velocity of the disc and the velocity of its centre.

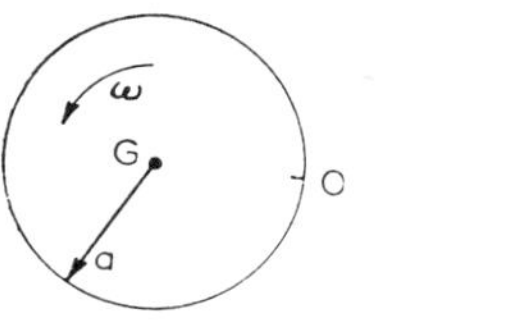

Fig. 8.4

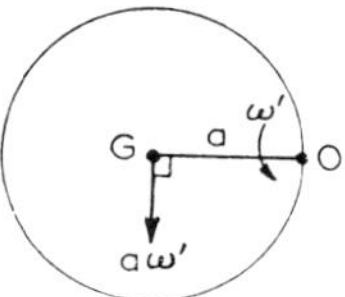

Fig. 8.5

Let O be the point in space coincident with the point on the rim that is to be fixed. As the centroid G is initially at rest, the moment of momentum of the system about O before fixing is $0+\frac{1}{2}ma^2\omega$, since $\frac{1}{2}ma^2$ is the moment of inertia of the disc about the normal to its plane through G.

Just after fixing, let ω' be the angular velocity of the disc about O. Then its centroid has velocity $a\omega'$ (Fig. 8.5). The moment of momentum about O is now $ma^2\omega'+\frac{1}{2}ma^2\omega'$ or $\frac{3}{2}ma^2\omega'$. Thus since there is no force moment about O,

$$\tfrac{3}{2}ma^2\omega' = \tfrac{1}{2}ma^2\omega$$

or

$$\omega' = \tfrac{1}{3}\omega\ , \quad a\omega' = \tfrac{1}{3}a\omega\ .$$

Hence immediately after fixing, the disc rotates with angular velocity $\frac{1}{3}\omega$ about O, the centroid moving with velocity $\frac{1}{3}a\omega$.

Example 2

A frame made of thin wire of line density ρ has the shape of a circle of radius a together with a diameter AB of the circle. The frame lies on a smooth horizontal table and is free to rotate about a fixed vertical axis through a point O on the circumference of the circle and on the diameter normal to AB. An insect of mass $\frac{1}{3}a\rho$ initially at rest at A begins to crawl along the diameter AB. If ϕ is the angle turned backwards by the frame in space after the insect has crawled a distance x along AB, show that

$$\{n^2a^2+(a-x)^2\}\dot{\phi} = a\dot{x}\ ,$$

where $n^2=12\pi+9$. Prove also that when the insect reaches B, the frame will have rotated through an angle $\dfrac{1}{n}\tan^{-1}\left(\dfrac{2n}{n^2-1}\right)$.

(L.U.)

Fig. 8.6 shows the system at time t after the insect has left A and before it reaches B. P is the position of the insect and C the centre of the circle.

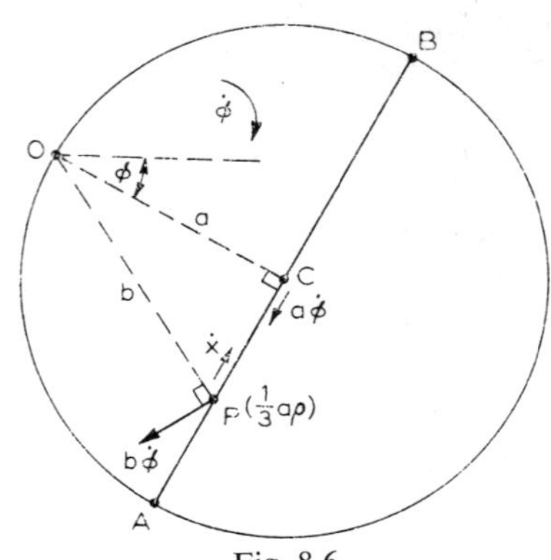

Fig. 8.6

The spin of the frame is $\dot{\phi}$, and so the velocity of C is $a\dot{\phi}$ along $\overline{CA}$, while the insect has a velocity $\dot{x}$ along $\overline{PC}$ together with $b\dot{\phi}$ along the direction perpendicular to OP in the sense of ϕ increasing ($b = OP$). Now the mass of the frame is $2a\rho(\pi+1)$ and its moment of inertia about C is

$$(2a\pi\rho)a^2 + \tfrac{1}{3}(2a\rho)(a^2) = 2a^3\rho(\pi+\tfrac{1}{3}) \ .$$

Hence the angular momentum of the frame about O in the direction of ϕ increasing is

$$a\{2a\rho(\pi+1).a\dot{\phi}\} + 2a^3\rho(\pi+\tfrac{1}{3})\dot{\phi} = 2a^3\rho\dot{\phi}(2\pi+\tfrac{4}{3}) \ .$$

The angular momentum of the particle in this direction is

$$\tfrac{1}{3}a\rho\{b^2\dot{\phi} - a\dot{x}\} = \tfrac{1}{3}a\rho\{(a-x)^2\dot{\phi} + a^2\dot{\phi} - a\dot{x}\} \ .$$

Since the moment about O of the external forces on the entire system comprising frame and particle is zero, the total angular momentum about this point is constant and equal to zero as the system is initially at rest.

$$\therefore \ 2a^3\rho\dot{\phi}(2\pi+\tfrac{4}{3}) + \tfrac{1}{3}a\rho\{(a-x)^2 + a^2\}\dot{\phi} = \tfrac{1}{3}a^2\rho\dot{x} \ ,$$

i.e.

$$\underline{a\dot{x} = \{n^2a^2 + (a-x)^2\}\dot{\phi} \ .}$$

This last relation gives

$$d\phi/dx = a\{n^2 + (a - x^2\}^{-1} \ .$$

Let $\phi = \phi_o$ when $x = 2a$. Then integrating between $x = 0$ and $x = 2a$,

$$\phi_o - 0 = \int_0^{2\alpha} \frac{a\,dx}{n^2 + (a-x)^2} \ ,$$

which leads to the stated result.

8.4 PROBLEMS ILLUSTRATING THE LAW OF CONSERVATION OF ENERGY

It is most important to understand the conditions under which this principle is applicable. Generally speaking the principle of conversation of energy cannot be applied to impact problems, because energy is dissipated as heat. Thus in Example 1 of the last section, dissipation occurs at the point of fixing. In fact the initial K.E. of the system is

$$\tfrac{1}{2}(\tfrac{1}{2}ma^2)\omega^2=\tfrac{1}{4}ma^2\omega^2$$

and the final value is

$$\tfrac{1}{2}m(a\omega')^2+\tfrac{1}{2}(\tfrac{1}{2}ma^2)\omega'^2=\tfrac{3}{4}ma^2\omega'^2=\tfrac{1}{12}ma^2\omega^2\ ,$$

showing that the net loss of K.E. is $\frac{1}{6}ma^2\omega^2$.

Example 2 of the last section is another case in which the principle is inapplicable because the insect is doing work on the system and no information is given concerning its total energy expenditure.

It should, however, be borne in mind that internal forces do no work when the distances between their points of application remain invariant (since action and reaction are equal and opposite). Also, normal reaction components do no work when a system is displaced because the reaction is perpendicular to the displacement. During the rolling of one surface on another without slipping, there are equal and opposite friction forces called into play on both surfaces and the points of application of these forces move equal distances in a small displacement and hence the total work done in a finite displacement is zero. This result is, however, no longer true for sliding. Thus in Example 1 of Section 8.2 of this chapter the principle of conservation of energy could be applied, but in Example 2 of the same section the principle is inapplicable whenever the hoop passes through slipping phases.

Example 1—The compound pendulum

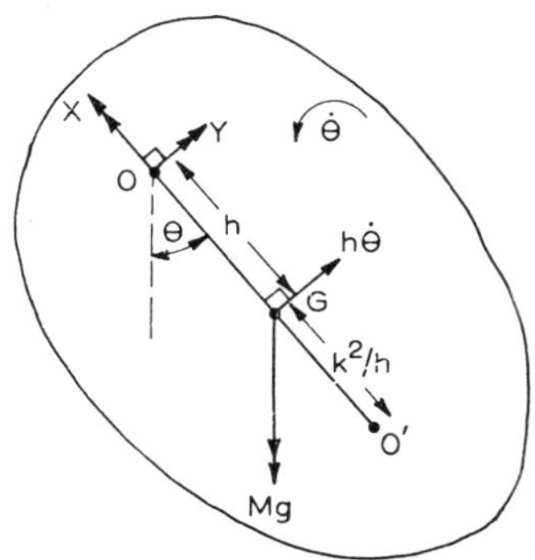

Fig. 8.7

Fig. 8.7 shows a rigid body of mass M rotating about a horizontal axis through the point O, the axis being distant h from the centroid G. Let k be the radius of gyration about the horizontal axis through G so that Mk^2 is the moment of inertia about this axis. At time t, suppose OG makes an angle θ with the downward vertical: then $\dot{\theta}$ is the angular velocity of the body in the sense of θ increasing and $h\dot{\theta}$ is the linear velocity of G in this sense. Thus the kinetic energy is

$$\tfrac{1}{2}M(h\dot{\theta})^2+\tfrac{1}{2}Mk^2\dot{\theta}^2=\tfrac{1}{2}M(h^2+k^2)\dot{\theta}^2\ .$$

The potential energy referred to the horizontal through O as zero level, is $-Mgh\cos\theta$. Since the sum of the K.E. and P.E. is constant, we have

$$\tfrac{1}{2}M(h^2+k^2)\dot{\theta}^2-Mgh\cos\theta=\text{const.} \tag{1}$$

Differentiating w.r.t. t and dividing through by $M\dot{\theta}$ gives

$$(h^2+k^2)\ddot{\theta}+gh\sin\theta=0\ . \tag{2}$$

The equation (2) could also be derived by taking moments about O.

When θ is small, $\sin\theta\simeq\theta$ and to first order (2) gives

$$\ddot{\theta}+gh(h^2+k^2)^{-1}\,\theta=0\ . \tag{3}$$

(3) is the equation of a S.H.M. of period

$$T=2\pi(l/g)^{\frac{1}{2}}\ ,$$

where l, the length of the simple equivalent pendulum, is given by

$$l=(h^2+k^2)/h=h+k^2/h\ . \tag{4}$$

The reader will see that replacement of h by k^2/h in (4) leaves the value of l unaltered. Hence if we take O' in OG so that $GO'=k^2/h$, the period of small oscillations about O' is the same as that about O. Further, from (4),

$$l=(\sqrt{h}-k/\sqrt{h})^2+2k\ ,$$

and so min.$l=2k$, when $h=k$.

Suppose now the pendulum has finite angular amplitude α so that $\dot{\theta}=0$ when $\theta=\alpha$. Then (1) gives

$$\tfrac{1}{2}M(h^2+k^2)\dot{\theta}^2-Mgh\cos\theta=-Mgh\cos\alpha\ ,$$

$$\dot{\theta}^2=2gh(\cos\theta-\cos\alpha)/(h^2+k^2) \tag{5}$$

and $$\ddot{\theta}=-gh\sin\theta/(h^2+k^2)\ . \tag{6}$$

To find the force exerted at O, suppose $[X, Y]$ are its components along $\overline{GO}$ and at right angles to $\overline{GO}$ as shown. Considering the motion along $\overline{GO}$,

$$X-Mg\cos\theta=Mh\dot{\theta}^2\ . \tag{7}$$

For the motion along the direction of Y,

$$Y-Mg\sin\theta=Mh\ddot{\theta}\ . \tag{8}$$

Thus we find

$$\begin{cases}X=Mg\{\cos\theta+2h^2(\cos\theta-\cos\alpha)/(h^2+k^2)\}\ ,\\ Y=Mgk^2\sin\theta/(h^2+k^2)\ .\end{cases}$$

Example 2

A uniform heavy solid hemisphere of radius a is held at rest with its base vertical and its curved surface in contact with a horizontal plane. If the hemisphere is released when the plane is rough enough to prevent slipping, show that the angle θ that the base makes with the horizontal at time t is such that

$$(d\theta/dt)^2 = 15g\cos\theta/\{a(28-15\cos\theta)\} \ .$$

(L.U.)

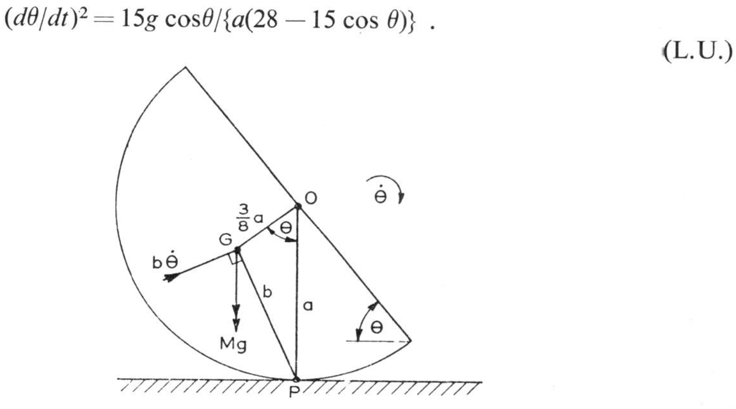

Fig. 8.8

Denote the mass of the hemisphere by M, the centroid by G, O the centre of the base, so that $OG = \frac{3}{8}a$. Let $PG = b$. Since $\dot{\theta}$ is the angular velocity of the hemisphere in the sense of θ increasing and P is the instantaneous centre (Fig. 8.8), $b\dot{\theta}$ is the velocity of G (perpendicular to PG) and so the K.E. of the body is

$$T = \tfrac{1}{2}M(b\dot{\theta})^2 + \tfrac{1}{2}I_G\dot{\theta}^2 \ .$$

Now
$$I_G = I_o - M(\tfrac{3}{8}a)^2$$
$$= \tfrac{2}{5}Ma^2 - M(\tfrac{3}{8}a)^2 = 83Ma^2/320 \ .$$

Since
$$b^2 = a^2 + (\tfrac{3}{8}a)^2 - 2a(\tfrac{3}{8}a\cos\theta)$$
$$= a^2(73-48\cos\theta)/64 \ ,$$
$$T = Ma^2\dot{\theta}^2(28-15\cos\theta)/40 \ .$$

The P.E. referred to the horizontal through P as zero level, is

$$V = Mga(1-\tfrac{3}{8}\cos\theta) \ .$$

Now $T+V =$ constant, and so

$$a\dot{\theta}^2(28-15\cos\theta) + 40g(1-\tfrac{3}{8}\cos\theta) = \text{const.} = 40g \ ,$$

since $\dot{\theta} = 0$ when $\theta = \frac{1}{2}\pi$. Hence we obtain the stated result.

8.5 PROBLEMS ILLUSTRATING IMPULSIVE MOTION

When an impulse is applied to a rigid body, changes are produced in both the linear momentum and in the moment of momentum about any point. The linear momentum change is found simply by replacing the given distribution by a single particle of the same mass concentrated at its centroid and transferring the impulse parallel to itself to the centroid position.

The gain in moment of momentum about a point is simply equal to the moment of the impulse about that point, both being measured in the same sense.

Example 1

A uniform rod AB of mass M and length $2a$ lies at rest on a smooth horizontal table. An impulse J is applied at A in the plane of the table and perpendicular to the rod. Determine the velocity of the centroid and the angular velocity of the rod.

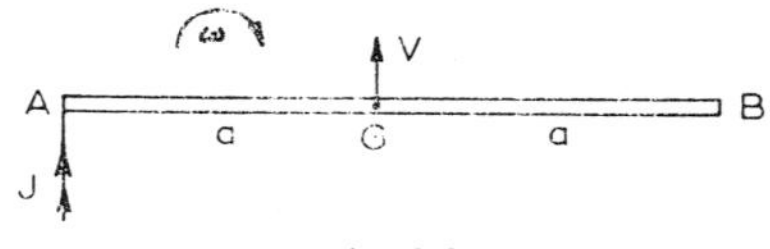

Fig. 8.9

Let V be the velocity of the centroid G (essentially perpendicular to the rod) and ω the acquired instantaneous angular velocity as shown in Fig. 8.9. The linear momentum gain in the direction of the applied impulse is MV so that

$$J = MV \ . \tag{1}$$

The moment of the impulse about G is Ja and the gain in moment of momentum about G in the same sense is $\frac{1}{3}Ma^2\omega$ so that

$$Ja = \tfrac{1}{3}Ma^2\omega \ . \tag{2}$$

From (1), (2), $V = J/M$, $\omega = 3J/(Ma)$.
N.B. Taking moments about A would give

$$\tfrac{1}{3}Ma^2\omega - MVa = 0 \ ,$$

which follows from (1), (2).

Example 2

A uniform rod AB of mass $2m$ is freely jointed at B to a second rod BC of mass m. The rods lie on a smooth horizontal plane at right angles to each other and an impulse I is applied to AB at A in a direction parallel to BC. Find the initial velocity of BC and prove that the kinetic energy generated is $\frac{5}{6}I^2/m$. (L.U.)

Let B have velocity components u, v along $\overline{AB}$, $\overline{BC}$ and let ω_1, ω_2 be the angular velocities of the rods in the senses shown (Fig. 8.10). The centroids of AB, BC have velocity components $[u, v + a\omega_1]$, $[u + b\omega_2, v]$. The lengths of AB, BC are $2a$, $2b$ respectively.

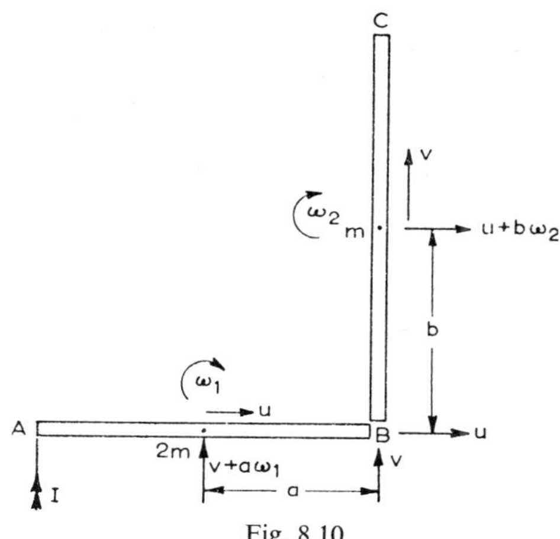

Fig. 8.10

The total linear momentum of the system is conserved along $\overline{AB}$.

$$\therefore\ 2mu+m(u+b\omega_2)=0\ ,$$

or

$$3u+b\omega_2=0\ . \tag{1}$$

The total linear momentum produced by I in the direction $\overline{BC}$ is $2m(v+a\omega_1)+mv=m(3v+2a\omega_1)$ and so

$$m(3v+2a\omega_1)=I\ . \tag{2}$$

An impulsive action is set up at B. By taking moments about B for the two rods we need not introduce this impulse. Thus, for the rod AB, the moment of momentum generated about B is $2m(v+a\omega_1)a+\frac{1}{3}(2ma^2\omega_1)$ $=m(2va+\frac{8}{3}a^2\omega_1)$ and the moment of I about B is $2aI$ in the same sense. Equating these gives

$$m(v+\tfrac{4}{3}a\omega_1)=I\ . \tag{3}$$

For the rod BC, the moment of momentum about B is $mb(u+b\omega_2)+$ $+\frac{1}{3}mb^2\omega_2$.
Equating this to zero gives

$$u+\tfrac{4}{3}b\omega_2=0\ . \tag{4}$$

From (1), (4), $u=0$, $\omega_2=0$ and from (2), (3),

$$v=-\tfrac{1}{3}I/m,\ a\omega_1=I/m\ .$$

Hence the initial velocity of BC is $\frac{1}{3}I/m$ in the direction $\overline{CB}$.

The K.E. generated is

$$\tfrac{1}{2}(2m)(v+a\omega_1)^2+\tfrac{1}{2}(2m)(a^2/3)\omega_1{}^2+\tfrac{1}{2}mv^2=(\tfrac{4}{9}+\tfrac{1}{3}+\tfrac{1}{18})I^2/m=\underline{\tfrac{5}{6}I^2/m}\ .$$

N.B. Since the linear momentum gain of BC is $\frac{1}{3}I$ along $\overline{CB}$, the impulsive action of the hinge at B on BC is $\frac{1}{3}I$ along $\overline{CB}$ and that on AB is equal and opposite.

8.6 FURTHER EXAMPLES

The following are rather harder examples and illustrate the applications of several dynamical principles.

Example 1

A pefectly rough circular hoop of radius a rolls on a horizontal floor with velocity U towards an inelastic step of height $h(<\frac{1}{2}a)$, the plane of the hoop being vertical and perpendicular to the edge of the step. Prove that the hoop can mount the step without losing contact at any stage if

$$4a^2hg < U^2(2a-h)^2 < 4a^2(a-h)g \ .$$

On the other hand, if the hoop rolls with velocity V towards the edge on the top of the step in a vertical plane perpendicular to the edge, prove that it can roll down the step without losing contact if $V^2 < (a-2h)g$.

(L.U.)

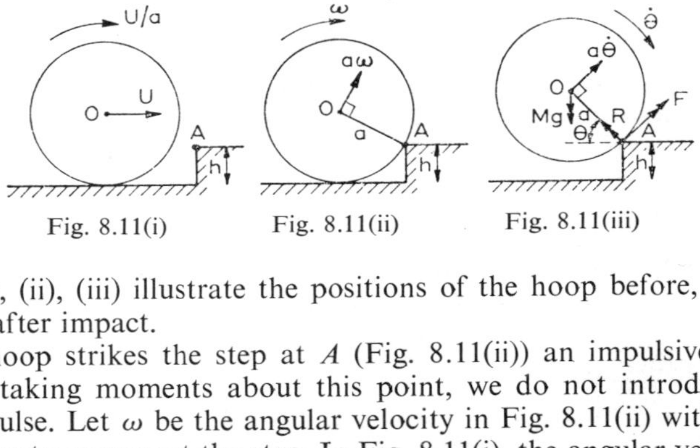

Fig. 8.11(i) Fig. 8.11(ii) Fig. 8.11(iii)

Figs. 8.11(i), (ii), (iii) illustrate the positions of the hoop before, during, and at time t after impact.

When the hoop strikes the step at A (Fig. 8.11(ii)) an impulsive action is set up. By taking moments about this point, we do not introduce this unwanted impulse. Let ω be the angular velocity in Fig. 8.11(ii) with which the hoop begins to surmount the step. In Fig. 8.11(i), the angular velocity is U/a. Hence, since the moment of momentum is conserved about A during the phases represented by Figs. 8.11(i), (ii), we have, taking M to be the mass of the hoop,

$$MU(a-h) + Ma^2(U/a) = Ma^2\omega + Ma^2\omega \ ,$$

or

$$a\omega = U(1-h/2a) \ . \qquad (1)$$

During the subsequent phase of mounting the step, energy is conserved since no slipping occurs at A in the motion from Fig. 8.11(ii) to Fig. 8.11(iii). Taking the horizontal through A as zero level of P.E., since the sum of the K.E. and P.E. is constant,

$$\frac{1}{2}Ma^2\dot{\theta} + \frac{1}{2}Ma^2\dot{\theta}^2 + Mga\sin\theta = \text{const.}$$

$$\therefore\ a\dot{\theta}^2 + g\sin\theta = \text{const.} = (U^2/a)(1-h/2a)^2 + g(1-h/a) \ . \qquad (2)$$

The value of $a\dot{\theta}^2$ when $\theta = \frac{1}{2}\pi$ is to be >0 for the hoop to surmount the step. Thus (2) gives

$$(U^2/a)(1-h/2a)^2 - gh/a > 0$$

or

$$4a^2hg < U^2(2a-h)^2 \ . \tag{3}$$

The equation of motion of the hoop along AO in Fig. 8.11(iii), taking R to be the reaction component of the step in this direction, is

$$Mg \sin\theta - R = Ma\dot{\theta}^2 \ . \tag{4}$$

From (2), (4), we find

$$R = 2Mg\sin\theta - (MU^2/a)(1-h/2a)^2 - Mg(1-h/a) \ .$$

The smallest value of $\sin\theta$ is $1-h/a$.

$$\therefore \ \min.R = Mg(1-h/a) - (MU^2/a)(1-h/2a)^2 \ .$$

Contact is never broken if $\min.R > 0$,

i.e. if

$$Mg(1-h/a) - (MU^2/a)(1-h/2a)^2 > 0 \ ,$$

or

$$U^2(2a-h)^2 < 4a^2g(a-h) \ . \tag{5}$$

From (3), (5),

$$\underline{4a^2hg < U^2(2a-h)^2 < 4a^2g(a-h) \ .}$$

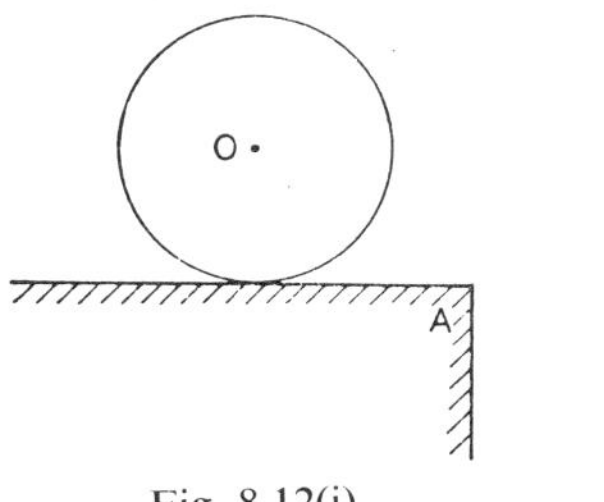

Fig. 8.12(i)

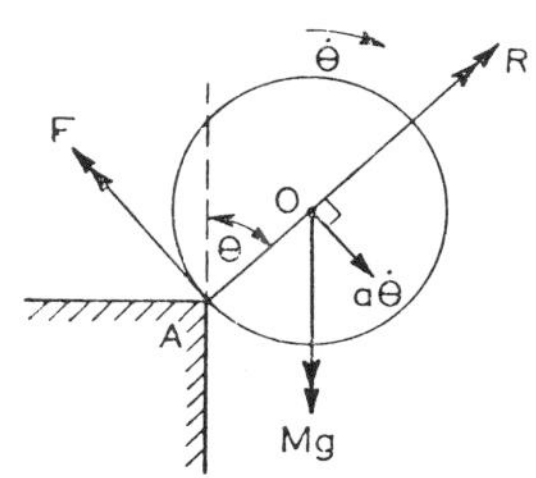

Fig. 8.12(ii)

Consider the motion down the step as depicted in Figs. 8.12(i), 8.12(ii). There is no impact at A, and so energy is conserved. Hence taking the horizontal through A as the zero level of P.E., the energy equation applied to Fig. 8.12(ii) is

$$\tfrac{1}{2}Ma^2\dot{\theta}^2 + \tfrac{1}{2}Ma^2\dot{\theta}^2 + Mga\cos\theta = \text{const.}$$

$$\therefore \ a\dot{\theta}^2 + g\cos\theta = \text{const.} = V^2/a + g \ . \tag{6}$$

If R denote the reaction component along $\overline{AO}$, the radial equation of motion is

$$Mg\cos\theta - R = Ma\dot{\theta}^2 \ . \tag{7}$$

From (6), (7),

$$R = M(2g\cos\theta - V^2/a - g) \ . \qquad (8)$$

R is a minimum when θ is a maximum. In this case,

$$\text{min. } \cos\theta = 1 - h/a \ .$$
$$\therefore \text{ min. } R = M[g(1 - 2h/a) - V^2/a] \ .$$

R does not vanish if min. $R > 0$, i.e. if

$$g(1 - 2h/a) - V^2/a > 0 \ ,$$

or

$$\underline{V^2 < g(a - 2h)} \ .$$

The condition $h < \frac{1}{2}a$ ensures the possibility of this inequality.

Example 2

A uniform rod AB of length $2a$ can turn freely about one end A, and at time $t = 0$ is hanging from A in equilibrium under gravity. The end A is then set in motion in a horizontal straight line so that, when $t \geqslant 0$, $OA = vt + \frac{1}{2}ft^2$, where O is a fixed point in the line and v, f are constants. Show that the initial angular velocity of the rod is $3v/4a$ and it will make complete revolutions about A if

$$3v^2 > 8a[g + (g^2 + f^2)^{\frac{1}{2}}] \ . \qquad \text{(L.U.)}$$

(i) *Initial Motion* ($t = 0$).

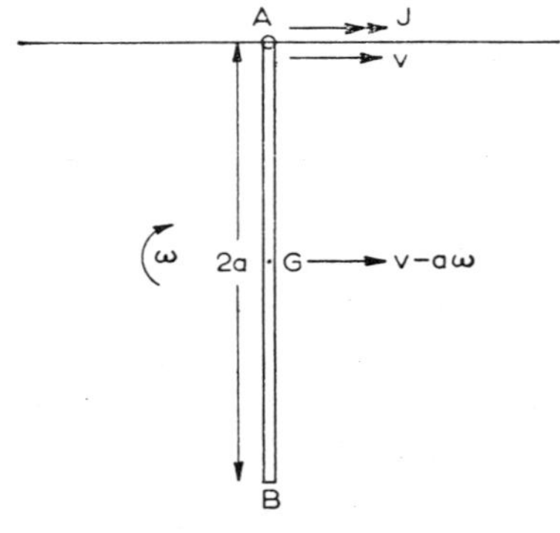

Fig. 8.13

Let J be the impulse applied at A, v the velocity of A, ω the angular velocity of the rod and m its mass (Fig. 8.13). Since $v - a\omega$ is the velocity of the centroid G in the direction of J, the linear momentum equation in this direction is

$$J = m(v - a\omega) \ . \qquad (1)$$

Taking moments about G,

$$Ja = \tfrac{1}{3}ma^2\omega \ . \qquad (2)$$

From (1), (2),
$$\underline{\omega = 3v/4a} \ .$$

(ii) *Subsequent motion* (*time* $t > 0$).

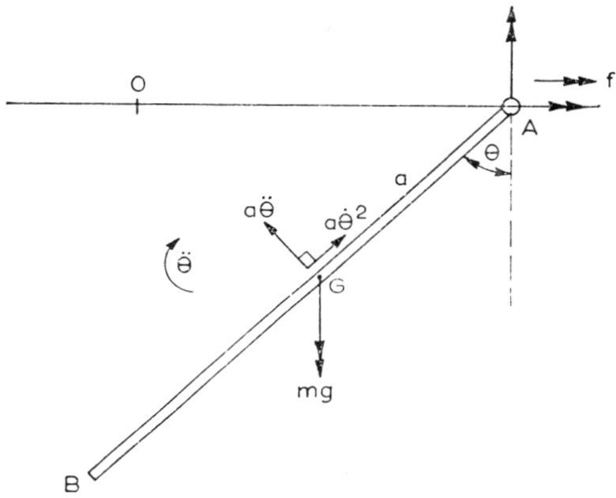

Fig. 8.14

There is an unknown reaction at A. To avoid introducing it, take moments about A. A has an acceleration f along $\overline{OA}$ and G has acceleration components *relative to A* of $a\dot{\theta}^2$ along $\overline{GA}$ and $a\ddot{\theta}$ in the perpendicular direction, where θ is the inclination of the rod to the downward vertical at time t (Fig. 8.14). Thus the total acceleration of G has components $[a\dot{\theta}^2 + f\sin\theta,\ a\ddot{\theta} - f\cos\theta]$ along $\overline{GA}$ and at right angles to it and the total moment of the rate of change of momentum about A is

$$\tfrac{1}{3}ma^2\ddot{\theta} + am(a\ddot{\theta} - f\cos\theta) \ .$$

The moment of mg about A in the same sense is $-mga\sin\theta$, and so equating the two gives

$$\tfrac{4}{3}a\ddot{\theta} = f\cos\theta - g\sin\theta \ . \qquad (3)$$

Since $\ddot{\theta} = \dot{\theta}\,d\dot{\theta}/d\theta$, integrating (3) and using $\dot{\theta} = 3v/4a$ when $\theta = 0$,

$$\tfrac{2}{3}a[\dot{\theta}^2 - (3v/4a)^2] = f\sin\theta + g\cos\theta - g \ .$$

$$\therefore\ \tfrac{2}{3}a\dot{\theta}^2 = f\sin\theta + g\cos\theta - g + 3v^2/8a$$

$$= (f^2 + g^2)^{\frac{1}{2}}\sin(\theta + \alpha) - g + 3v^2/8a \ ,$$

where $\cos\alpha = f(f^2 + g^2)^{-\frac{1}{2}}$, $\sin\alpha = g(f^2 + g^2)^{-\frac{1}{2}}$.

$$\therefore\ \tfrac{2}{3}a\dot{\theta}^2 \geqslant -(f^2 + g^2)^{\frac{1}{2}} - g + 3v^2/8a \ .$$

$\therefore$ If $3v^2 > 8a[g + (f^2 + g^2)^{\frac{1}{2}}]$, then $\tfrac{2}{3}a\dot{\theta}^2 > 0$, i.e. $\dot{\theta}^2 > 0$ for all θ and the rod will make complete revolutions.

EXERCISE 8

1. A uniform rod of mass m and length $2a$ falls from rest in a vertical position with one end fixed on a table which is so rough that slipping never occurs. Show that when the rod is inclined to the vertical at an angle θ the angular velocity is $(3g/a)^{\frac{1}{2}} \sin(\theta/2)$ and that the rod will leave the table when the inclination to the vertical is $\cos^{-1}(\frac{1}{3})$.

2. A uniform rod is placed on a horizontal table with two thirds of its length hanging over the edge of the table. If the rod is at right angles to the edge and is released show that it will begin to slip when the rod has turned through an angle of $\tan^{-1}(\frac{1}{2}\mu)$, where μ is the coefficient of friction between rod and table.

3. A flywheel with an axle of radius r, total weight W and radius of gyration k, rolls with its axle on two parallel horizontal rails, rough enough to prevent slip. Motion is caused by a weight w attached to a light cord wrapped round the axle, with w hanging freely. Show that motion is possible in which the hanging portion of the cord makes a steady angle α with the vertical given by

$$\sin\alpha/(1-\sin\alpha)^2 = wr^2/W(r^2+k^2) ,$$

and that the acceleration of the flywheel is then $g\tan\alpha$. (L.U.)

4. A uniform rod of mass m and length $2a$ is held inclined to the vertical at an angle α with its lower end in contact with a smooth horizontal table and is released from rest. Prove that when the inclination of the rod to the vertical is θ, the reaction of the table is

$$mg(4-6\cos\alpha\cos\theta+3\sin^2\theta)(1+3\sin^2\theta)^{-2} .$$

Prove that the rod can never leave the table.

5. A thin uniform rod of length $2a$ is held so that it stands upright on a smooth horizontal plane and in contact with a vertical wall. When the rod is released, the upper end moves away from the wall and the rod falls in a vertical plane at right angles to the wall. Show that the lower end of the rod leaves the wall when the rod has turned through an angle $\cos^{-1}(\frac{2}{3})$, and prove that when the rod strikes the horizontal plane the velocity of its mid-point is $\sqrt{(14ga/9)}$. (L.U.)

6. A compound pendulum consists of a bar AB of length $2a$ and mass M and a thin uniform square plate, coplanar with the bar, of side $a/8$ and mass $64M$, fixed so that the intersection of the diagonals of the plate lies on the bar at a distance x from A. The pendulum performs small oscillations about a fixed horizontal axis through A perpendicular to the plane of the plate. If the period is $3/2\sqrt{2}$ times the period when the bar swings alone, find the values of x. (L.U.)

7. Show that the moment of inertia of a uniform solid right circular cone, of mass M, height h, and vertical angle 2α, about a diameter of its base is $Mh^2(3\tan^2\alpha+2)/20$.

The cone can turn freely about a horizontal hinge attached to a diameter of its base and is released from rest with its axis horizontal. Show that, when the axis is vertical the cone exerts on the hinge a force of magnitude $\frac{3}{2}Mg(2+\cos^2\alpha)/(3-\cos^2\alpha)$. (L.U.)

8. With the usual notation, prove the formula $\frac{1}{2}M(V^2+k^2\omega^2)$ for the kinetic energy of a lamina moving in its plane.

A particle of mass m is fixed to a point P of the rim of a uniform circular disc, centre O, mass m, and radius a. The disc is held, with its plane vertical, its lowest point in contact with a perfectly rough horizontal table, and with OP inclined at 60° to the upward vertical and is then released. If the subsequent motion continues in the same vertical plane, show that, when OP makes an angle θ with the upward vertical,

$$a(7+4\cos\theta)\dot{\theta}^2=2g(1-2\cos\theta)\ .$$

Show also that when OP is first horizontal the acceleration of O is $18g/49$, and find the reaction between the disc and table at this instant. (L.U.)

9. A conservative system with one degree of freedom has kinetic energy $T=f(\theta)\dot{\theta}^2$ and potential energy $V=h(\theta)$. If $\theta=\alpha$ is a position of equilibrium of the system, show that $h'(\alpha)=0$ and that the period of small oscillations about $\theta=\alpha$ is $2\pi\{2f(\alpha)/h''(\alpha)\}^{\frac{1}{2}}$, provided $h''(\alpha)>0$.

A uniform rod AB, of mass $2m$ and length a, is smoothly hinged to a fixed point at A and has a particle of mass m attached to B, which is connected by a light elastic string of modulus mg and natural length $a/2$ to a point C vertically below A. If $AC=2a$, show that the period of small oscillations about the position of equilibrium with B vertically below A is $\pi(5a/3g)^{\frac{1}{2}}$. (L.U.)

10. A wheel, of mass M and centre O, consists of a uniform circular disc of radius a in which have been cut four circular holes. The radius of each hole is $a/4$, and their centres, which are at a distance $a/2$ from O, form the vertices of a square. Show that the moment of inertia of the wheel about an axis through O normal to the plane of the wheel is $55Ma^2/96$.

The wheel is free to rotate about this axis. A light elastic string of modulus $2Mg$ and natural length a joins P, a point on the circumference of the wheel, to a fixed point A in the plane of the wheel. If $AO=3a$, show that the length of the equivalent simple pendulum for small oscillations of the wheel about its stable position of equilibrium is $55a/288$. (L.U.)

11. State the principles of conservation of energy and of angular momentum.

A cube of edge $2a$ and mass m is free to rotate about a vertical edge. A smooth horizontal groove is cut across the upper face along the diagonal which does not intersect the axis of rotation, and a particle of mass $2m/3$ is placed at the midpoint of the groove. The cube is given an initial angular velocity Ω. If the velocity of the particle relative to the groove is V when the particle is about to leave the groove and the angular velocity of the cube at the same instant is ω, show that

$$V\sqrt{2}=4a(2\omega-\Omega)$$

and that ω satisfies the equation

$$6\omega^2 - 6\omega\Omega + \Omega^2 = 0 \ . \qquad \text{(L.U.)}$$

12. A uniform rod AB of mass M and length $2a$ rests in equilibrium on a rough fixed horizontal circular cylinder of radius a, with AB perpendicular to the axis of the cylinder. A particle of mass m is fixed to each end of the rod. Show that, if the rod rolls on the cylinder in a vertical plane, its kinetic energy is given by

$$a^2\dot{\theta}^2\{3\theta^2(M+2m)+(M+6m)\}/6 \ ,$$

where θ is the inclination of the rod to the horizontal. Show that the period of small oscillations of the rod about its position of stable equilibrium is

$$2\pi\{a(M+6m)/g(3M+6m)\}^{\frac{1}{2}} \ . \qquad \text{(L.U.)}$$

13. A wire bent in the form of the curve of intrinsic equation $s=\frac{1}{4}a \tan \psi$ is fixed in a vertical plane with its vertex uppermost. A uniform disc of radius a rolls down the wire, the plane of the disc being vertical throughout the motion. If the motion starts from rest at the vertex, show that the disc leaves the wire when the common normal makes an angle $\pi/3$ with the vertical. (L.U.)

14. A uniform solid circular cylinder of radius a moves on the inside of a fixed straight tube of inner radius $c+a$ whose axis is horizontal. The cylinder starts by rolling from the lowest position in the tube with an angular velocity $(ngc)^{\frac{1}{2}}/a$. Prove that, if the cylinder is to roll completely round the tube, $n > 11/3$, and the coefficient of friction between the cylinder and the tube $> \{(3n-11)(3n+3)\}^{-\frac{1}{2}}$. (L.U.)

15. A uniform thin hemispherical shell is held with its curved surface in contact with a smooth vertical wall and its rim parallel to the wall and touching a smooth horizontal table. It is then released and the upper edge of the rim begins to move towards the wall. Show that the shell will remain in contact with the wall until the rim is horizontal. Prove that subsequently the maximum inclination of the rim to the horizontal is $\cos^{-1}(3/8)$. (L.U.)

16. A uniform solid rough sphere of radius a and mass m rests on a uniform board of mass M which lies on a smooth horizontal table. If the board is projected with a velocity v along the table, show that the sphere will slip on the board for a time $2Mv/\{\mu(7M+2m)g\}$, where μ is the coefficient of friction between the sphere and the board. Find the velocity of the board at this instant and the velocity with which the sphere starts to roll. (L.U.)

17. A uniform rod AB of length $2a$ is placed vertically on a rough diving board with A above B and B in contact with the edge of the board. The rod is slightly disturbed by turning about B and it leaves the board when the reaction along the rod vanishes. Show that when this occurs the rod has turned through an acute angle θ where $\cos\theta = 3/5$. Find the angular velocity of the rod at this instant.

If the rod strikes the water vertically after rotating through the angle π from its initial position, show that the height of the board above the surface of the water is

$$a(24\phi + 25\phi^2 + 12)/30 \ ,$$

where $\phi = \pi - \phi$. (L.U.)

18. A uniform circular disc of radius a and mass m is free to rotate in a vertical plane about a smooth horizontal axis through its centre C. An insect A of mass $m/20$ is at the lowest point of the disc and the whole system is at rest. The insect suddenly starts to crawl along the rim of the disc with uniform speed V relative to the disc. Show that the initial spin of the disc is $V/(11a)$. If in the subsequent motion θ is the angle between CA and the downward vertical, prove that

$$11a\ddot{\theta} + g \sin \theta = 0 \ .$$

Show that the component of the force exerted by the insect on the disc in the direction perpendicular to the radius is $(mg/22)\sin \theta$. (L.U.)

19. A hollow circular cylinder of radius b is made to rotate about its axis, which is horizontal, with variable angular velocity ω. A uniform solid circular cylinder of radius $a(<b)$ rolls without slipping inside the hollow circular cylinder, the axes of the two cylinders being parallel and the plane through them making an angle θ with the downward vertical. Prove that at any time

$$3(b-a)\ddot{\theta} = b\dot{\omega} - 2g \sin \theta \ .$$

If $\dot{\omega}$ is constant and equal to kg/b and the system starts from rest with the smaller cylinder in its lowest position, find the normal component R of the reaction between the two cylinders in any position. Show that, if $2k>7$, both $\dot{\theta}$ and R increase steadily in the subsequent motion.

Show also that, if the smaller cylinder rolls completely round inside the larger cylinder for a value of $k<2$, it is necessary that $\dot{\theta}$ should exist when $\theta = \pi - \sin^{-1}(k/2)$ and R should be >0 when $\theta = \pi - \sin^{-1}(2k/7)$. (L.U.)

20. A hollow circular cylinder of internal radius a is fixed with its axis horizontal. A light hoop of radius $b(<a)$ has a heavy particle attached to its rim and can roll on the inside surface of the cylinder in a vertical plane perpendicular to the axis. Show that the position of equilibrium in which the hoop is in contact with the lowest point of the cylinder and the particle is at the highest point of the hoop is unstable if $2b<a$.

If $b=a/4$, and the hoop is slightly displaced from this position of equilibrium, show that in the subsequent motion the plane containing the axis of the cylinder and the centre of the hoop makes an angle θ with the downward vertical, where

$$9a\dot{\theta}^2(1+\cos 4\theta) = 4g(3 \cos \theta - \cos 3\theta - 2) \ .$$

Deduce that θ cannot exceed $\cos^{-1}\{(\sqrt{3}-1)/2\}$. (L.U.)

21. A rigid body of mass M is in the shape of a thin cylindrical shell whose normal cross section is a circle of radius a. Owing to non-uniformity of density, the centre of mass G of the shell is at a distance ka from the axis of symmetry A of the shell, which rolls on a perfectly rough horizontal table. Show that the velocity v of A is given by $v^2 = ag(\lambda - \nu)/\nu$, where λ is a constant, $\nu = 1 - k\cos\theta$, and θ is the angle between AG and the downward vertical.

Furthermore, show that the normal reaction of the shell on the plane is $Mg[\nu + \frac{1}{2}\lambda\{(1-k^2)/\nu^2 - 1\}]$. (L.U.)

22. A uniform solid sphere, balanced on the top of a fixed rough sphere, not necessarily of the same radius, is slightly disturbed. Show that the initial motion is one of rolling and that slipping commences before the spheres separate.

If θ is the inclination of the line of centres of the spheres to the upward vertical, and the coefficient of friction between the spheres is 1/5, show that the slipping commences when $\theta = \cos^{-1}(340/389)$. (L.U.)

23. A uniform solid sphere of radius a, rotating about a horizontal diameter, is gently placed on the interior surface of a fixed rough circular cylinder of radius $a(1+\sqrt{2})$ whose axis is horizontal and parallel to the axis of rotation of the sphere. The coefficient of friction between the sphere and the cylinder is $1/\sqrt{2}$ and the angle made by the plane of the axis of the cylinder and the centre of the sphere with the downward vertical is θ. If the initial value of θ is $\pi/6$, and the sense of rotation is such as to tend to make the sphere roll upwards, show that the centre of the sphere does begin to rise and that while slipping continues

$$2a\dot{\theta}^2/g = 2\sin\theta - \exp\{(\theta - \pi/6)\sqrt{2}\} \ . \qquad \text{(L.U.)}$$

24. A thin, uniform, hollow cylinder of mass M and radius b rolls on a rough horizontal table, and inside it is a rough, uniform, solid cylinder of mass m and radius a. The axes of the cylinders are parallel and their centres of gravity lie in the same cross section. Show that, when the inclination to the downward vertical of the plane through the axes is θ, the kinetic energy of the system is

$$Mv^2 + \tfrac{1}{4}m(u+v)^2 + \tfrac{1}{2}m(u^2 + 2uv\cos\theta + v^2) \ ,$$

where v is the speed of the axis of M and $u = (b-a)\dot{\theta}$.

Prove also that, for small oscillations of the system, the length of the simple equivalent pendulum is

$$6M(b-a)/(4M+3m) \ .$$

(L.U.)

25. A uniform solid circular cylinder of mass M and height $2a$ rests with its axis vertical on a smooth horizontal table. The end A of a uniform rod AB of length $2a$ and mass m is smoothly connected to a point on the perimeter of the base of the cylinder; the other end B is connected by a light elastic string of natural length a and modulus $mg/2$ to a point on the perimeter of the upper plane face of the cylinder. The rod, string and axis

of the cylinder all lie in the same vertical plane. If θ is the angle between the rod and the vertical and the system is released from rest, show that, provided the string is always taut and the cylinder remains vertical,

$$\{4(M+m)-3m\cos^2\theta\}a\dot{\theta}^2-6(M+m)g(\cos\theta+2\sin\theta/2)=\text{const.}$$

Deduce that $\theta=\pi/3$ is a position of equilibrium and that the period of small oscillations about this position is

$$2\pi\{a(16M+13m)/9g(M+m)\}^{\frac{1}{2}}\ .$$

(L.U.)

26. A uniform rod PQ of length $2a$ and mass m is freely hinged at P to a point on the circumference of a uniform circular disc of radius a, centre O and mass m. The system rests on a smooth horizontal table with O, P, Q collinear in this order. An impulse I is applied at Q in a direction perpendicular to PQ. Show that the initial angular velocity of the disc is of magnitude $4I/7ma$. Show further that if the rod is rigidly fixed to the disc at P instead of being hinged, the angular velocity is $12I/17ma$. (L.U.)

27. Two rods AB, BC, each of length a, are freely jointed at B and placed on a horizontal table. The end A is freely pivoted to a fixed point on the table and the angle ABC is $\pi/3$. The system is then set in motion in such a way that the end C begins to move with velocity V and acceleration V^2/a, both in the direction $\overline{CB}$. Find the initial angular velocities of the rods and show that their initial angular accelerations are $8V^2/3\sqrt{3}a^2$ and $10v^2/3\sqrt{3}a^2$.

If the rods are uniform and each of mass m, find the initial kinetic energy generated. (L.U.)

28. Two uniform rods AB, AC, each of mass m and length $2c$, are freely hinged at A and held in a vertical plane with A above BC and each rod inclined at α to the downward vertical. The system is then released from rest and, after falling through a height h, B and C strike a smooth inelastic horizontal table. Show from energy considerations, or otherwise, that neither rod rotates before the impact. Show also that immediately after the impact each rod has an angular velocity of $3(2gh\sin^2\alpha)^{\frac{1}{2}}/4a$ and find the impulsive reaction at the joint A. (L.U.)

29. A uniform rod AB of length $2a$ and mass m is freely hinged to a fixed point at A and to an equal rod BC at B. The system is moving in a vertical plane about A and, when ABC is vertically downwards, the angular velocity of BC is 2ω and that of AB is ω in the same sense. At this instant a horizontal impulse J, opposing the motion, is applied to BC at a point $2ax$ below B, where $x<1$. If the ratio of the angular velocities is unchanged by the blow, show that $x=1/2$.

Show that the initial kinetic energy of the system is $28ma^2\omega^2/3$ and find J if this is unchanged by the impulse. (L.U.)

30. A uniform rod AB, length $2l$ and mass m, rests on a smooth horizontal table, and the end B is constrained to move along a small smooth straight groove on the table. Initially the rod is at rest parallel to the groove. A horizontal impulse J is now applied at the end A perpendicular to its length.

Show that the angular velocity of the rod immediately after the impulse is $3J/2lm$. Show further that when AB is at right angles to the direction of the groove, the angular velocity of the rod is $3J/lm$, and find the reaction between the rod and the groove at this instant. (L.U.)

31. A uniform solid circular cylinder of radius a rolls without slipping down a plane inclined at angle α to the horizontal, the axis of the cylinder being always horizontal. After rolling from rest through a distance l it strikes a perfectly rough inelastic bar which is fixed in a horizontal position parallel to the plane and at a perpendicular distance h from it. Show that the cylinder will surmount the obstacle if

$$[1-ah/(a^2+k^2)]^2 l \sin\alpha > a[1-\sin(\alpha+\beta)] ,$$

where k is the radius of gyration of the cylinder about its axis and $\sin\beta = 1-h/a$, it being supposed that $\alpha+\beta<\frac{1}{2}\pi$. (L.U.)

32. A uniform circular hoop in a vertical plane is projected with velocity v down a rough plane inclined at an angle α to the horizontal. The hoop has an initial angular velocity ω tending to make it roll up the plane. Show that the hoop will turn back if

$$a\omega(\mu-\tan\alpha) > v\mu ,$$

where $\mu(>\tan\alpha)$ is the coefficient of friction and a is the radius of the hoop.

Show that the least value of ω for which the hoop will return to the point of projection satisfies the equation

$$\omega^2 - 2\omega\frac{v\mu}{a(\mu-\tan\alpha)} + \frac{v^2}{a^2} = 0 .$$

(L.U.)

Chapter 9

Some Three-dimensional Rigid Body Dynamics

9.1 INTRODUCTION

The last chapter illustrates the applications of the principles of rigid body dynamics to two-dimensional problems. In this chapter, we extend the principles so far developed and proceed to apply them to three-dimensional problems in particle and rigid body dynamics. Their applications to the latter resemble the corresponding applications in two-dimensional problems, but it is often more convenient to formulate the three-dimensional problems in vectorial form before solving them.

9.2 EULER'S DYNAMICAL EQUATIONS FOR THE MOTION OF A RIGID BODY ABOUT A FIXED POINT

Suppose a rigid body turns about a point O fixed in both body and space. Let OX, OY, OZ be the three principal axes through O. These axes are essentially mutually orthogonal and fixed in the body. Let **i, j, k** stipulate the unit vectors in $\overline{OX}$, $\overline{OY}$, $\overline{OZ}$. At any time t, the body has an angular velocity $\boldsymbol{\omega}=[\omega_1, \omega_2, \omega_3]$, say, referred to these axes, and so the moment of momentum about O is instantaneously

$$\mathbf{H}=A\omega_1\mathbf{i}+B\omega_2\mathbf{j}+C\omega_3\mathbf{k} \qquad (1)$$

where A, B, C are the principal moments of inertia about these axes.

Using the results of Chapter 2, the rate of change of moment of momentum about O is

$$d\mathbf{H}/dt=\partial\mathbf{H}/\partial t+\boldsymbol{\omega}_\wedge\mathbf{H} \ ,$$

since $\boldsymbol{\omega}$ is the spin of the frame, $\partial\mathbf{H}/\partial t$ denoting the rate of change of angular momentum *relative to the frame.*

$$\therefore \ d\mathbf{H}/dt=A\dot{\omega}_1\mathbf{i}+B\dot{\omega}_2\mathbf{j}+C\dot{\omega}_3\mathbf{k}+\boldsymbol{\omega}_\wedge\mathbf{H} \ .$$

Since O is fixed,

$$d\mathbf{H}/dt=\mathbf{L} \ , \qquad (2)$$

where $\mathbf{L}=[L_1, L_2, L_3]$ is the vector moment of the external forces about O. Equating components gives

$$\left.\begin{aligned} L_1 &= A\dot{\omega}_1 - (B-C)\omega_2\omega_3 \ , \\ L_2 &= B\dot{\omega}_2 - (C-A)\omega_3\omega_1 \ , \\ L_3 &= C\dot{\omega}_3 - (A-B)\omega_1\omega_2 \ . \end{aligned}\right\} \tag{3}$$

These equations (3) are known as *Euler's dynamical equations.*

The special case of motion about O under no forces is of importance. The dynamical equations become

$$\left.\begin{aligned} A\dot{\omega}_1 - (B-C)\omega_2\omega_3 &= 0 \ , \\ B\dot{\omega}_2 - (C-A)\omega_3\omega_1 &= 0 \ , \\ C\dot{\omega}_3 - (A-B)\omega_1\omega_2 &= 0 \ . \end{aligned}\right\} \tag{4}$$

Multiplying these three equations by ω_1, ω_2, ω_3 and adding gives

$$A\omega_1\dot{\omega}_1 + B\omega_2\dot{\omega}_2 + C\omega_3\dot{\omega}_3 = 0 \ ,$$

whence, on integration,

$$A\omega_1{}^2 + B\omega_2{}^2 + C\omega_3{}^2 = \text{const.} \tag{5}$$

Since the L.H.S. of (5) is $2T$, T being the kinetic energy, (5) expresses the constancy of the K.E. during the motion. This is obvious since no work is being done on the body, there being no external forces.

Another integral of equations (4) can be obtained by multiplying them by $A\omega_1$, $B\omega_2$, $C\omega_3$ and adding to give

$$A^2\omega_1\dot{\omega}_1 + B^2\omega_2\dot{\omega}_2 + C^2\omega_3\dot{\omega}_3 = 0$$

which integrates to give

$$A^2\omega_1{}^2 + B^2\omega_2{}^2 + C^2\omega_3{}^2 = \text{const.} \tag{6}$$

Equation (6) expresses the constancy of the magnitude of the angular momentum H during the motion. Physically this is attributable to the zero moment of forces about O.

Example

A rigid body is free to rotate about its centroid G, the principal moments of inertia at which are 7, 25, 32 units respectively. The body is given an angular velocity Ω about a line through G whose direction ratios are 4:0:3. Show that after time t the components of angular velocity about the principal axes of inertia at G are

$$\tfrac{4}{5}\Omega \cos \phi, \qquad \tfrac{4}{5}\Omega \sin \phi, \qquad \tfrac{3}{5}\Omega \cos \phi$$

where $\tan(\phi/2) = \tanh(3\,\Omega t/10)$.

Deduce that ultimately the body rotates about the principal axis of intermediate moment of inertia. (L.U.)

Since the body rotates about its centroid, there is zero moment about this point. Taking $A=7$, $B=25$, $C=32$ and $[\omega_1, \omega_2, \omega_3]$ to be the angular velocities about the P.As. at time t, Euler's dynamical equations (4) become on simplification

$$\dot{\omega}_1+\omega_2\omega_3=0\ , \tag{7}$$
$$\dot{\omega}_2-\omega_3\omega_1=0\ , \tag{8}$$
$$16\dot{\omega}_3+9\omega_1\omega_2=0\ , \tag{9}$$

the initial values of the angular velocities being $\omega_{10}=\frac{4}{5}\Omega$, $\omega_{20}=0$, $\omega_{30}=\frac{3}{5}\Omega$.

Multiplying (7), (8) by ω_1, ω_2 respectively, adding, integrating and using initial values gives

$$\omega_1{}^2+\omega_2{}^2=16\Omega^2/25\ . \tag{10}$$

Multiplying (8), (9) by $9\omega_2$, ω_3 respectively, adding, integrating and using initial values gives

$$9\omega_2{}^2+16\omega_3{}^2=144\Omega^2/25\ . \tag{11}$$

From (8), (10), (11), we have $dt=\mathrm{d}\omega_2/(\omega_3\omega_1)=\frac{4}{3}d\omega_2/[(\frac{4}{5}\Omega)^2-\omega_2{}^2]$, which on integration gives

$$t=(5/3\Omega)\mathrm{th}^{-1}(5\omega_2/4\Omega)\ ,$$

or

$$\omega_2=(4\Omega/5)\mathrm{th}(3\Omega t/5)=\tfrac{4}{5}\Omega\sin\phi\ . \tag{12}$$

Substitution of (12) into (10), (11), then gives the stated values of ω_1, ω_3.

Further, as $t\rightarrow\infty$, $\mathrm{th}(3\Omega t/10)\rightarrow 1$, and so $\phi/2\rightarrow\pi/4$ or $\phi\rightarrow\pi/2$. Thus, $\omega_1\rightarrow 0$, $\omega_2\rightarrow\frac{4}{5}\Omega$, $\omega_3\rightarrow 0$, showing that ultimately rotation about the axis of intermediate moment of inertia ensues.

9.3 FURTHER PROPERTIES OF RIGID BODY MOTION UNDER NO FORCES

In the last section, we saw that the scalars, T, H are constant for rotation about a fixed point under no forces. Further, since $d\mathbf{H}/dt=\mathbf{0}$, $\mathbf{H}$ is a constant vector and so $\hat{\mathbf{H}}$ is also constant. Thus the direction of $\mathbf{H}$ is invariant.

We further observe

$$\mathbf{H}.\boldsymbol{\omega}=2T\ ,$$

and so

$$\boldsymbol{\omega}.\hat{\mathbf{H}}=2T/H=\text{const.},$$

showing that the projection of $\boldsymbol{\omega}$ on the constant direction $\hat{\mathbf{H}}$ is constant. Thus, if O is the point of rotation and if $\overline{OP}\equiv\boldsymbol{\omega}$, then the locus of P is an *invariable plane* perpendicular to $\mathbf{H}$.

Further, we observe that the locus in space of the instantaneous axis of rotation is a cone (not necessarily right circular) having O as vertex. This locus is called a *herpolhode*.

The locus of the instantaneous axis relative to the body is also a cone of vertex O and is called a *polhode*. Its equation may be found as follows.

Referred to the moving frame of principle axes **i**, **j**, **k**, the equation of the instantaneous axis of rotation is

$$x/\omega_1 = y/\omega_2 = z/\omega_3 \ .$$

Further,

$$A^2\omega_1{}^2 + B^2\omega_2{}^2 + C^2\omega_3{}^2 = H^2 = \text{const.},$$
$$A\omega_1{}^2 + B\omega_2{}^2 + C\omega_3{}^2 \quad = 2T = \text{const.},$$

and from these three equations we derive

$$2T(A^2x^2 + B^2y^2 + C^2z^2) = H^2(Ax^2 + By^2 + Cz^2) \ .$$

This is a cone of the second degree since it is homogeneous of the second degree in x, y, z.

The polhode and herpolhode touch along the instantaneous axis of rotation. Thus *rigid body motion about a fixed point is equivalent to the rolling of one cone on another.*

We now establish the theorem that *the motion of a rigid body about a fixed point may be represented by the rolling of an ellipsoid fixed in the body upon a plane fixed in space.*

Proof

Take current co-ordinates (x, y, z) with respect to the P.As. **i**, **j**, **k** fixed n the body. Consider the ellipsoid

$$Ax^2 + By^2 + Cz^2 = 2T \ .$$

The instantaneous axis

$$x/\omega_1 = y/\omega_2 = z/\omega_3 = r/\omega$$

meets this ellipsoid at $r = \omega$, i.e. at the point $(\omega_1, \omega_2, \omega_3)$.

The equation of the tangent plane to the ellipsoid at $(\omega_1, \omega_2, \omega_3)$ is

$$A\omega_1 x + B\omega_2 y + C\omega_3 z = 2T \ .$$

which is at a distance p from the origin (point of rotation) where

$$p = 2T/\sqrt{(A^2\omega_1{}^2 + B^2\omega_2{}^2 + C^2\omega_3{}^2)} = 2T/H = \text{const.}$$

Thus the tangent plane at $(\omega_1, \omega_2, \omega_3)$ is the invariable plane and so the ellipsoid touches the invariable plane.

The point of contact of the ellipsoid and the invariable plane lies on the instantaneous axis of rotation and so there can be no slipping at this point. Thus the ellipsoid rolls on the invariable plane.

9.4 SOME PROBLEMS OF GENERAL THREE-DIMENSIONAL RIGID BODY MOTION

The general motion of a rigid body may be described by stipulating the motion of its centroid together with the instantaneous angular velocity and acceleration (Chapter 2). The basic principles of general rigid body dynamics have been built up and have already been applied in Chapter 8 to some two-dimensional problems. We now apply these principles to three dimensional problems. It is convenient to formulate these problems vectorially.

Example 1—Non-rotating axes.

Set up the general equations of motion of a uniform sphere on a fixed rough horizontal plane under a given system of external forces through its centre. Show that if the sphere rolls without slipping its translation is equivalent to that of a particle of equal mass on a smooth horizontal plane subject to external forces in the same directions but having magnitudes equal to 5/7 of those acting on the sphere.

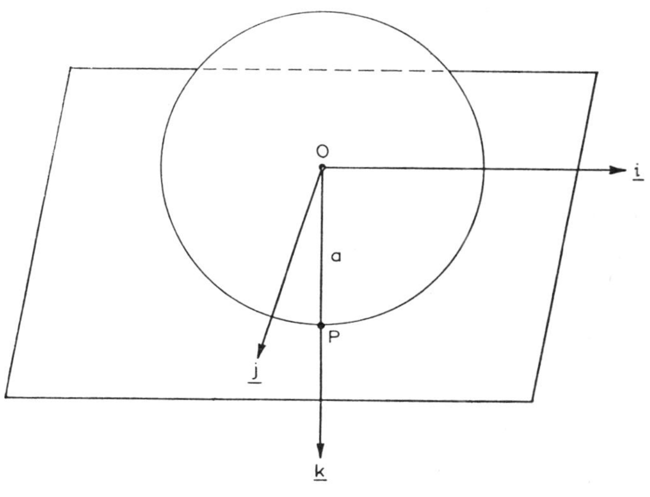

Fig. 9.1

Fig. 9.1 shows the sphere, centre O, radius a moving generally on the fixed horizontal plane. The axis $\mathbf{k}$ is taken vertically downwards and $\mathbf{i}$, $\mathbf{j}$, are chosen in the horizontal plane through O so that $\mathbf{i}\wedge\mathbf{j}=\mathbf{k}$, etc.,

Suppose the externally applied force through O is $X\mathbf{i}+Y\mathbf{j}+Z\mathbf{k}$, and that the friction at the contact point P is $F_1\mathbf{i}+F_2\mathbf{j}$. Let $\boldsymbol{\omega}=[\omega_1, \omega_2, \omega_3]$ be the angular velocity at time t. The moment of momentum about O is

$$\mathbf{H}=\tfrac{2}{5}Ma^2(\omega_1\mathbf{i}+\omega_2\mathbf{j}+\omega_3\mathbf{k}) \ ,$$

since the unit vectors specify principal axes for the sphere. Since the axes are *not* rotating, we have

$$d\mathbf{H}/dt=\tfrac{2}{5}Ma^2(\dot{\omega}_1\mathbf{i}+\dot{\omega}_2\mathbf{j}+\dot{\omega}_3\mathbf{k}) \ .$$

The moment of the external forces about O is

$$a\mathbf{k}_\wedge(F_1\mathbf{i}+F_2\mathbf{j})=(-F_2\mathbf{i}+F_1\mathbf{j})a \ .$$

Since O is the centroid, it may be treated as a fixed point, and so

$$a\mathbf{k}_\wedge(F_1\mathbf{i}+F_2\mathbf{j})=\tfrac{2}{5}Ma^2(\dot{\omega}_1\mathbf{i}+\dot{\omega}_2\mathbf{j}+\dot{\omega}_3\mathbf{k}) \ ,$$

or $$\dot{\omega}_1=-5F_2/2Ma,\ \dot{\omega}_2=5F_1/2Ma,\ \dot{\omega}_3=0 \ . \qquad (1)$$

$$\therefore\ \omega_3=n=\text{const.}$$

Let $-R\mathbf{k}$ be the normal reaction at P. Then the total force acting on the sphere is

$$(X+F_1)\mathbf{i}+(Y+F_2)\mathbf{j}+(Mg-R)\mathbf{k}$$

so that if the displacement of O in time t is $x\mathbf{i}+y\mathbf{j}$, the acceleration is $\ddot{x}\mathbf{i}+\ddot{y}\mathbf{j}$. Hence

$$(X+F_1)\mathbf{i}+(Y+F_2)\mathbf{j}+(Mg-R)\mathbf{k}=M(\ddot{x}\mathbf{i}+\ddot{y}\mathbf{j}) \ ,$$

whence

$$X+F_1=M\ddot{x},\quad Y+F_2=M\ddot{y},\quad R=Mg \ . \qquad (2)$$

The condition for rolling is that P be instantaneously at rest. The velocity of P is $(\dot{x}\mathbf{i}+\dot{y}\mathbf{j})+\boldsymbol{\omega}_\wedge a\mathbf{k}$, and so we obtain

$$\dot{x}+a\omega_2=0 \ , \quad \dot{y}-a\omega_1=0 \ .$$

Differentiating these gives

$$\ddot{x}+a\dot{\omega}_2=0 \ , \quad \ddot{y}-a\dot{\omega}_1=0 \ . \qquad (3)$$

From (1), (3),

$$F_1=-\tfrac{2}{5}M\ddot{x} \ , \quad F_2=-\tfrac{2}{5}M\ddot{y} \ ,$$

so that (2) gives

$$X=M\ddot{x}-F_1=\tfrac{7}{5}M\ddot{x} \ ,$$
$$Y=M\ddot{y}-F_2=\tfrac{7}{5}M\ddot{y} \ .$$

Example 2—Non-rotating axes.

A uniform solid sphere rolls without slipping on a rough horizontal plane which is rotating with uniform angular velocity about a vertical axis. If there are no forces acting on the sphere save its weight and the friction at the contact, prove that the locus of the centre of the sphere is a circle.

The notation is the same as in the previous example, but the plane now has uniform angular velocity $\Omega\mathbf{k}$ (Fig. 9.2).

The axes $\mathbf{i}$, $\mathbf{j}$, $\mathbf{k}$ through the moving point O are again fixed in direction. Let $\boldsymbol{\omega}=[\omega_1,\ \omega_2,\ \omega_3]$ be the angular velocity of the sphere.

Taking moments about O we obtain as before

$$F_1=\tfrac{2}{5}Ma\dot{\omega}_2,\quad F_2=-\tfrac{2}{5}Ma\dot{\omega}_1,\quad \dot{\omega}_3=0 \ . \qquad (1)$$

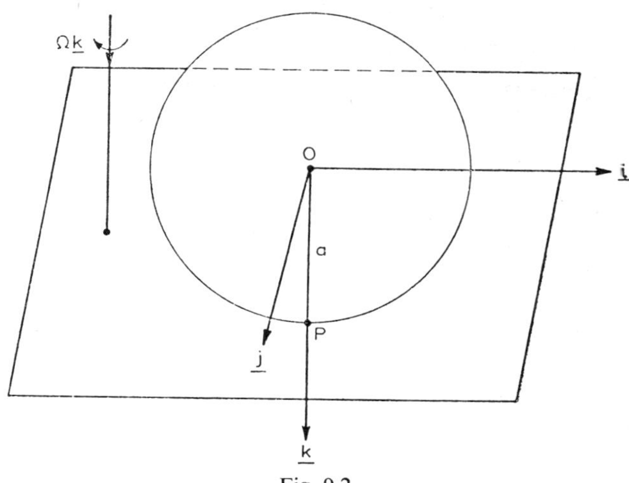

Fig. 9.2

The linear motion of the sphere is given by

$$F_1\mathbf{i}+F_2\mathbf{j}=M(\ddot{x}\mathbf{i}+\ddot{y}\mathbf{j}) ,$$

and so $$F_1=M\ddot{x} , \quad F_2=M\ddot{y} . \tag{2}$$

Since P is the point (x, y, a), its velocity on the plane is

$$\Omega\mathbf{k}_\wedge(x\mathbf{i}+y\mathbf{j}+a\mathbf{k})=\Omega(-y\mathbf{i}+x\mathbf{j}) ,$$

taking the origin of Cartesian co-ordinates to be at the point on the axis of rotation of the plane at a height a above the plane. Also, the velocity of P on the sphere is

$$\dot{x}\mathbf{i}+\dot{y}\mathbf{j}+\boldsymbol{\omega}_\wedge a\mathbf{k}=(\dot{x}+a\omega_2)\mathbf{i}+(\dot{y}-a\omega_1)\mathbf{j} .$$

Thus the conditions for no slipping at P are

$$\dot{x}+a\omega_2=-\Omega y , \quad \dot{y}-a\omega_1=\Omega x . \tag{3}$$

From (1), (2),

$$\dot{\omega}_1=-5\ddot{y}/2a , \quad \dot{\omega}_2=5\ddot{x}/2a . \tag{4}$$

Differentiating equations (3) and using equations (4),

$$\ddot{x}+\tfrac{2}{7}\Omega\dot{y}=0 , \quad \ddot{y}-\tfrac{2}{7}\Omega\dot{x}=0 . \tag{5}$$

To the first of equations (5), add i multiplied by the second, getting

$$D(D-i\tfrac{2}{7}\Omega)(x+iy)=0 \quad (D\equiv d/dt) .$$

The general solution of this is

$$x+iy=A+B\exp(i\tfrac{2}{7}\Omega t) ,$$

where A and B are complex constants. Taking $A=\alpha+i\beta$, $B=\lambda+i\mu$, where α, β, λ, μ are all real constants, we have

$$x=\alpha+\lambda\cos(\tfrac{2}{7}\Omega t)-\mu\sin(\tfrac{2}{7}\Omega t)\ ,$$
$$y=\beta+\mu\cos(\tfrac{2}{7}\Omega t)+\lambda\sin(\tfrac{2}{7}\Omega t)\ .$$

From these equations we obtain the circle

$$(x-\alpha)^2+(y-\beta)^2=\lambda^2+\mu^2\ ,\quad z=0$$

for the locus of O.

Example 3—Rotating axes.

A uniform sphere rolls without slipping on the rough interior of a fixed vertical cylinder of greater radius. Discuss the motion of the sphere.

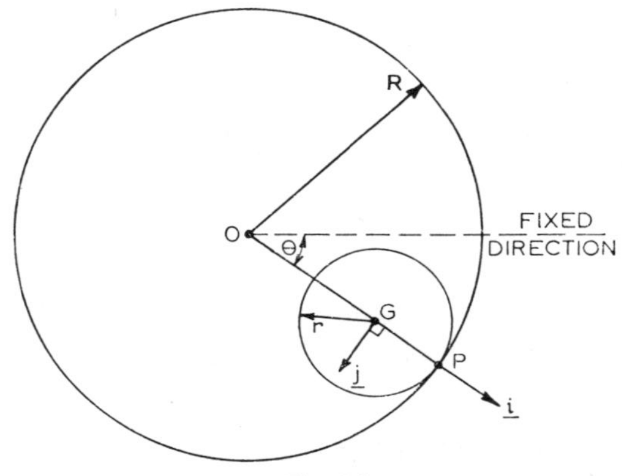

Fig. 9.3

Let r be the radius of the sphere, R that of the cylinder, $\mathbf{i}$ the unit vector in $\overline{OG}$ where O and G are the centres of the horizontal sections of the cylinder and the sphere through the contact point P (Fig. 9.3), $\mathbf{k}$ the unit vector in the downward vertical and $\mathbf{j}=\mathbf{k}_\wedge\mathbf{i}$, etc. Suppose $\boldsymbol{\omega}=[\omega_1,\ \omega_2,\ \omega_3]$ is the angular velocity of the sphere. Since the axes are principal axes, the moment of momentum about O is

$$\mathbf{H}=\tfrac{2}{5}Mr^2(\omega_1\mathbf{i}+\omega_2\mathbf{j}+\omega_3\mathbf{k})\ ,$$

M being the mass of the sphere. If OG makes an angle θ with a fixed horizontal direction, then $\dot{\theta}\mathbf{k}$ is the angular velocity of the frame $\mathbf{i}$, $\mathbf{j}$, $\mathbf{k}$ and so

$$\begin{aligned} d\mathbf{H}/dt &= \partial\mathbf{H}/\partial t+\dot{\theta}\mathbf{k}_\wedge\mathbf{H} \\ &= \tfrac{2}{5}Mr^2[\dot{\omega}_1-\omega_2\dot{\theta})\mathbf{i}+(\dot{\omega}_2+\omega_1\dot{\theta})\mathbf{j}+\dot{\omega}_3\mathbf{k}]\ . \end{aligned}$$

If $d=R-r$, the acceleration of G is $[-d\dot{\phi}^2,\ d\ddot{\phi},\ \ddot{z}]$, z being the depth of G below some fixed horizontal plane.

The moment of the rate of change of momentum about P is

$$d\mathbf{H}/dt + (-r\mathbf{i})_\wedge M(-d\dot{\theta}^2\mathbf{i} + d\ddot{\theta}\mathbf{j} + \ddot{z}\mathbf{k})$$
$$= \tfrac{2}{5}Mr^2[(\dot{\omega}_1 - \omega_2\dot{\theta})\mathbf{i} + (\dot{\omega}_2 + \omega_1\dot{\theta})\mathbf{j} + \dot{\omega}_3\mathbf{k}] + Mr\ddot{z}\mathbf{j} - Mrd\ddot{\theta}\mathbf{k} \ .$$

The moment of the forces about P is

$$-r_\wedge Mg\mathbf{k} = Mgr\mathbf{j} \ .$$

Hence equating the two expressions, we obtain

$$\dot{\omega}_1 - \omega_2\dot{\theta} = 0 \ , \tag{1}$$
$$\tfrac{2}{5}r(\dot{\omega}_2 + \omega_1\dot{\theta}) + \ddot{z} = g \ , \tag{2}$$
$$\tfrac{2}{5}r\dot{\omega}_3 - d\ddot{\theta} = 0 \ . \tag{3}$$

The velocity of G is $d\dot{\theta}\mathbf{j} + \dot{z}\mathbf{k}$ and the velocity of P relative to G is

$$(\omega_1\mathbf{i} + \omega_2\mathbf{j} + \omega_3\mathbf{k})_\wedge r\mathbf{i} = r\omega_3\mathbf{j} - r\omega_2\mathbf{k} \ ,$$

and so the total velocity of P is $(d\dot{\theta} + r\omega_3)\mathbf{j} + (\dot{z} - r\omega_2)\mathbf{k}$. Since P is instantaneously at rest during pure rolling, we have

$$d\dot{\theta} + r\omega_3 = 0 \ , \tag{4}$$
$$\dot{z} - r\omega_2 = 0 \ . \tag{5}$$

As the equations of linear motion would involve the unknown friction forces and reaction at P we shall not construct them.

We now obtain from (1)–(5) an equation expressing the co-ordinate z and its derivatives.

From (3), (4), $\dot{\omega}_3 = 0$, $\ddot{\theta} = 0$, and so

$$\omega_3 = n \ , \quad \dot{\theta} = -rn/d \ , \tag{6}$$

where n is a constant. From (1), (5), (6),

$$\dot{\omega}_1 = \dot{\theta}\omega_2 = -n\dot{z}/d \ ,$$
$$\therefore \ \omega_1 = c - nz/d \ , \tag{7}$$

where c is a constant. Thus (2) becomes

$$\tfrac{2}{5}\{\ddot{z} + (c - nz/d)(-r^2n/d)\} + \ddot{z} = g$$

or

$$\tfrac{7}{5}\ddot{z} + \tfrac{2}{5}(rn/d)^2 z = g + \tfrac{2}{5}r^2nc/d = \text{const.} \tag{8}$$

Equation (8) shows that the vertical motion of the centre of the sphere is a S.H.M. of period $2\pi(7/2)^{\frac{1}{2}}(R - r)/rn$.

Example 4—Rotating axes.

A hoop, radius a, rolls without slipping on a rough horizontal surface. Obtain the equations of its motion in the form

$$3\ddot{\theta} - \dot{\phi}^2 \sin\theta\cos\theta + 4\omega\dot{\phi}\sin\theta = -2(g/a)\cos\theta \ ,$$
$$(d/dt)(\dot{\phi}\sin^2\theta) - 2\omega\dot{\theta}\sin\theta = 0 \ ,$$
$$2\dot{\omega} - \dot{\theta}\dot{\phi}\sin\theta = 0 \ ,$$

where θ is the angle the plane of the hoop makes with the horizontal, $\dot{\phi}$ is the angular velocity of the vertical plane containing the diameter through the point of contact, and ω is the angular velocity of the hoop about its axis of symmetry.

If the hoop rolls along a straight line with its plane vertical, show that the motion is stable if

$$\omega > \tfrac{1}{2}(g/a)^{\frac{1}{2}} .$$

(L.U.)

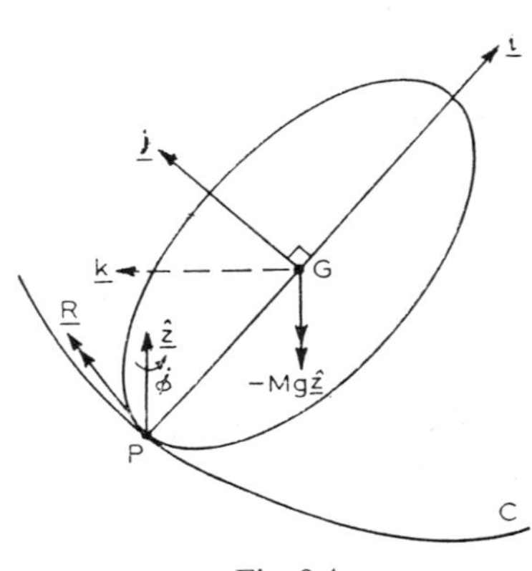

Fig. 9.4

Fig. 9.4 shows the hoop of mass M rolling along a curve C in the horizontal plane. $\mathbf{i}$ is the unit vector through the centroid G in the direction $\overline{PG}$, $\mathbf{j}$ the unit vector normal to the hoop so that $\mathbf{i}_\wedge\mathbf{j}=\mathbf{k}$, etc. Let $\hat{\mathbf{z}}$ be the unit vector in the upward vertical direction, $\mathbf{R}$ the total reaction at the contact point P. Then $\hat{\mathbf{z}}=\sin\theta\mathbf{i}+\cos\theta\mathbf{k}$ and the resultant spin of the frame is $\mathbf{\Omega}$ where

$$\mathbf{\Omega}=\dot{\phi}\hat{\mathbf{z}}-\dot{\theta}\mathbf{j}=[\dot{\phi}\sin\theta, \quad -\dot{\theta}\ , \quad \dot{\phi}\cos\theta] .$$

The angular velocity of the hoop is $\boldsymbol{\omega}$ where

$$\boldsymbol{\omega}=[\dot{\phi}\sin\theta\ , \quad -\dot{\theta}\ , \quad \omega] .$$

Since P is instantaneously at rest, the velocity of the centroid is

$$\mathbf{v}_G=\boldsymbol{\omega}_\wedge a\mathbf{i}=a(\omega\mathbf{j}+\dot{\theta}\mathbf{k}) . \tag{1}$$

The equation of translation of the hoop is

$$\mathbf{R}-Mg\hat{\mathbf{z}}=Md\mathbf{v}_G/dt . \tag{2}$$

The moment of momentum about G is

$$\mathbf{H}_G=\tfrac{1}{2}Ma^2(\dot{\phi}\sin\theta\mathbf{i}-\dot{\theta}\mathbf{j}+2\omega\mathbf{k}) ,$$

since $\mathbf{i}$, $\mathbf{j}$, $\mathbf{k}$ are in the directions of principal axes of the hoop and the corresponding principal moments of inertia are $\frac{1}{2}Ma^2$, $\frac{1}{2}Ma^2$, Ma^2. Thus

$$d\mathbf{H}_G/dt = \partial\mathbf{H}_G/\partial t + \mathbf{\Omega}_\wedge\mathbf{H}_G$$
$$= \tfrac{1}{2}Ma^2\{\ddot{\phi}\sin\theta + \dot{\phi}\,\dot{\theta}\cos\theta)\mathbf{i} - \ddot{\theta}\mathbf{j} + 2\dot{\omega}\mathbf{k}\}$$
$$+\tfrac{1}{2}Ma^2\begin{vmatrix} \mathbf{i} & \mathbf{j} & \mathbf{k} \\ \dot{\phi}\sin\theta & -\dot{\theta} & \dot{\phi}\cos\theta \\ \dot{\phi}\sin\theta & -\dot{\theta} & 2\omega \end{vmatrix}$$
$$= \tfrac{1}{2}Ma^2\{(\ddot{\phi}\sin\theta + 2\dot{\theta}\dot{\phi}\cos\theta - 2\omega\dot{\theta})\mathbf{i} - (\ddot{\theta} + 2\omega\dot{\phi}\sin\theta\cos\theta - \dot{\phi}^2\sin\theta\cos\theta)$$
$$\mathbf{j} + 2\dot{\omega}\mathbf{k}\}\ .$$

Since G may be treated as a fixed point, we have, on taking moments about it,

$$d\mathbf{H}_G/dt = -a\mathbf{i}_\wedge\mathbf{R}\ . \tag{3}$$

From (2), (3)

$$d\mathbf{H}_G/dt = -a\mathbf{i}_\wedge(Mg\hat{\mathbf{z}} + Md\mathbf{v}_G/dt)\ . \tag{4}$$

Now $d\mathbf{v}_G/dt = \partial\mathbf{v}_G/\partial t + \mathbf{\Omega}_\wedge\mathbf{v}_G$

$$= a(\dot{\omega}\mathbf{j} + \ddot{\theta}\mathbf{k}) + a\begin{vmatrix} \mathbf{i} & \mathbf{j} & \mathbf{k} \\ \dot{\phi}\sin\theta & -\dot{\theta} & \dot{\phi}\cos\theta \\ 0 & \omega & \dot{\theta} \end{vmatrix}$$
$$= a\{-(\dot{\theta}^2 + \omega\dot{\phi}\cos^2\theta)\mathbf{i} + (\dot{\omega} - \dot{\phi}\dot{\theta}\sin\theta)\mathbf{j} + (\ddot{\theta} + \dot{\phi}\omega\sin\theta)\mathbf{k}\}\ .$$

Thus $-a\mathbf{i}_\wedge(Mg\hat{\mathbf{z}} + Md\mathbf{v}_G/dt)$

$$= Ma[\{g\cos\theta + a(\ddot{\theta} + \omega\dot{\phi}\sin\theta)\}\mathbf{j} - a(\dot{\omega} - \dot{\phi}\dot{\theta}\sin\theta)\mathbf{k}]\ .$$

The vector equation (4) yields the three scalar equations

$$\ddot{\phi}\sin\theta + 2\dot{\theta}\dot{\phi}\cos\theta - 2\omega\dot{\theta} = 0\ , \tag{5}$$
$$3\ddot{\theta} + 4\omega\dot{\phi}\sin\theta - \dot{\phi}^2\sin\theta\cos\theta = -2(g/a)\cos\theta\ , \tag{6}$$
$$2\dot{\omega} - \dot{\phi}\dot{\theta}\sin\theta = 0\ . \tag{7}$$

On multiplying (5) through by $\sin\theta$, we obtain the form

$$d/dt(\dot{\phi}\sin^2\theta) - 2\omega\dot{\theta}\sin\theta = 0\ . \tag{5'}$$

To test stability of the hoop when rolling along the straight line with its plane vertical, allow the hoop to become slightly perturbed from this motion so that

$$\theta = \pi/2 + \epsilon\ , \quad \omega = \omega_0 + \eta\ ,$$

where ϵ, η, ϕ and their derivates are small, and ω_0 is the steady value of ω for $\theta = \pi/2$. Then to the first order of smallness, (5), (6) give

$$\ddot{\phi} - 2\omega_0\dot{\epsilon} = 0\ , \tag{8}$$
$$3\ddot{\epsilon} + 4\omega_0\dot{\phi} = 2g\epsilon/a\ . \tag{9}$$

From (8), $\dot{\phi} = 2\omega_0\epsilon$, so that (9) gives

$$3\ddot{\epsilon} + (8\omega_0{}^2 - 2g/a)\epsilon = 0\ . \tag{10}$$

(10) describes a S.H.M. if $\omega_0^2 > g/4a$, which is the required stability condition.

9.5 THE ROTATING EARTH

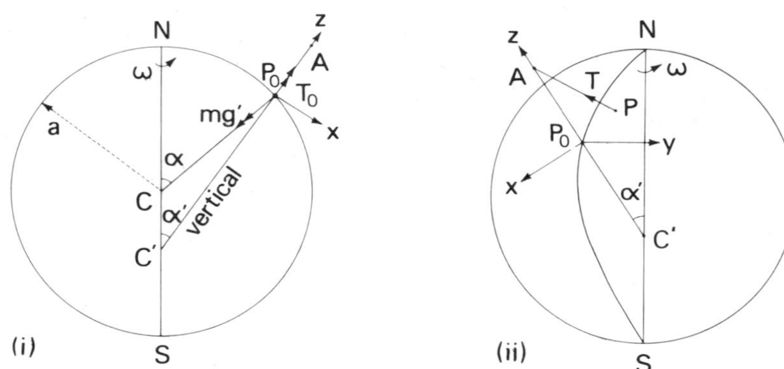

Fig. 9.5—(i) Relative rest. (ii) Relative motion.

In most problems in dynamics, the weight of a body is taken as the force with which the Earth attracts it when there is no relative motion between it and the Earth. It is measured by the tension in a supporting string and the direction of gravity is determined by that of the string which sets itself in the direction of the centre of the Earth when the Earth is taken to be a uniformly attracting sphere. However, this ignores the rotation of the Earth. As the Earth is rotating about its axis, with a uniform angular velocity, the particle will have an acceleration towards the centre of the horizontal circle which it describes and so the vertical or plumb line will not pass through the Earth's centre.

Figs. 9.5(i) and (ii) depict the Earth as a uniform sphere of radius a and centre C and rotating positively about the axis SN with uniform angular velocity ω, S being the South Pole and N the North Pole. A pendulum of length l is suspended from A above the Earth its bob P of mass m being close to the Earth's surface. In Fig. 9.5(i) there is relative rest between the bob and the Earth: in Fig 9.5(ii) the bob is set in motion by giving it a small displacement from the position of equilibrium. Let us first consider the equilibrium configuration of Fig. 9.5(i). The tension $\mathbf{T}_0$ in the string P_0A is equal and opposite to the apparent weight of the bob. $\overline{AP_0}$ is the direction of the plumb line at A and it intersects SN at C' and not at C. The force $\mathbf{T}_0$ that would be measured were a small spring balance inserted in the string would be equal and opposite to the apparent weight $m\mathbf{g}$ of the particle. The actual gravitational attraction at P_0 is $m\mathbf{g}'$, where $\mathbf{g}'$ is directed along the radius $\overline{P_0C}$. Thus the mass m at P_0 is subjected to a force of magnitude mg along $\overline{P_0A}$ and one of magnitude mg' along $\overline{P_0C}$. If $\angle P_0CN = \alpha$, $\angle P_0CN = \alpha'$, then the equation of motion of m at P_0 in the direction taken perpendicularly to SN is

$$mg' \sin \alpha - mg \sin \alpha' = ma \sin \alpha\ \omega^2 \tag{1}$$

and that in the direction of SN is

$$mg' \cos \alpha - mg \cos \alpha' = 0 \tag{2}$$

and from (1), (2) we readily obtain the results:

$$\sin(\alpha - \alpha') = (a\omega^2/2g)\sin 2\alpha \ , \tag{3}$$

$$g' \sin(\alpha - \alpha') = a\omega^2 \sin \alpha \cos \alpha' \ , \tag{4}$$

$$g' \cos(\alpha - \alpha') = g + a\omega^2 \sin \alpha \sin \alpha' \ . \tag{5}$$

Equation (3) gives $(\alpha - \alpha')$, the deviation of the plumb line from the radial direction. Equations (4), (5) express the gravitational force components per unit mass at P_0 perpendicularly to and along P_0C'.

Now suppose the bob is given a small displacement from its equilibrium position P_0 so that at time t its position is specified by P, Fig. 9.5(ii). Through P_0 introduce the right-handed orthogonal co-ordinate frame P_0x, P_0y, P_0z where $\overline{P_0z}$ is taken along $\overline{C'P_0}$, $\overline{P_0y}$ is perpendicular to the meridional plane SNP_0 and in the easterly direction through P_0 and $\overline{P_0x}$ is perpendicular to $\overline{P_0y}$ and $\overline{P_0z}$ and in the sense of a positive turn from the former to the latter. Then this frame is fixed relatively to the Earth and moves with it with vector angular velocity $\boldsymbol{\omega} = \omega(-\sin \alpha\mathbf{i} + \cos \alpha\mathbf{k})$. The velocity of P_0 is $\mathbf{v}_0 = a\omega \sin \alpha\mathbf{j}$ and its acceleration is $\mathbf{f}_0 = a\omega^2 \sin \alpha(-\cos \alpha'\mathbf{i} - \sin \alpha'\mathbf{k})$. Let $\overline{P_0P} \equiv \mathbf{r} = xi + y\mathbf{j} + z\mathbf{k}$, so that (x, y, z) are the co-ordinates of P in the moving frame. Then the velocity of P relative to P_0 is

$$\begin{aligned}\frac{d\mathbf{r}}{dt} &= \frac{\partial \mathbf{r}}{\partial t} + \boldsymbol{\omega}_\wedge \mathbf{r} \\ &= \dot{x}\mathbf{i} + \dot{y}\mathbf{j} + \dot{z}\mathbf{k} + (-\omega \sin \alpha\mathbf{i} + \omega \cos \alpha\mathbf{k})_\wedge(x\mathbf{i} + y\mathbf{j} + z\mathbf{k}) \\ &= (\dot{x} - \omega y \cos \alpha)\mathbf{i} + (\dot{y} + \omega x \cos \alpha + \omega z \sin \alpha)\mathbf{j} + (\dot{z} - \omega y \sin \alpha)\mathbf{k}\end{aligned}$$

The acceleration of P relative to P_0 is

$$\begin{aligned}\frac{d^2\mathbf{r}}{dt^2} &= \frac{\partial}{\partial t}\left(\frac{d\mathbf{r}}{dt}\right) + \boldsymbol{\omega}_\wedge \frac{d\mathbf{r}}{dt} \\ &= (\ddot{x} - \omega\dot{y}\cos \alpha)\mathbf{i} + (\ddot{y} + \omega\dot{x}\cos \alpha + \omega\dot{z}\sin \alpha)\mathbf{j} + (\ddot{z} - \omega\dot{\mathbf{j}}\sin \alpha)\mathbf{k} \\ &\quad + \begin{vmatrix} \mathbf{i} & \mathbf{j} & \mathbf{k} \\ -\omega \sin \alpha & 0 & \omega \cos \alpha \\ (\dot{x} - \omega y \cos \alpha) & (\dot{y} + \omega x \cos \alpha + \omega z \sin \alpha) & (\dot{z} - \omega y \sin \alpha) \end{vmatrix} \\ &= (\ddot{x} - 2\omega\dot{y} \cos \alpha - \omega^2 x \cos^2\alpha - \omega^2 z \sin \alpha \cos \alpha)\mathbf{i} \\ &\quad + (\ddot{y} + 2\omega\dot{z} \sin \alpha + 2\omega\dot{x} \cos \alpha - \omega^2 y)\mathbf{j} \\ &\quad + (\ddot{z} - 2\omega\dot{y} \sin \alpha - \omega^2 x \sin \alpha \cos \alpha - \omega^2 z \sin {}^2\alpha)\mathbf{k} \ .\end{aligned} \tag{6}$$

Thus the total acceleration of P (referred to a stationary fixed frame) is

$$\mathbf{f}_P = \mathbf{f}_0 + d^2\mathbf{r}/dt^2 \ . \tag{7}$$

The equation of motion of the bob in Fig. 9.5(ii) is simply

$$m\mathbf{f}_P = m\mathbf{g}' + \mathbf{T} \tag{8}$$

where $\mathbf{T}$ is the tension in the string and $\mathbf{g}'$, the gravitational attraction at P may, since the displacement is small and $|\mathbf{r}| << a$, be taken to be the same as that at P_0 so that

$$\begin{aligned}\mathbf{g}' &= -g' \sin(\alpha - \alpha')\mathbf{i} - g'\cos(\alpha - \alpha')\mathbf{k} \\ &= -a\omega^2 \sin \alpha \cos \alpha' \mathbf{i} - (g + a\omega^2 \sin \alpha \sin \alpha')\mathbf{k} \ ,\end{aligned} \tag{9}$$

using (4), (5). Since

$$\mathbf{f}_0 = -a\omega^2 \sin \alpha(\cos \alpha' \mathbf{i} + \sin \alpha' \mathbf{k}) \ , \tag{10}$$

the equations (6), (7), (8), (9), (10) give

$$\begin{aligned}\mathbf{T}/M = &(\ddot{x} - 2\omega\dot{y} \cos \alpha - \omega^2 x \cos^2\alpha - \omega^2 z \sin \alpha \cos \alpha)\mathbf{i} \\ &+ (\ddot{y} + 2\omega\dot{z} \sin \alpha + 2\omega\dot{x} \cos \alpha - \omega^2 y)\mathbf{j} \\ &+ (\ddot{z} - 2\omega\dot{y} \sin \alpha - \omega^2 x \sin \alpha \cos \alpha - \omega^2 z \sin^2\alpha + g)\mathbf{k} \ .\end{aligned} \tag{11}$$

In Fig. 9.5(ii), A has co-ordinates $(0, 0, l)$ and P has co-ordinates (x, y, z) so that the unit vector in $\overline{PA}$ is $[-x\mathbf{i} - y\mathbf{j} + (l - z)\mathbf{k}]/[x^2 + y^2 + (l - z)^2]^{\frac{1}{2}}$ and this approximates to $(-x\mathbf{i} - y\mathbf{j} + l\mathbf{k})/l$, if $l \gg x$, y and z. Furthermore ω is small—calculation shows it to be approximately 0.73 radians per second since the Earth rotates through 2π radians in 24 hours. Hence in equation (11) second-order terms such as $\omega^2 x$, $\omega^2 z$, $\omega\dot{x}$, $\dot{z}$ etc., may be neglected and on resolving into the three directions, $\mathbf{i}$, $\mathbf{j}$ and $\mathbf{k}$, we obtain the approximations

$$\ddot{x} - 2\omega\dot{y} \cos \alpha = -Tx/ml \ , \tag{12}$$

$$\ddot{y} + 2\omega\dot{x} \cos \alpha = -Ty/ml \ , \tag{13}$$

$$g - 2\omega\dot{y} \sin \alpha = T/m \ . \tag{14}$$

To equation (12), add i times equation (13) and write $\zeta = x + iy$. Then, denoting $D \equiv d/dt$, $n^2 = g/l$,

$$(D^2 + 2i\omega \cos \alpha D + n^2)\zeta = 0$$

or

$$[(D + i\omega \cos \alpha)^2 + \mu^2]\zeta = 0 \qquad (\mu^2 = n^2 + \omega^2 \cos^2\alpha) \ .$$

This has the general solution

$$\zeta = \exp(-i\omega t \cos \alpha)(A \cos \mu t + B \sin \mu t) \ ,$$

where A and B are complex integration constants. Write

$$Z = X + iY = \zeta \exp(i\omega t \cos \alpha) \ ,$$

$$A = c + id \ , \qquad B = e + if \ ,$$

where X, Y, c, d, e, f are all real. Then

$$\left.\begin{aligned} c \cos \mu t + e \sin \mu t = X \ , \\ d \cos \mu t + f \sin \mu t = Y \ , \end{aligned}\right\}$$

and so

$$\cos \mu t/(fX - eY) = \sin \mu t/(cY - dX) = 1/(cf - de) \ .$$

Elimination of t now gives

$$(fX - eY)^2 + (cY - dX)^2 = (cf - de)^2 \ . \tag{15}$$

Equation (15) is an ellipse, centre $X = 0 = Y$.

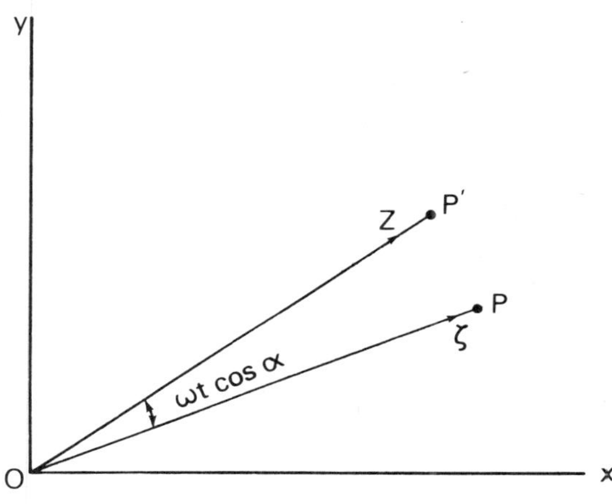

Fig. 9.6

Fig. 9.6 is the Argand diagram for the complex numbers $\overline{OP} \equiv \zeta$, $\overline{OP'} \equiv Z$. Then

$$\arg Z = \arg \zeta + \omega t \cos \alpha \ , \qquad |Z| = |\zeta| \ ,$$

and so $\angle POP' = \omega t \cos \alpha$, $OP' = OP$. We have seen that the locus of Z is an ellipse having O as centre. Since ζ lags behind Z an angular distance of $\omega t \cos \alpha$, the locus of ζ is an ellipse which revolves around the normal to the Argand plane with angular velocity $-\omega \cos \alpha$. Hence the locus of the projection of the bob of the pendulum on the horizontal plane through P_0 in Fig. 9.5(ii) is an ellipse revolving around the vertical axis $C'P_0$ with angular velocity $-\omega \cos \alpha$. Thus there is relative motion between the vertical plane of motion of the bob of the pendulum and the Earth. It is easy to measure the relative angular shift made by the plane of motion over a period of a few days, thereby confirming that the Earth has rotated.

The above device for detecting the Earth's rotation is called a *Foucault pendulum*. In practice, l is very large and x, y small. The Foucault pendulum in the South Kensington Science Museum is of length 24.08 m (79 ft) and carries a brass sphere containing lead shot of total weight 13.6 kg (30 lb).

EXERCISE 9

1. A uniform heavy square lamina, free to turn about its centre which is fixed, is set rotating with angular velocity Ω about an axis inclined at 45° to its plane. Show that, in the subsequent motion, the axis of figure

of the lamina describes, with uniform angular velocity, a right circular cone whose axis is inclined at $\tan^{-1}(1/3)$ with the invariable line. Find the semi-angle of the cone and the uniform angular velocity with which it is described. (L.U.)

2. A rigid lamina moves under no forces. The principal axes at the centre of mass O are Ox, Oy, Oz, the last being normal to the lamina, and the principal moments of inertia about Ox, Oy are A, B respectively ($B > A$). Prove that the resolutes of the angular velocity vector about Ox, Oy, Oz are respectively given by

$$\omega \cos \lambda \ , \qquad \omega \sin \lambda \ , \qquad \dot{\lambda} \ ,$$

where ω is a constant and λ satisfies the differential equation

$$2\ddot{\lambda} + \kappa^2 \sin 2\lambda = 0 \ , \qquad \kappa^2 = \omega^2(B-A)/(B+A) \ .$$

If the lamina is uniform and rectangular, and is set spinning about a diagonal with angular velocity ω, show that in the subsequent motion the lamina periodically spins about the other diagonal and that this first happens after a time

$$\frac{2}{\omega}(\sec 2\alpha)^{\frac{1}{2}} \int_0^{\pi/2} \frac{d\theta}{\sqrt{(1-\sin^2\alpha \sin^2\theta)}} \ ,$$

where 2α is the acute angle between the diagonals. (L.U.)

3. A uniform solid rectangular parallelepiped is free to turn about its centre of gravity which is fixed. The edges of the solid have lengths $2a$, $2a$, a. Initially the solid has angular velocity Ω about a diagonal. Find the components, parallel to the edges of the solid, of its angular velocity after a time t. Hence or otherwise show that the resolute of the angular velocity of the solid about the same diagonal after time t is

$$\tfrac{1}{9}\Omega[1 + 8 \cos (\Omega t/5)] \ .$$

(L.U.)

4. A rigid body is moving freely about a fixed point O under no external forces. Prove that, if two of the principal moments of inertia at O are equal, the instantaneous axis of rotation describes a circular cone, with vertex at O, fixed in space.

A rigid straight light rod, length $5a$, is rigidly and normally attached at one end to a fixed point O on the surface of a uniform solid sphere of radius $5a$. To the other end of the rod is fixed a particle of mass equal to that of the sphere. The system is free to turn about O and is given an angular velocity Ω about a line OA where A is a specified point on the sphere such that $OA = 8a$. Prove that at time t later, the component of angular velocity about OA is

$$(\Omega/25)\{16 + 9 \cos (2\Omega t/3)\} \ .$$

Show that the instantaneous axis of the system describes in space a cone of angle $2 \tan^{-1}(6/7)$. (L.U.)

5. A uniform rectangular lamina of sides $2a$, $3a$ is freely hinged to a

horizontal axis along one of its shorter edges. This axis is fixed to a vertical shaft which passes through the midpoint of the hinged edge, and the shaft is forced to rotate with constant angular velocity ω. If θ is the inclination of the plate to the downward vertical at time t, show that one of Euler's dynamical equations can be written

$$a\ddot{\theta} - a\omega^2 \sin\theta \cos\theta = -\tfrac{1}{2}g \sin\theta \ .$$

Show that if ω is sufficiently large there is a position of relative equilibrium in which the lamina makes an acute angle with the downward vertical. Prove also that this position is stable, and find the period of small oscillations when the lamina is disturbed slightly from this position. (R.U.)

6. Prove that the first component of the angular momentum of a rigid body rotating with one point fixed is $A\omega_1 - F\omega_2 - E\omega_3$ in the usual notation.

A uniform square lamina is rotating under no impressed forces about one corner O, which is fixed. If ω_1, ω_2, ω_3 are the components of the angular velocity about the principal axes at O (in the order of increasing moments of inertia), show that $\omega_1{}^2 + \omega_2{}^2$ and $3\omega_2{}^2 + 4\omega_3{}^2$ are constant. (R.U.)

7. An ellipsoid, free to move about its centre, is set in rotation at time $t = 0$ with an agular velocity $\mathbf{\Omega}$. If n, 0, $3n$ are the components of $\mathbf{\Omega}$ along the principal axes at the centre, and $6A$, $3A$, A the moments of inertia about these axes, find the components of the angular velocity at time t, and show that, for large values of t, the magnitude of the angular velocity is practically $n\sqrt{5}$. (R.U.)

8. Co-ordinate axes $O1$, $O2$, $O3$ are taken with origin O at the centre of a uniform rigid rectangular plate so that $O1$, $O2$ are parallel to the edges and $O3$ is normal to the plate. The moments of inertia about $O1$, $O2$, $O3$ are A, B, $A + B$, respectively, with $A > B$. The plate can rotate freely about O and is set spinning about an axis of spin through O having direction cosines $[0, \cos\alpha, \sin\alpha]$ with respect to the co-ordinate axes. If $\tan^2\alpha > \lambda^2 = (A - B)/(A + B)$, prove that the axis of spin rotates continually in the same direction relative to the plate.

If $\tan^2\alpha = \lambda^2$, prove that the plate will ultimately spin around the axis $O1$. (L.U.)

9. A plane lamina is pivoted at a point O and moves under the action of no forces other than the reaction at O. The principal axes in the plane of the lamina are Ox, and Oy and the corresponding principal moments of inertia are A and B ($B > A$). Initially the lamina is rotating with angular velocity Ω about a line (not necessarily in the plane of the lamina) which makes a small angle with Ox. Show that, during the subsequent motion, the direction of the component of angular velocity in the plane of the lamina oscillates about Ox with period approximately

$$(2\pi/\Omega)(B + A)^{\frac{1}{2}}/(B - A)^{\frac{1}{2}} \ .$$

(C.U.)

10. Establish Euler's equations of motion of a rigid body with one point fixed.

A body whose principal moments of inertia at its centre of mass G are A_1, A_2, A_3 spins about a fixed axis through G which has constant direction cosines $[l_1, l_2, l_3]$ with respect to the principal axes at G, being constrained by smooth bearings. Prove that the angular velocity is constant and show that the body exerts on the bearings a couple of moment

$$\omega^2\{(A_2-A_3)^2l_2^2l_3^2+(A_3-A_1)^2l_3^2l_1^2+(A_1-A_2)^2l_1^2l_2^2\}^{\frac{1}{2}} ,$$

where ω is the angular velocity.

(M.T.II)

11. A uniform solid of revolution is set rotating about its centre of mass O. The principal moments at O are A, A, C, and the total initial spin is Ω, while the initial component of spin about the axis of symmetry is n. There is a resisting couple k times the angular velocity. Prove that after time t the angle ϕ between the axes of spin and symmetry is given by

$$\tan\phi=\frac{\sqrt{(\Omega^2-n^2)}}{n}\exp\left\{-\left(\frac{C-A}{CA}\right)kt\right\} .$$

Hence show that if the solid is an oblate spheroid the resistance tends to reduce the wobble, and if it is prolate to increase the wobble.

(M.T.II)

12. A rigid body is free in space under no external forces and its motion consists of a pure rotation about an axis fixed in the body. The principal moments of inertia A, B, C at the centre of mass are all unequal. Establish that:

(i) the angular velocity is constant in magnitude;
(ii) the axis of rotation is also fixed in space;
(iii) the centre of mass lies on the axis of rotation;
(iv) the axis of rotation coincides with a principal axis.

Given that $A<B<C$, determine the principal axes about which the rotation is stable. (M.T.II)

13. A perfectly rough plane is made to rotate about a vertical axis in its plane with constant angular velocity Ω. A uniform solid sphere rolls on the plane under the action of gravity. Initially it is at rest relative to the plane. Prove that the centre never descends more than $5g/\Omega^2$ below its original level. (L.U.)

14. A uniform solid sphere rolls without slipping on an inclined plane rotating about a fixed axis normal to itself with angular velocity Ω. Show that if the centre of the sphere is initially at rest, it will in the ensuing motion always lie between two fixed horizontal planes distant $35g\sin^2\alpha/2\Omega^2$ apart, and move from one to the other in time $7\pi/2\Omega$, where α is the inclination of the plane to the horizontal. (L.U.)

15. A hollow cylinder of internal radius b is fixed with its axis OZ vertical. A uniform solid sphere of centre C and radius a moves under gravity in contact with the inner surface of the cylinder which is rough enough to

prevent slipping. Show that ω, the angular velocity of the plane OZC, is constant and that the height of C above the fixed point O is in simple harmonic motion. Find the period of this motion in terms of ω. (L.U.)

16. A uniform circular disc of radius a and mass m is free to turn about a fixed point A of its circumference in any way. It is rotating with angular velocity Ω about the vertical diameter through A. A moving particle, also of mass m, hits the disc at an end of the horizontal diameter and adheres to the disc. When the particle hits the disc it has a velocity u perpendicular to the disc and is moving in the same direction as the point where it hits. Find the velocity of the particle after impact, and show that the impulsive reaction at A is $m(u-a\Omega)/29$. (L.U.)

17. A uniform solid sphere of radius a rolls on the outside of a fixed rough cylinder of the same radius whose axis is horizontal. Initially the sphere stands on top of the cylinder and spins about its vertical axis with angular velocity Ω. When the normal at the point of contact makes an angle θ with the vertical, show that $7a\dot{\theta}^2=5g(1-\cos\theta)$. Show also that the angular velocity of the sphere about the normal is then $\Omega\cos n\theta$, where $7n^2=2$. (L.U.)

18. A uniform rectangular plate $ABCD$ of mass M, for which $AB=2a$, $BC=2b$, is at rest and is then given a blow of impulse J perpendicular to the plate at C. Show that, if A is fixed but the plate is free to turn about it, the kinetic energy generated by the blow is $12J^2/7M$.

If the plate is completely free find the kinetic energy generated by the blow and the velocities of B and C. (L.U.)

19. A lamina is rotating with uniform angular velocity Ω about a vertical axis in the lamina through its centre of gravity G. The axis is freely supported at G and at a point P at a distance l from G. GX is the line in the plane of the lamina perpendicular to GP. The moments of inertia of the lamina about GX and GP are A and B respectively, and the product of inertia of the lamina with respect to GX and GP is H. Show that the horizontal reactions at G and P are of magnitude $H\Omega^2/l$ and state their directions.

If now the support P is removed, show that P has no initial velocity but has an initial acceleration. Find this acceleration in terms of l, Ω, A, B, H, and state its direction. (L.U.)

20. A uniform solid right circular cone of height h, mass M and semi-vertical angle $\pi/4$ moves freely about its vertex O, which is fixed, under no forces other than the reaction at O. If the cone is set spinning with angular velocity Ω about a generator Ol, find the components of the angular velocity at any subsequent time t in the right-handed orthogonal frame Oa, Ob, Oc fixed in the cone, where Oc is the axis of the cone and Ol bisects the angle between Oa and Oc.

Find the reaction at O and show that its magnitude is $3\sqrt{29}\ Mh\Omega^2/40$. [The principal moments of inertia at O are $3Mh^2/4$, $3Mh^2/4$, $3Mh^2/10$.] (L.U.)

21. Taking axes Ox, Oy, Oz, due east, due north and vertically upwards at a point O in latitude λ on the earth's surface, show that the component accelerations of a particle close to O are

$$\ddot{x}+2\omega\dot{z}\cos\lambda-2\omega\dot{y}\sin\lambda,\quad \ddot{y}+2\omega\dot{x}\sin\lambda,\quad \ddot{z}-2\omega\dot{x}\cos\lambda\ ,$$

neglecting the square of the earth's angular velocity ω.

A train is travelling through this point along a straight line in a given horizontal direction with uniform speed V. Find the effective value of gravity inside a carriage of the train, and show that a plumb line will be inclined at $2\omega V\sin\lambda/g$ to the vertical. (R.U.)

22. Obtain the expression $\ddot{\mathbf{r}}+2\boldsymbol{\omega}_\wedge\dot{\mathbf{r}}+\dot{\boldsymbol{\omega}}_\wedge\mathbf{r}+\boldsymbol{\omega}_\wedge(\boldsymbol{\omega}_\wedge\mathbf{r})$ for the acceleration of a particle referred to a fixed frame of reference S in terms of the acceleration $\ddot{\mathbf{r}}$ and the velocity $\dot{\mathbf{r}}$ of the particle referred to a frame of reference rotating with angular velocity $\boldsymbol{\omega}$ with respect to S and having the same origin as S.

A projectile is fired eastward in a northern latitude λ with speed v at an angle of elevation α. Neglecting terms in ω^2, where ω is the angular speed of the earth's diurnal rotation, show that the projectile will strike the horizontal plane through the point of projection at a point whose distance from the plane of projection is

$$(4\omega v^3\sin^2\alpha\cos\alpha\sin\lambda)/g^2$$

towards the south. (L.U.)

23. A particle is projected northward with speed V at an elevation α from a point of the earth's surface in north latitude λ. Find approximately the east-west deviation of the particle from the vertical plane of projection after time t. [The gravitational field may be assumed uniform and air resistance may be neglected.] (M.T.II)

Chapter 10

Generalized Co-ordinates

10.1 NOTE ON DYNAMICAL METHODS

It is evident that two seemingly different approaches have been used for the solution of the problems considered so far. In some cases the laws of motion (linear and/or rotational) have been invoked; such an approach often leads to many equations containing unknown and unwanted internal forces of the system. In other cases, however, where the forces have been conservative in kind, we have used the principle of conservation of energy, thereby avoiding the introduction of unwanted internal forces.

The first method is essentially *vectorial* and derives directly from the work of Newton. In the second method, *scalar* quantities, i.e. potential and kinetic energies, are calculated and an energy equation is set up. Of the two methods, the first is the more cumbersome, but it is the more general, since many problems involve non-conservative fields of force or else impulsive motions to which the principle of conservation of energy is not applicable. The two approaches (where both are applicable), are not really distinct, since the conservation of energy principle is established using the motion equations. However, the energy method is prefereable in certain cases and often leads to the desired result much more quickly than does the vectorial method.

In 1788, Joseph Louis Lagrange's celebrated *Mécanique Analytique* was published in Paris. The object of this treatise was to show how all dynamical problems could be solved from a knowledge of the kinetic energy, the forces and couples, and the co-ordinates of the components of the given dynamical system at any time. In many cases Lagrange's analytical method saves labour in that unwanted internal forces are not introduced. The analytical method, which we develop in this chapter, has proved a most powerful tool in engineering dynamics, quantum mechanics and other branches of mathematical physics.

10.2 PRELIMINARY NOTIONS

A *dynamical system* is one comprised of particles; it may also include rigid bodies since these are made up of particles. We suppose the system to be made up of N particles of masses $m_i(i=1, 2, \ldots, N)$ and that at any time t the position of each particle may be specified by means of n independent variables $q_j(j=1, 2, \ldots, n)$. These variables are termed the *generalized co-ordinates* of the system and in number they equal the number of degrees of freedom of the system.

To illustrate the above notions, consider the motion of a two-particle system. Three Cartesian co-ordinates specify the position of one particle and another three those of the other particle, making a total of six. There are six degrees of freedom for such a system requiring six generalized co-ordinates.

Now suppose the two particles are connected by a rigid weightless rod of length l, so that they remain a constant distance apart. If (x, y, z) denote the Cartesian co-ordinates of the first particle and if the second particle has spherical polar co-ordinates (r, θ, ϕ) with respect to the first, then $r=l$, so that the rigid connection has reduced the number of degrees of the two-particle system from six to five.

The first two-particle system is said to be *unconstrained*; the second is *constrained*, the constraint being the rigid connection.

Let us extend these ideas to the case of N particles where $N \geqslant 3$. Firstly, if the system is unconstrained, then $3N$ co-ordinates are required for its specification. Now suppose all pairs of particles are at invariant distance apart so that the system constitutes a rigid body. Let d_N be the number of invariable distances between the pairs of particles of the system. If we add another $(N+1)$th particle to the system, then for a rigid body of $(N+1)$ particles, the distances of this $(N+1)$th particle from any three particles of the original system of N must be specified as constant. Thus

$$d_{N+1}=d_N+3\ , \qquad N \geqslant 3\ .$$

This, together with $d_3=3$, can easily be seen to lead to $d_N=3N-6$. Hence a rigid body comprising N particles requires $3N-6$ constraints and so the number of degrees of freedom is six. This result has been obtained in another way in Section 2.4.

However, it is important to distinguish between different kinds of constraints. In the case of the rigidly connected two-particle system the constraint was expressed by the equation $r=l$, or alternatively by $l^2=(x_2-x_1)^2 +(y_2-y_1)^2+(z_2-z_1)^2$ if (x_1, y_1, z_1); (x_2, y_2, z_2) denote the Cartesian co-ordinates of the particles. Thus the single constraint requires but one equation specifying that constant. In the case of a particle moving on a wire there are two constraints since the wire may be thought of as the curve of intersection of two surfaces, each of which requires an equation for its specification. In this case there is one degree of freedom of the particle and a suitable co-ordinate for specification of the particle is its arcual distance of travel along the curve from a fixed point on it.

Reverting again to the rigidly connected two-particle system, we may choose x_1, y_1, z_1, x_2, y_2 to specify it, since z_2 is derivable from the equation of constraint. With such a selection, it is quite obvious that five arbitrary variations

δx_1, δy_1, δz_1, δx_2, δy_2 in these co-ordinates may be made. If for a dynamical system specified by the n generalized co-ordinates $q_i(i=1, 2, \ldots, n)$, we can make independent variations $\delta q_i(i=1, 2, \ldots, n)$ in all the co-ordinates, then the system is said to be *holonomic*: otherwise it is said to be *non-holonomic*. The rigidly connected two-particle system specified by the five co-ordinates mentioned is a holonomic system.

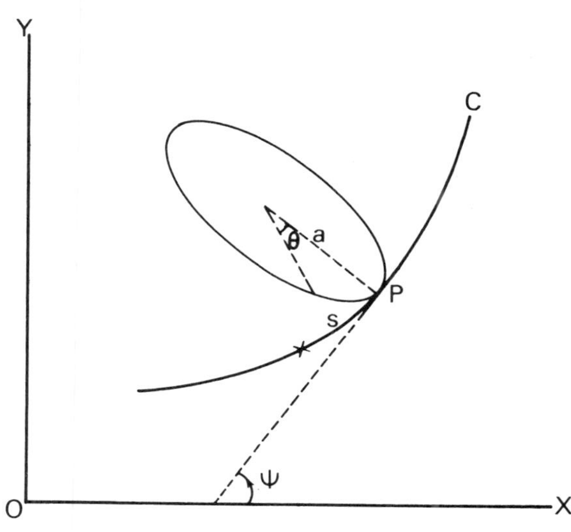

Fig. 10.1

To illustrate the case of a non-holonomic system, consider the motion of a vertical wheel of radius a on a rough horizontal plane specified by co-ordinate axes OX, OY (Fig. 10.1). The contact point P of the wheel moves along a curve C in the plane. The motion is presumed to be one of rolling without slipping. Let ψ be the angle the tangent at P to C makes with OX when P has co-ordinates (x, y) and let θ be the angular rotation of the wheel when P has travelled a distance s along C. Then the four co-ordinates x, y, θ, ψ specify the position of the wheel at any instant. Now suppose the position of the wheel is varied from the state defined by (x, y, θ, ψ) to $(x+\delta x, y+\delta y, \theta+\delta\theta, \psi+\delta\psi)$. Using $s=a\theta$ so that $\delta s=a\delta\theta$, we obtain

$$\delta x=\delta s \cos\psi=a\cos\psi\delta\theta \; ; \qquad \delta y=\delta s \sin\psi=a\sin\psi\delta\theta \; .$$

These relations show quite clearly that the variations δx, δy in x and y depend upon the variation $\delta\theta$ in θ. Thus the system (x, y, θ, ψ) is non-holonomic. Another feature of it is that the differential equations

$$dx=a\cos\psi d\theta \; ; \qquad dy=a\sin\psi d\theta$$

are non-integrable since ψ is arbitrary. It is a feature of non-holonomic systems that constraints are expressible as non-integrable relations.

10.3 GENERALIZED VELOCITIES

Let the dynamical system be comprised of N particles of masses $m_i(i=1, 2, \ldots, N)$ and at time t suppose each particle is specified by the n generalized co-ordinates $q_j(j=1, 2, \ldots, n)$. Then the n quantities $\dot{q}_j = dq_j/dt$ $(j=1, 2, \ldots, n)$ are called the *generalized velocities* of the system.

Let $\mathbf{r}_i$ be the position vector of m_i at time t so that

$$\mathbf{r}_i = \mathbf{r}_i(q_1, q_2, \ldots, q_n; t) .$$

We shall throughout this chapter denote total differentiations w.r.t. time t by a dot. Then

$$\dot{\mathbf{r}}_i = \frac{d\mathbf{r}_i}{dt} = \frac{\partial \mathbf{r}_i}{\partial q_1}\dot{q}_1 + \frac{\partial \mathbf{r}_i}{\partial q_2}\dot{q}_2 + \ldots + \frac{\partial \mathbf{r}_i}{\partial q_j}\dot{q}_j + \ldots + \frac{\partial \mathbf{r}_i}{\partial q_n}\dot{q}_n + \frac{\partial \mathbf{r}_i}{\partial t} .$$

Now regard the new independent variables as $\dot{q}_1, \dot{q}_2, \ldots, \dot{q}_n, t$. Then

$$\partial \dot{\mathbf{r}}_i/\partial \dot{q}_j = \partial \mathbf{r}_i/\partial q_j (j=1, 2, \ldots, n) .$$

10.4 VIRTUAL WORK AND GENERALIZED FORCES

Suppose the particles of a dynamical system undergo a small instantaneous displacement independent of time, consistent with the constraints of the system and such that all internal and external forces remain unchanged in magnitude and direction during the displacement. Such a displacement is said to be *virtual* because of its hypothetical nature. Although purely fictitious the notion of virtual displacement is essential to the elucidation of the principles of analytical dynamics.

Let the ith particle m_i at position $\mathbf{r}_i$ at time t undergo a virtual displacement to position $\mathbf{r}_i + \delta\mathbf{r}_i$. Let $\mathbf{F}_i$, $\mathbf{F}_i'$ be the external and internal forces acting on m_i. Then the *virtual work* done on m_i in the displacement is $(\mathbf{F}_i + \mathbf{F}_i').\delta\mathbf{r}_i$ and so the total virtual work done on all particles of the system when similar displacements are made is

$$\delta W = \sum_{i=1}^{N} (\mathbf{F}_i + \mathbf{F}_i').\delta\mathbf{r}_i = \sum_{i=1}^{N} \mathbf{F}_i.\delta\mathbf{r}_i + \sum_{i=1}^{N} \mathbf{F}_i'.\delta\mathbf{r}_i .$$

Now $\sum_{i=1}^{N} \mathbf{F}_i'.\delta\mathbf{r}_i$ is the total work done by the internal forces of the system. In many cases this is zero, e.g. when the particles of the system are connected by rigid constraints. In future we shall, unless otherwise stated, assume the internal forces $\mathbf{F}_i'(i=1, \ldots, n)$ to be of the type which do no work in displacement (see Chapter 8, Section 8.4), though we shall certainly meet cases (such as elastic connections) where the internal forces, though equal and opposite, do work in displacement. When the internal forces do no work in a virtual displacement,

$$\delta W = \sum_{i=1}^{N} \mathbf{F}_i.\delta\mathbf{r}_i = \sum_{i=1}^{N} (X_i\,\delta x_i + Y_i\delta y_i + Z_i\delta z_i)\ ,$$

where $\mathbf{F}_i = [X_i,\ Y_i,\ Z_i]$, $\delta\mathbf{r}_i = [\delta x_i,\ \delta y_i,\ \delta z_i]$. δW is termed the *virtual work function* and we note that the coefficients in it of δx_i, δy_i, etc. are the *external* force components X_i, Y_i, etc.

Now suppose the system is holonomic and specified by the n generalized co-ordinates $q_j(j=1, \ldots, n)$. Then we can change q_j to $q_j + \delta q_j$ without making attendant changes in the other $(n-1)$ co-ordinates. Let this virtual displacement take effect instantaneously and suppose the corresponding work done on the dynamical system to be $Q_j\delta q_j$. Then

$$Q_j\delta q_j = \sum_{i=1}^{N} \mathbf{F}_i.\delta\mathbf{r}_i.$$

If we now make similar variations in each $q_j(j=1, \ldots, n)$, then

$$\delta W = \sum_{j=1}^{n} Q_j\delta q_j = \sum_{i=1}^{N} \mathbf{F}_i.\delta\mathbf{r}_i.$$

This last relationship shows that Q_j is the coefficient of $\delta q_j(j=1, \ldots, n)$ in the virtual work function δW which has been constructed *solely from the external forces acting on the system.* Since the coefficient of δx_i in δW is the force X_i, we call the coefficient of the generalized virtual displacement δq_j, the *generalized force* Q_j associated with co-ordinate $q_j(j=1, \ldots, n)$.

10.5 DERIVATION OF LAGRANGE'S EQUATIONS FOR A HOLONOMIC SYSTEM

We now establish Lagrange's equations for a holonomic dynamical system in the form

$$\frac{d}{dt}\left(\frac{\partial T}{\partial \dot{q}_j}\right) - \frac{\partial T}{\partial q_j} = Q_j(j=1, \ldots, n)$$

where T is the kinetic energy of the system at time t when the system is specified by the n generalized co-ordinates $q_j(j=1, \ldots, n)$ and Q_j $(j=1, \ldots, n)$ are the generalized forces.

Proof

The equation of motion of the ith particle of mass m_i is

$$\mathbf{F}_i + \mathbf{F}_i' = m_i\ddot{\mathbf{r}}_i\ , \tag{1}$$

in the notation previously defined, dots denoting *total* time differentiations. Now

$$\ddot{\mathbf{r}}_i.(\partial\mathbf{r}_i/\partial q_j) = (d/dt)\{\dot{\mathbf{r}}_i.(\partial\mathbf{r}_i/\partial q_j)\} - \dot{\mathbf{r}}_i.(\partial\dot{\mathbf{r}}_i/\partial q_j)\ ,$$

since $(d/dt)(\partial\mathbf{r}_i/\partial q_j)=\partial\dot{\mathbf{r}}_i/\partial q_j$. We have shown in Section 10.3, however, that $\partial\mathbf{r}_i/\partial q_j=\partial\dot{\mathbf{r}}_i/\partial\dot{q}_j$ and so

$$\ddot{\mathbf{r}}_i.(\partial\mathbf{r}_i/\partial q_j)=(d/dt)\{\tfrac{1}{2}\partial\dot{\mathbf{r}}_i^2/\partial\dot{q}_j\}-\tfrac{1}{2}\partial\dot{\mathbf{r}}_i^2/\partial q_j\ . \tag{2}$$

Scalar multiply both sides of (1) by $\partial\mathbf{r}_i/\partial q_j$ and use (2). Then

$$(d/dt)\{\partial(\tfrac{1}{2}m_i\dot{\mathbf{r}}_i^2)/\partial\dot{q}_j\}-\partial(\tfrac{1}{2}m_i\dot{\mathbf{r}}_i^2)/\partial q_j=(\mathbf{F}_i+\mathbf{F}_i').\partial\mathbf{r}_i/\partial q_j\ . \tag{3}$$

Now sum both sides of (3) from $i=1$ to N. The total K.E. of the system is $T=\frac{1}{2}\Sigma m_i\dot{\mathbf{r}}_i^2$, and so

$$(d/dt)(\partial T/\partial q_j)-\partial T/\partial q_j=\sum_{i=1}^{N}(\mathbf{F}_i+\mathbf{F}_i').\partial\mathbf{r}_i/\partial q_j\ . \tag{4}$$

When all generalized co-ordinates are constant save q_j, the virtual work function is

$$\delta W=Q_j\delta q_j=\sum_{i=1}^{N}\mathbf{F}_i.\delta\mathbf{r}_i=\sum_{i=1}^{N}(\mathbf{F}_i+\mathbf{F}_i').\delta\mathbf{r}_i\ .$$

and so $$Q_j=\sum_{i=1}^{N}(\mathbf{F}_i+\mathbf{F}_i').\partial\mathbf{r}_i/\partial q_j\ .$$

Thus (4) becomes

$$(d/dt)(\partial T/\partial\dot{q}_j)-\partial T/\partial q_j=Q_j\ . \tag{5}$$

Allowing j to vary from 1 to n generates the required set of n equations of Lagrange.

Résumé

Lagrange's equations for a holonomic dynamical system specified by the n generalized co-ordinates $q_j(j=1,\ldots,n)$ are

$$(d/dt)(\partial T/\partial\dot{q}_j)-\partial T/\partial q_j=Q_j(j=1,\ldots,n)\ ,$$

where T is the K.E. of the system and Q_j is the coefficient of δq_j in the virtual work function

$\delta W=\sum_{j=1}^{n}Q_j\delta q_j$ *constructed solely from the forces which do work in displacement, when we allow the generalized co-ordinates $q_j(j=1,\ldots,n)$ to undergo instantaneous virtual increments $\delta q_j(j=1,\ldots,n)$.*

10.6 WORKED EXAMPLES

Example 1—Planetary motion.

Fig. 10.2 shows a planet P of mass m orbiting round the Sun S (considered fixed) under the inverse square law of attraction $\mu m/r^2$. Let (r,θ) be the

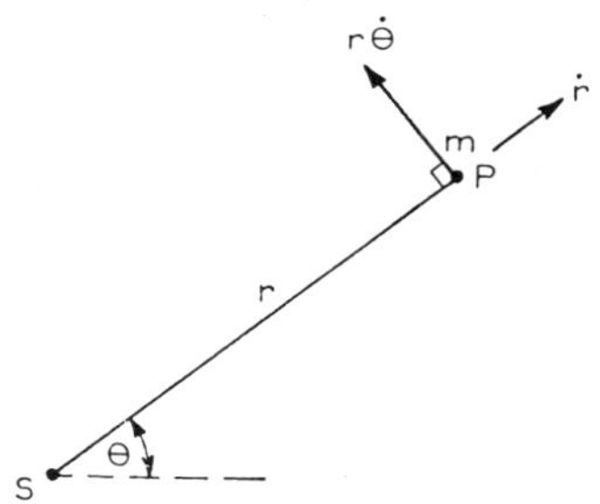

Fig. 10.2

polar co-ordinates of P with respect to S at time t. Then these are the generalized co-ordinates of the system and the K.E. is

$$T = \tfrac{1}{2}m(\dot{r}^2 + r^2\dot{\theta}^2) \ . \tag{1}$$

r, θ may be varied independently here to give radial and transverse displacements of P of amounts δr, $r\delta\theta$ corresponding to virtual displacements δr, $\delta\theta$. Thus the system is holonomic and the virtual work function is

$$\delta W = (-\mu m/r^2)\delta r + 0(r\delta\theta) = -(\mu m/r^2)\delta r \ . \tag{2}$$

Lagrange's equations for the system are

$$\left.\begin{aligned}(d/dt)(\partial T/\partial \dot{r}) - \partial T/\partial r &= Qr \ ,\\ (d/dt)(\partial T/\partial \dot{\theta}) - \partial T/\partial \theta &= Q_\theta,\end{aligned}\right\} \tag{3}$$

where Q_r, Q_θ are the coefficients of δr, $\delta\theta$ in δW: thus

$$Q_r = -\mu m/r^2 \ , \qquad Q_\theta = 0 \ . \ . \ .$$

Now

$$\partial T/\partial r = mr\dot{\theta}^2, \quad \partial T/\partial\theta = 0, \quad (d/dt)(\partial T/\partial\dot{r}) = m\ddot{r}, \quad (d/dt)(\partial T/\partial\dot{\theta}) = d/dt(mr^2\dot{\theta}),$$

and so Lagrange's equations become

$$\left.\begin{aligned}m\ddot{r} - mr\dot{\theta}^2 &= -\mu m/r^2 \ ,\\ (d/dt)(mr^2\dot{\theta}) &= 0 \ .\end{aligned}\right\}$$

These are familiar results.

Example 2—Pulley system.

It is required to find the motion of the system in Fig. 10.3 the pulley wheels having negligible masses and moments of inertia, their axles being frictionless.

Let the axle of the moveable pulley be x below that of the fixed one and let the mass of 4 units be y below the axle of the moveable pulley. Then the 5, 2, 4 masses are at distances $(l_1 - x)$, $(x + l_2 - y)$, $(x + y)$ below the axles of the upper pulley, l_1 and l_2 being the constant lengths of the two strings minus their overlaps round the wheels. Clearly x and y are the generalized co-ordinates of

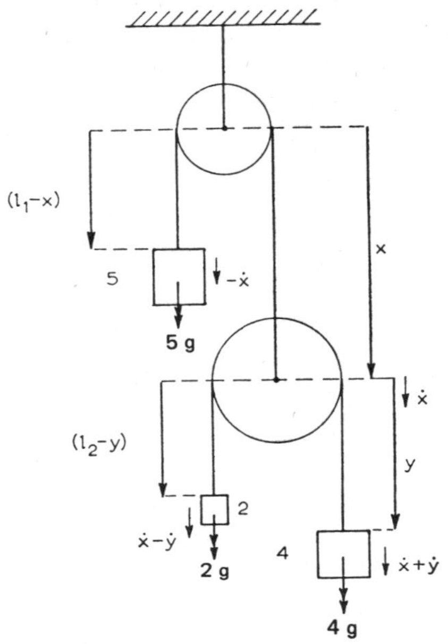

Fig. 10.3

this 3-particle system and since x, y can be varied independently, the system is holonomic.

The downward velocities of the masses are $-\dot{x}$, $(\dot{x}-\dot{y})$, $(\dot{x}+\dot{y})$ and so the K.E. is

$$T=\tfrac{1}{2}\{5\dot{x}^2+2(\dot{x}-\dot{y})^2+4(\dot{x}+\dot{y})^2\} \ .$$

Allowing x, y to increase by virtual displacements δx, δy, the virtual work function constructed from the external forces is

$$\begin{aligned}\delta W &= 5g\delta(l_1-x)+2g\delta(x+l_2-y)+4g\delta(x+y)\\ &= g(\delta x+2\delta y) \ ,\end{aligned}$$

whence the generalized forces are $Q_x=g$, $Q_y=2g$.

Lagrange's equations are

$$\left.\begin{aligned}(d/dt)(\partial T/\partial\dot{x})-\partial T/\partial x &= Q_x \ ,\\ (d/dt)(\partial T/\partial\dot{y})-\partial T/\partial y &= Q_y \ ,\end{aligned}\right\}$$

i.e.

$$\left.\begin{aligned}11\ddot{x}+2\ddot{y} &= g \ ,\\ 2\ddot{x}+6\ddot{y} &= 2g \ ,\end{aligned}\right\}$$

and so $\ddot{x}=g/31$, $\ddot{y}=10g/31$. Thus the downward accelerations of the particles of masses 5, 2, 4 units are respectively

$$\left.\begin{aligned}-\ddot{x} &= -g/31 \ ;\\ \ddot{x}-\ddot{y} &= -9g/31 \ ;\\ \ddot{x}+\ddot{y} &= 11g/31 \ .\end{aligned}\right\}$$

Note that the tensions in the string are internal forces which do no work in displacement and so they are not introduced in δW.

Example 3

A horizontal circular wire has radius R, centre C and is free to rotate about a vertical axis through a point O in its plane distant d from C. The wire carries a smooth particle P and $\angle OCP = \theta$ at time t. If ω is the angular velocity of the wire show that

$$R\ddot{\theta} + \dot{\omega}(R - d\cos\theta) = d\,\omega^2 \sin\theta \ .$$

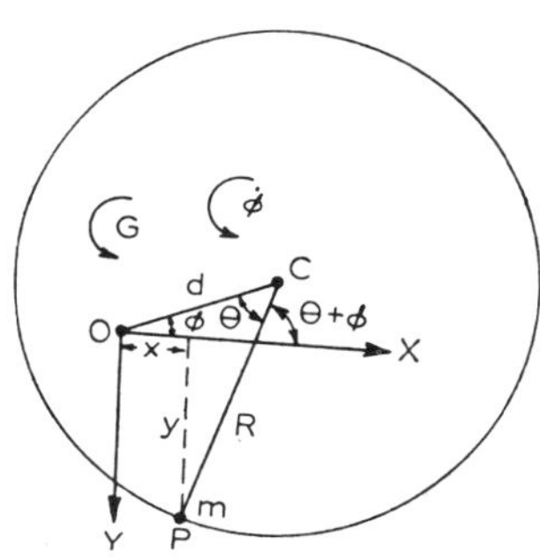

Fig. 10.4

At time t, let OC make an angle ϕ with the fixed line OX, so that PC makes an angle $(\theta + \phi)$ with OX. The generalized co-ordinates are ∂, ϕ. Let P have co-ordinates x, y referred to OX, OY (Fig. 10.4). Then

$$x = d\cos\phi - R\cos(\theta + \phi) \ , \qquad y = R\sin(\theta + \phi) - d\sin\phi \ ,$$
$$\dot{x} = -d\dot{\phi}\sin\phi + R(\dot{\theta} + \dot{\phi})\sin(\theta + \phi) \ ,$$
$$\dot{y} = R(\dot{\theta} + \dot{\phi})\cos(\theta + \phi) - d\dot{\phi}\cos\phi \ .$$

Hence, taking m to be the mass of the particle P, the K.E. is

$$T = \tfrac{1}{2}m(\dot{x}^2 + \dot{y}^2)$$
$$= \tfrac{1}{2}m\{R^2(\dot{\theta} + \dot{\phi})^2 + d^2\dot{\phi}^2 - 2Rd\dot{\phi}(\dot{\theta} + \dot{\phi})\cos\theta\} \ .$$

Let G be the couple applied in the direction of ϕ increasing. Then the virtual work function is

$$\delta W = G\delta\phi \ ,$$

and so the generalized forces are $Q_\theta = 0$, $Q_\phi = G$.

Lagrange's equation for co-ordinate θ is

$$(d/dt)(\partial T/\partial\dot{\theta}) - \partial T/\partial\theta = Q_\theta \ ,$$

or

$$(d/dt)\{R^2(\dot{\theta} + \dot{\phi}) - Rd\dot{\phi}\cos\theta\} - Rd\dot{\phi}(\dot{\theta} + \dot{\phi})\sin\theta = 0 \ .$$
$$\therefore\ R(\ddot{\theta} + \ddot{\phi}) - d\ddot{\phi}\cos\theta + d\dot{\phi}\dot{\theta}\sin\theta - d\dot{\phi}(\dot{\theta} + \dot{\phi})\sin\theta = 0 \ .$$

Putting $\dot{\phi}=\omega$, $\ddot{\phi}=\omega$ leads to the required result.

Example 4

Two uniform rods AB, AC, each of mass m and length $2a$, are smoothly hinged together at A and move on a horizontal plane. At time t the mass-centre of the rods is at the point (ξ, η) referred to fixed perpendicular axes Ox, Oy in the plane, and the rods make angles $\theta \pm \phi$ with Ox. Prove that the kinetic energy of the system is

$$m[\dot{\xi}^2+\dot{\eta}^2+(\tfrac{1}{3}+\sin^2\phi)a^2\dot{\theta}^2+(\tfrac{1}{3}+\cos^2\phi)a^2\dot{\phi}^2] \ ,$$

and derive Lagrange's equations of motion for the system if an external force with components $[X, Y]$ along the axes acts at X. (L.U.)

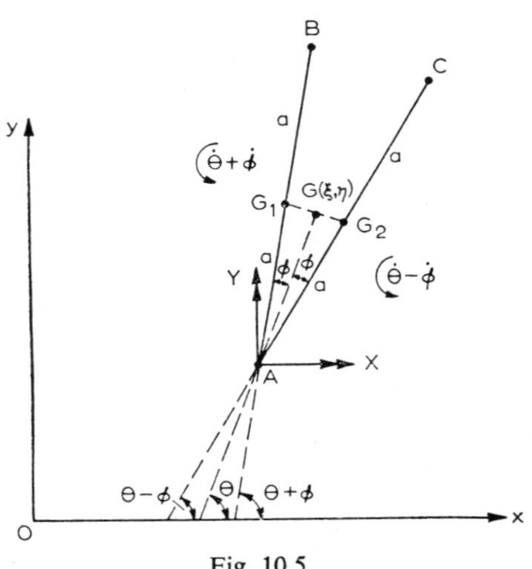

Fig. 10.5

In Fig. 10.5, G_1, G_2 are the centroids of the rods AB, AC. The co-ordinates of G_1, G_2 are respectively $(\xi \mp a \sin\phi \sin\theta,\ \eta \pm a \sin\phi\cos\theta)$, and their velocity components are

$$[\dot{\xi} \mp a\,\dot{\phi}\cos\phi\sin\theta \mp a\,\dot{\theta}\sin\phi\cos\theta\ ,\quad \dot{\eta} \pm a\,\dot{\phi}\cos\phi\cos\theta \mp a\,\dot{\theta}\sin\phi\sin\theta].$$

Since the angular velocities of AB, AC are $(\dot{\theta}+\dot{\phi})$, $(\dot{\theta}-\dot{\phi})$ respectively, the total K.E. is given by

$$\begin{aligned}T=\tfrac{1}{2}m\{(&\dot{\xi}-a\,\dot{\phi}\cos\phi\sin\theta-a\dot{\theta}\sin\phi\cos\theta)^2\\ &+(\dot{\eta}+a\dot{\phi}\cos\phi\cos\theta-a\dot{\theta}\sin\phi\sin\theta)^2\}+\tfrac{1}{6}ma^2(\dot{\theta}+\dot{\phi})^2\\ &+\tfrac{1}{2}m\{(\dot{\xi}+a\dot{\phi}\cos\phi\sin\theta+a\dot{\theta}\sin\phi\cos\theta)^2\\ &+(\dot{\eta}-a\dot{\phi}\cos\phi\cos\theta+a\dot{\theta}\sin\phi\sin\theta)^2\}+\tfrac{1}{6}ma^2(\dot{\theta}-\dot{\phi})^2\\ &=\underline{m\{\dot{\xi}^2+\dot{\eta}^2+(\tfrac{1}{3}+\sin^2\phi)a^2\dot{\theta}^2+(\tfrac{1}{3}+\cos^2\phi)a^2\dot{\phi}^2\}}\ .\end{aligned}$$

The virtual work function is

$$\begin{aligned}\delta W &= X\delta x_A + Y\delta y_A \\ &= X\delta(\xi - a\cos\phi\cos\theta) + Y\delta(\eta - a\cos\phi\sin\theta) \\ &= X(\delta\xi + a\delta\phi\sin\phi\cos\theta + a\delta\theta\cos\phi\sin\theta) \\ &\quad + Y(\delta\eta + a\delta\phi\sin\phi\sin\theta - a\delta\theta\cos\phi\cos\theta)\ .\end{aligned}$$

Thus the generalized components of force associated with co-ordinates ξ, η, θ, ϕ are respectively

$$Q_\xi = X\ , \qquad Q_\eta = Y\ ,$$
$$Q_\theta = a\cos\phi(X\sin\theta - Y\cos\theta)\ , \quad Q_\phi = a\sin\phi(X\cos\theta + Y\sin\theta).$$

Lagrange's equations for the generalized co-ordinates ξ, η, θ, ϕ are $(d/dt)(\partial T/\partial\dot{\xi}) - \partial T/\partial\xi = Q_\xi$, etc. These become on simplification

$$\begin{aligned}&2m\ddot{\xi} = X\ , \\ &2m\ddot{\eta} = Y\ , \\ &2ma[(\tfrac{1}{3} + \sin^2\phi)\ddot{\theta} + \dot{\theta}\dot{\phi}\sin 2\phi] = \cos\phi(X\sin\theta - Y\cos\theta)\ , \\ &2ma[(\tfrac{1}{3} + \cos^2\phi)\ddot{\phi} - 2\dot{\phi}^2\sin\phi\cos\phi] - ma(\dot{\theta}^2 - \dot{\phi}^2)\sin 2\phi \\ &\qquad = \sin\phi(X\cos\theta + Y\sin\theta)\ .\end{aligned}$$

Example 5—Flyball governor.

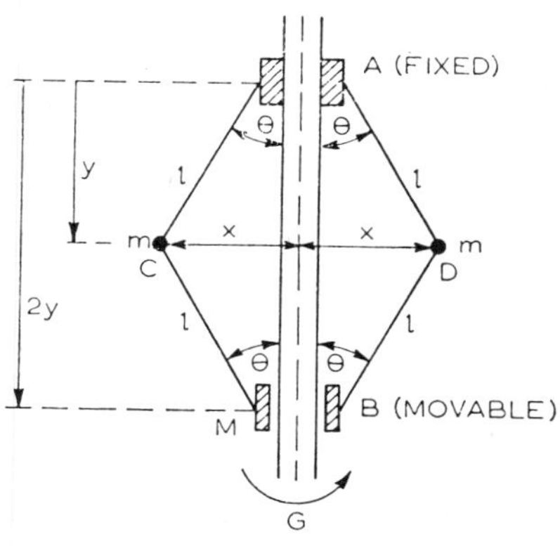

Fig. 10.6

Fig. 10.6 shows a flyball governor, the balls C, D each being of mass m and attached by light rods AC, AD; BC, BD to heavy collars A and B, A being rigidly attached to a vertical shaft and B, of mass M, free to slide up and down the shaft.

At time t, the rods are inclined at θ to the vertical and we suppose that under the action of a torque of moment G, the system has rotated through an angle ϕ. Let x denote the horizontal displacement of either mass m

from the axis of the shaft, y the depth of each below the horizontal through A, so that $2y$ is the depth of the mass M below this level. Then $x = l \sin \theta$, $y = l \cos \theta$. The velocity components of each m are $[\dot{x}, \dot{y}, x\dot{\phi}]$ in the directions of x, y, ϕ increasing, i.e. $[l\dot{\theta} \cos \theta, \ -l\dot{\theta} \sin \theta, \ l\dot{\phi} \sin \theta]$, and the downward velocity of M is $2\dot{y}$ or $-2l\dot{\theta} \sin \theta$. Neglecting moments of inertia, the K.E. of the system is thus

$$T = 2 \times \tfrac{1}{2} m(\dot{x}^2 + \dot{y}^2 + x^2\dot{\phi}^2) + \tfrac{1}{2} M(2\dot{y})^2$$
$$= \underline{ml^2(\dot{\theta}^2 + \dot{\phi}^2 \sin^2 \theta) + 2Ml^2\dot{\theta}^2 \sin^2 \theta} \ .$$

The virtual work function is

$$\delta W = 2mg \ \delta y + Mg\delta(2y) + G\delta\phi$$
$$= \underline{-2(m+M)gl \sin \theta \ \delta\theta + G \ \delta\phi} \ .$$

Thus, taking θ, ϕ as generalized co-ordinates, the generalized forces associated with them are

$$Q_\theta = -2(m+M)gl \sin \theta \ , \qquad Q_\phi = G \ .$$

Lagrange's equations are

$$\left. \begin{aligned} (d/dt)(\partial T/\partial\dot{\theta}) - \partial T/\partial\theta &= Q_\theta \ , \\ (d/dt)(\partial T/\partial\dot{\phi}) - \partial T/\partial\phi &= Q_\phi \ , \end{aligned} \right\}$$

i.e. $\quad (d/dt)\{2l^2\dot{\theta}(m + 2M \sin^2 \theta)\} - 2ml^2\dot{\phi}^2 \sin \theta \cos \theta - 4Ml^2\dot{\theta}^2 \sin \theta \cos \theta$

$$= -2(m+M)gl \sin \theta \ ,$$
$$d/dt\{2ml^2\dot{\phi} \sin^2 \theta\} = G \ .$$

When rotation is steady and $\dot{\phi} = \omega$ (= const.) these give

$$l\ddot{\theta}(m + 2M \sin^2 \theta) + 2Ml \sin \theta \cos \theta \ \dot{\theta}^2 - m\omega^2 l \sin \theta \cos \theta = -(m+M)g \sin \theta, \tag{1}$$

$$4ml^2\omega\dot{\theta} \sin \theta \cos \theta = G \ . \tag{2}$$

Let $\theta = \alpha$(= const.) define a steady condition of 'dynamic equilibrium' so that $\dot{\theta} = 0 = \ddot{\theta}$. Then (2) shows $G = 0$ and (1) gives, assuming $\alpha \neq 0$,

$$\underline{\cos \alpha = (M+m)g/m\omega^2 l, \text{ if } \omega^2 > (M+m)g/ml} \ . \tag{3}$$

To study small oscillations about $\theta = \alpha$, put $\theta = \alpha + \epsilon$ in (1) to give

$$[m + 2M \sin^2(\alpha + \epsilon)]l\ddot{\epsilon} + Ml \sin(2\alpha + 2\epsilon)\dot{\epsilon}^2 - \tfrac{1}{2} ml\omega^2 \sin(2\alpha + 2\epsilon)$$
$$= -(M+m)g \sin(\alpha + \epsilon).$$

Retaining only first order terms,

$$(m + 2M \sin^2 \alpha)l\ddot{\epsilon} + [(M+m)g \cos \alpha - ml\omega^2 \cos 2\alpha] \ \epsilon = 0 \ .$$

Using the value of ω^2 given by (3), we find on simplification that

$$\ddot{\epsilon} = -n^2\epsilon \ ,$$

where $n^2 = \dfrac{(M+m)g \sin^2 \alpha}{(m + 2M \sin^2 \alpha)l \cos \alpha}$. This shows the motion is S.H.M. of period $2\pi/n$.

10.7 CASE OF CONSERVATIVE FORCES

When the forces are *conservative* and the system is specified by the generalized co-ordinates $q_j (j=1, \ldots, n)$, we can find a potential function $V=V(q_1, \ldots, q_n)$ and such that $\delta W=-\delta V$, whence

$$Q_1\delta q_1+\ldots+Q_j\delta q_j+\ldots+Q_n\delta q_n=-(\delta q_1\partial V/\partial q_1+\ldots+\delta q_j\partial V/\partial q_j+\ldots+\delta q_n\partial V/\partial q_n)\ .$$

Hence $Q_j=-\partial V/\partial q_j (j=1, \ldots, n)$, so that Lagrange's equations for a conservative holonomic dynamical system become

$$(d/dt)(\partial T/\partial \dot{q}_j)-\partial T/\partial q_j=-\partial V/\partial q_j \qquad (j=1, \ldots, n)$$

i.e.

$$(d/dt)(\partial L/\partial \dot{q}_j)-\partial L/\partial q_j=0 \qquad (j=1, \ldots, n)\ , \qquad (1)$$

where

$$L=T-V\ . \qquad (2)$$

L is called the *Lagrangian* or the *kinetic potential.*

Example

A uniform rod OA of length $2a$ and mass m is smoothly pivoted at one end to a fixed point O. The rod makes an angle θ with the downward vertical OZ and the plane AOZ makes an angle ϕ with a fixed vertical plane. A bead of mass λm slides on the smooth rod and is connected to O by a light elastic string of modulus nmg and natural length a. Show that the kinetic energy T of the system is given by

$$2T=\tfrac{4}{3}ma^2(\dot{\theta}^2+\dot{\phi}^2\sin^2\theta)+\lambda m(\dot{x}^2+x^2\dot{\theta}^2+x^2\dot{\phi}^2\sin^2\theta)$$

where x is the stretched length of the string, and derive the equations of motion.

Show that, if a steady motion with $\theta=\pi/3$ and $x=4a/3$ is possible, then $\dot{\phi}^2=3g/2a$ and $n=6\lambda$. (L.U.)

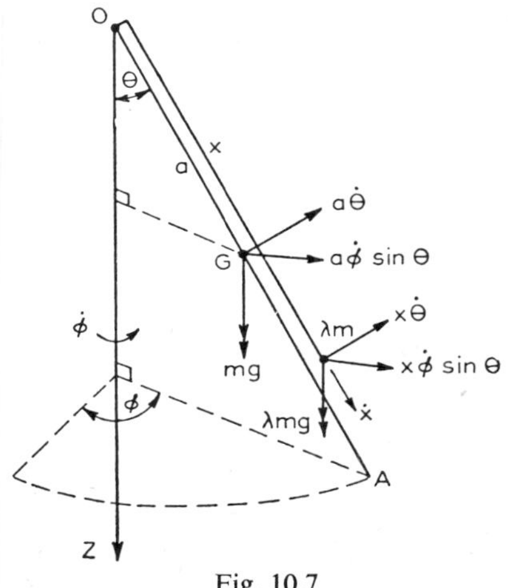

Fig. 10.7

K.E. of rod $=\frac{1}{2}m(a^2\dot{\theta}^2+a^2\dot{\phi}^2\sin^2\theta)+\frac{1}{2}(\frac{1}{3}ma^2)[\dot{\theta}^2+(\dot{\phi}\sin\theta)^2]$, since the rod has spin components $[\dot{\theta}, \dot{\phi}\sin\theta, \dot{\phi}\cos\theta]$ about the P.As. and the corresponding principal M.Is. are $\frac{1}{3}ma^2$, $\frac{1}{3}ma^2$, 0.

K.E. of particle $=\frac{1}{2}\lambda m[\dot{x}^2+(x\dot{\phi}\sin\theta)^2+(x\dot{\theta})^2]$.

$$\therefore\ \underline{T=\tfrac{2}{3}ma^2(\dot{\theta}^2+\dot{\phi}^2\sin^2\theta)+\tfrac{1}{2}\lambda m(\dot{x}^2+x^2\dot{\theta}^2+x^2\dot{\phi}^2\sin^2\theta)}\ .$$

The gravitational P.E. (referred to the horizontal through O as zero level) $=-mga\cos\theta-\lambda mgx\cos\theta$.
Since the stretch of the string is $(x-a)$, its elastic P.E. $=nmg(x-a)^2/2a$.
Thus the total P.E. is

$$\underline{V=-mga\cos\theta-\lambda mgx\cos\theta+nmg(x-a)^2/2a}\ .$$

$$\begin{aligned}\therefore\ L&=T-V\\&=\tfrac{2}{3}ma^2(\dot{\theta}^2+\dot{\phi}^2\sin^2\theta)+\tfrac{1}{2}\lambda m(\dot{x}^2+x^2\dot{\theta}^2+x^2\dot{\phi}^2\sin^2\theta)\\&\quad+mga\cos\theta+\lambda mgx\cos\theta-nmg(x-a)^2/2a\ .\end{aligned}$$

Lagrange's equations for the generalized co-ordinates x, θ, ϕ are

$$d/dt(\partial L/\partial\dot{q})-\partial L/\partial q=0 \qquad (q=x,\theta,\phi)\ ,$$

i.e.

$$\lambda\ddot{x}-\lambda x(\dot{\theta}^2+\dot{\phi}^2\sin^2\theta)-\lambda g\cos\theta+ng(x-a)/a=0\ , \qquad (1)$$

$$\tfrac{4}{3}a^2\ddot{\theta}+\lambda x^2\ddot{\theta}+2\lambda x\dot{x}\dot{\theta}-\tfrac{4}{3}a^2\dot{\phi}^2\sin\theta\cos\theta-\lambda x^2\dot{\phi}^2\sin\theta\cos\theta$$
$$+ga\sin\theta+\lambda gx\sin\theta=0\ , \qquad (2)$$

$$d/dt\{\tfrac{4}{3}a^2\dot{\phi}\sin^2\theta+\lambda x^2\dot{\phi}^2\sin^2\theta\}=0\ . \qquad (3)$$

For steady motion with $\theta=\pi/3$, $x=4a/3$, (3) is satisfied and (1), (2) give

$$-\lambda a\dot{\phi}^2-\tfrac{1}{2}\lambda g+\tfrac{1}{3}ng=0\ , \qquad (4)$$

$$-\tfrac{2}{3}a\dot{\phi}^2-\tfrac{8}{9}\lambda a\dot{\phi}^2+g+\tfrac{4}{3}\lambda g=0\ . \qquad (5)$$

From (5), $\dot{\phi}^2=3g/2a$ and (4) then gives $n=6\lambda$.
(*N.B.* In this problem, the internal elastic forces do work in displacement).

10.8 GENERALIZED COMPONENTS OF MOMENTUM AND IMPULSE

In this section, we introduce the notions of generalized impulses and momenta to complete the process which has already been undertaken for displacements, velocities and forces.

To generalize our notions of momentum, let us first consider the translation of a rigid body without rotation. If the centroid is at (x, y, z) at time t, then taking M to be the mass,

$$T=\tfrac{1}{2}M(\dot{x}^2+\dot{y}^2+\dot{z}^2)\ .$$

Defining $p_x=\partial T/\partial\dot{x}$, etc., we have

$$p_x=\partial T/\partial\dot{x}=M\dot{x}\ ;\quad p_y=\partial T/\partial\dot{y}=M\dot{y}\ ;\quad p_z=\partial T/\partial\dot{z}=M\dot{z}\ .$$

Thus $[p_x, p_y, p_z]$ are the components of linear momentum in the x-, y-, z- directions.

When a rigid body rotates about an axis l through its centroid which is fixed, if I is the moment of inertia about l and θ the angle turned through in time t, then $T=\frac{1}{2}I\dot{\theta}^2$. Defining $p_\theta=\partial T/\partial\dot{\theta}$, we have

$$p_\theta=\partial T/\partial\dot{\theta}=I\dot{\theta} \ ,$$

i.e. p_θ is the angular momentum about l.

If we now have a holonomic system specified by generalized co-ordinates $q_j(j=1, \ldots, n)$ at time t, we can evaluate the kinetic energy $T=T(q_1 \ldots, q_n; \dot{q}_1, \ldots, \dot{q}_n; t)$ and then form the n quantities $p_j=\partial T/\partial\dot{q}_j$ $(j=1, \ldots, n)$. These are called the *generalized components of momentum* of the system.

Now suppose $[X, Y, Z]$ to be the components of force acting at a point of a body. Then the impulses of these forces over a time interval $0<t<\tau$ are

$$\int_o^\tau X\,dt \ , \qquad \int_o^\tau Y\,dt \ , \qquad \int_o^\tau Z\,dt \ .$$

Further, suppose the forces $[X, Y, Z]$ become very large and the time interval τ very small in such a way that

$$\lim_{\substack{x\to\infty,\\ \tau\to 0}} \int_o^\tau X\,dt=J_x=\text{finite, etc.}$$

Then J_x is termed an *impulse*. By analogy, if for generalized forces $Q_j(j=1, \ldots, n)$ of a system,

$$\lim_{\substack{Q_j\to\infty,\\ \tau\to 0}} \int_o^\tau Q_j dt=J_j=\text{finite } (j=1, \ldots, n) \ ,$$

then the n quantities $J_j(j=1, \ldots, n)$ are termed *generalized impulses*. Since

$$\delta W=\sum_{j=1}^{n} Q_j\delta q_j \ ,$$

$$\int_o^\tau \delta W\,dt=\sum_{j=1}^{n}\left\{\delta q_j \int_o^\tau Q_j\,dt\right\} ,$$

if the above double limit exists, then

$$\delta U=\sum_{j=1}^{n} J_j\,\delta q_j \ ,$$

where δU is the *impulsive virtual work function* given by

$$\delta U=\lim_{Q_i\to\infty,\ \tau\to 0} \int_o^\tau \delta W\,dt \qquad (j=1, \ldots, n) \ .$$

δU is constructed from those impulses which do impulsive virtual work in a virtual displacement. Impulses which do no such impulsive work need not be included in δU.

10.9 LAGRANGE'S EQUATIONS FOR IMPULSIVE FORCES

Since $p_i = \partial T/\partial \dot{q}_i$, Lagrange's equations for a holonomic system are

$$dp_i/dt - \partial T/\partial q_i = Q_i \qquad (j=1,\ldots,n)\ .$$

Integrating these equations from $t=0$ to $t=\tau$ we obtain

$$(p_j)_{t=\tau} - (p_j)_{t=0} = \int_o^\tau \frac{\partial T}{\partial q_j}\,dt + \int_o^\tau Q_j dt \qquad (j=1,\ldots,n)\ .$$

Now let $Q_j \to \infty$, $\tau \to 0$ in such a way that $\int_o^\tau Q_j dt \to J_j =$ finite $(j=1,\ldots,n)$.

As the co-ordinates q_j do not change abruptly,

$$\lim_{\tau\to 0} \int_0^\tau \partial T/\partial q_j\, dt = 0 \qquad (j=1,\ldots,n)\ .$$

Writing $\Delta p_j = \lim_{\tau\to 0}[(p_j)_{t=\tau} - (p_j)_{t=0}]$, we thus obtain *Lagrange's equations in impulsive form*:

$$\Delta p_j = J_j \qquad (j=1,\ldots,n)\ .$$

The statement of these equations is: *the generalized momentum increment is equal to the generalized impulsive force associated with each generalized co-ordinate.*

Example 1

Two uniform rods AB, BC of masses m_1, m_2 and lengths $2a$, $2b$ are smoothly hinged at B and initially they lie at rest on a smooth table and in a straight line. AB receives a blow of impulse I at A perpendicular to AB. Construct the equations of motion of the system just after impact.

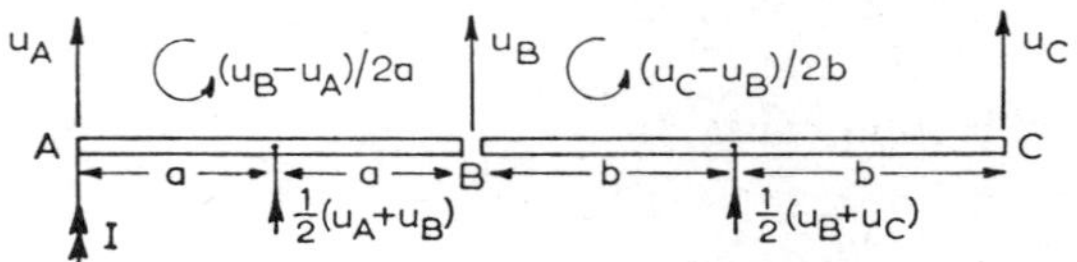

Fig. 10.8

Let u_A, u_B, u_C be the initial velocities of A, B, C perpendicular to the rods. The velocities of the centroids of AB, BC are $\frac{1}{2}(u_A+u_B)$, $\frac{1}{2}(u_B+u_C)$, and the anti-clockwise angular velocities of the rods are $(u_B-u_A)/2a$, $(u_C-u_B)/2b$. Hence the K.E. of the system generated by the blow is

$$T=\tfrac{1}{2}m_1\{\tfrac{1}{2}(u_A+u_B)\}^2+\tfrac{1}{2}(\tfrac{1}{3}m_1a^2)\{(u_B-u_A)/2a\}^2$$
$$+\tfrac{1}{2}m_2\{\tfrac{1}{2}(u_B+u_C)\}^2+\tfrac{1}{2}(\tfrac{1}{3}m_2b^2)\{(u_C-u_B)/2b\}^2$$
$$=\tfrac{1}{6}m_1(u_A{}^2+u_Au_B+u_B{}^2)+\tfrac{1}{6}m_2(u_B{}^2+u_Bu_C+u_C{}^2)\ .$$

Let A, B, C undergo virtual displacements δx_A, δx_B, δx_C in the direction of I just after its application so that $u_A=\dot{x}_A$, etc. The impulsive virtual work function is

$$\delta U=I\delta x_A=J_A\delta x_A+J_B\delta x_B+J_C\delta x_C\ ,$$

where J_A, J_B, J_C are the generalized impulses associated with co-ordinates x_A, x_B, x_C. Thus $J_A=I$, $J_B=0=J_C$.

The generalized momenta are

$$\left.\begin{aligned} p_A&=\partial T/\partial u_A=\tfrac{1}{6}m_1(2u_A+u_B)\ ,\\ p_B&=\partial T/\partial u_B=\tfrac{1}{6}m_1(u_A+2u_B)+\tfrac{1}{6}m_2(2u_B+u_C)\ ,\\ p_C&=\partial T/\partial u_C=\tfrac{1}{6}m_2(u_B+2u_C)\ .\end{aligned}\right\}$$

Lagrange's equations for impulsive motion are

$$\Delta p_A=J_A,\ \text{etc.}$$

i.e.

$$\left.\begin{aligned} &\tfrac{1}{6}m_1(2u_A+u_B)=I\ ,\\ &\tfrac{1}{6}m_1(u_A+2u_B)+\tfrac{1}{6}m_2(2u_B+u_C)=0\ ,\\ &\tfrac{1}{6}m_2(u_B+2u_C)=0\ .\end{aligned}\right\}$$

Thus the velocities are determinable. Observe that the impulsive action at B has not been introduced.

Example 2

AB, BC, CD are three equal rods freely jointed together to form the three sides of a square, the rods lying on a smooth horizontal table. The end A is pivoted to the table. If the end D receives a blow in the direction $\overline{AD}$, show that the initial angular velocities of the rods are as 1 : 0 : 11.

Take axes AX, AY as in Fig. 10.9 and suppose AB, BC, CD make angles θ, ϕ, ψ with $\overline{AX}$. Let m be the mass and $2a$ the length of each rod. The centroids of AB, BC, CD have co-ordinates $(a\cos\theta,\ a\sin\theta)$, $(2a\cos\theta+a\cos\phi,\ 2a\sin\theta+a\sin\phi)$, $(2a\cos\theta+2a\cos\phi-a\cos\psi,\ 2a\sin\theta+2a\sin\phi-a\sin\psi)$ and so their velocity components are respectively

$$[-a\dot{\theta}\sin\theta,\ a\dot{\theta}\cos\theta],\quad [-2a\dot{\theta}\sin\theta-a\dot{\phi}\sin\phi,\ 2a\dot{\theta}\cos\theta+a\dot{\phi}\cos\phi]\ ,$$
$$[-2a\dot{\theta}\sin\theta-2a\dot{\phi}\sin\phi+a\dot{\psi}\sin\psi,\ 2a\dot{\theta}\cos\theta+2a\dot{\varphi}\cos\phi-a\dot{\phi}\cos\psi]\ .$$

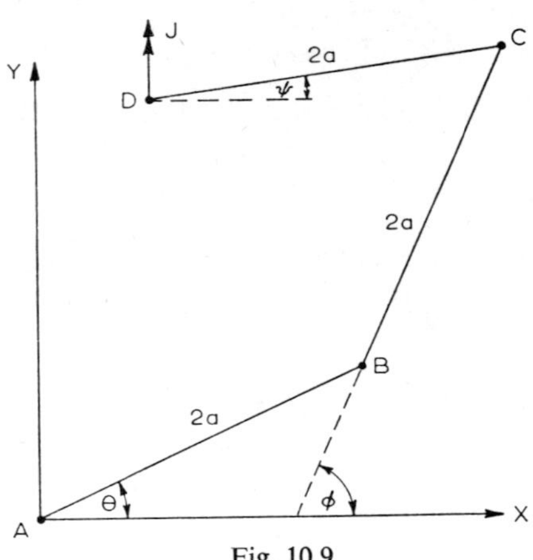

Fig. 10.9

Thus the K.E. of the system is

$$T=\tfrac{1}{2}ma^2\dot{\theta}^2+\tfrac{1}{2}m\{(-2a\dot{\theta}\sin\theta-a\dot{\phi}\sin\phi)^2+(2a\dot{\theta}\cos\theta+a\dot{\phi}\cos\phi)^2\}$$
$$+\tfrac{1}{2}m\{(-2a\dot{\theta}\sin\theta-2a\dot{\phi}\sin\phi+a\dot{\psi}\sin\psi)^2+(2a\dot{\theta}+2a\dot{\phi}\cos\phi-$$
$$a\dot{\psi}\cos\psi)^2\}+\tfrac{1}{6}ma^2(\dot{\theta}^2+\dot{\phi}^2+\dot{\psi}^2)\ ;$$
$$\therefore\ T=\tfrac{2}{3}ma^2\{7\dot{\theta}^2+4\dot{\phi}^2+\dot{\psi}^2+9\dot{\theta}\dot{\phi}\cos(\phi-\theta)-3\dot{\theta}\dot{\psi}\cos(\psi-\theta)$$
$$-3\dot{\phi}\dot{\psi}\cos(\psi-\phi)\}\ .$$

Thus the generalized momenta are

$$\left.\begin{aligned}p_\theta&=\partial T/\partial\dot{\theta}=\tfrac{2}{3}ma^2\{14\dot{\theta}+9\dot{\phi}\cos(\phi-\theta)-3\dot{\psi}\cos(\psi-\theta)\}\ ,\\ p_\phi&=\partial T/\partial\dot{\phi}=\tfrac{2}{3}ma^2\{8\dot{\phi}+9\dot{\theta}\cos(\phi-\theta)-3\dot{\psi}\cos(\psi-\phi)\}\ ,\\ p_\psi&=\partial T/\partial\dot{\psi}=\tfrac{2}{3}ma^2\{2\dot{\psi}-3\dot{\theta}\cos(\psi-\theta)-3\dot{\phi}\cos(\psi-\phi)\}\ .\end{aligned}\right\}$$

The y-co-ordinate of D is $2a(\sin\theta+\sin\phi-\sin\psi)$ and the impulsive virtual work function is

$$\begin{aligned}\delta U&=J\delta\{2a(\sin\theta+\sin\phi-\sin\psi)\}\\&=2aJ(\cos\theta\,\delta\theta+\cos\phi\,\delta\phi-\cos\psi\,\delta\psi)\ .\end{aligned}$$

Hence the generalized impulsive forces are

$$J_\theta=2aJ\cos\theta\ ,\quad J_\phi=2aJ\cos\phi\ ,\quad J_\psi=-2aJ\cos\psi\ .$$

Lagrange's equations of impulsive motion are

$$\Delta p_\theta=J_\theta,\ \text{etc.,}$$

with $\theta=0$, $\phi=\tfrac{1}{2}\pi$, $\psi=0$ and $\dot{\theta}=\omega_1$, $\dot{\phi}=\omega_2$, $\dot{\psi}=\omega_3$. Then

$$\left.\begin{aligned}\tfrac{2}{3}ma^2(14\omega_1-3\omega_3)&=2aJ\ ,\\ \tfrac{2}{3}ma^2(8\omega_2)&=0\ ,\\ \tfrac{2}{3}ma^2(2\omega_3-3\omega_1)&=-2aJ\ .\end{aligned}\right\}$$

Solving we find

$$\omega_1=-\tfrac{3}{19}aJ\ ,\quad \omega_2=0\ ,\quad \omega_3=-\tfrac{33}{19}aJ\ .$$

10.10 KINETIC ENERGY AS A QUADRATIC FUNCTION OF VELOCITIES

If at time t the position vector of the ith particle (mass m_i) of a holonomic system is $\mathbf{r}_i$, then the K.E. is

$$T=\tfrac{1}{2}\sum_{i=1}^{N}m_i\dot{\mathbf{r}}_i^{\,2}\ ,$$

where N is the number of particles. Suppose the system to be holonomic and specified by the n generalized co-ordinates $q_j(j=1,\ldots,n)$. Then $\mathbf{r}_i=\mathbf{r}_i(q_1,q_2,\ldots,q_n;\ t)\ \ (i=1,\ldots,N)$, and $\dot{\mathbf{r}}_i=\dot{q}_1\partial\mathbf{r}_i/\partial q_1+\dot{q}_2\partial\mathbf{r}_i/\partial q_2+\ldots+\dot{q}_n\partial\mathbf{r}_i/\partial q_n+\partial\mathbf{r}_i/\partial t\ (i=1,\ldots,N)$.

$$\begin{aligned}\therefore\ T&=\tfrac{1}{2}\sum_{i=1}^{N}m_i(\dot{q}_1\partial\mathbf{r}_i/\partial q_1+\dot{q}_2\partial\mathbf{r}_i/\partial q_2+\ldots+\dot{q}_n\partial\mathbf{r}_i/\partial q_n+\partial\mathbf{r}_i/\partial t)^2\\ &=\tfrac{1}{2}[(a_{11}\dot{q}_1^{\,2}+a_{22}\dot{q}_2^{\,2}+\ldots+a_{nn}\dot{q}_n^{\,2}+2a_{12}\dot{q}_1\dot{q}_2+\ldots)\\ &\quad+2(a_1\dot{q}_1+a_2\dot{q}_2+\ldots+a_n\dot{q}_n)+a]\ ,\end{aligned}$$

where

$$\left.\begin{aligned}a_{rs}&=\sum_{i=1}^{N}m_i(\partial\mathbf{r}_i/\partial q_r).(\partial\mathbf{r}_i/\partial q_s)\ (s\geqslant r)\ ,\\ a_r&=\sum_{i=1}^{N}m_i(\partial\mathbf{r}_i/\partial q_r).(\partial\mathbf{r}_i/\partial t)\ ,\\ a&=\sum_{i=1}^{N}m_i(\partial\mathbf{r}_i/\partial t)^2\ .\end{aligned}\right\}$$

Thus we see that T is a *quadratic function of the generalized velocities.*

The case when time t is not explicitly involved is of considerable importance. We now have $\partial\mathbf{r}_i/\partial t=\mathbf{0}\ (i=1,\ldots,N)$. Thus the K.E. assumes the form of a *homogeneous quadratic function of the generalized velocities*:

$$\begin{aligned}T&=\tfrac{1}{2}(a_{11}\dot{q}_1^{\,2}+a_{22}\dot{q}_2^{\,2}+\ldots+a_{nn}\dot{q}_n^{\,2}+2a_{12}\dot{q}_1\dot{q}_2+\ldots)\ .\\ &=\tfrac{1}{2}\sum_{s=1}^{n}\sum_{r=1}^{n}a_{rs}q_rq_s\ ,\ \text{where } a_{rs}=a_{sr}\ .\end{aligned}$$

In this case we have, using Euler's theorem for homogeneous functions,

$$\dot{q}_1 \partial T/\partial \dot{q}_1 + \dot{q}_2 \partial T/\partial \dot{q}_2 + \ldots + \dot{q}_n \partial T/\partial \dot{q}_n = 2T$$

or

$$\dot{q}_1 p_1 + \dot{q}_2 p_2 + \ldots + \dot{q}_n p_n = 2T \; .$$

Example

The ends of a uniform rigid rod of mass m are moving with velocities $\mathbf{u}$, $\mathbf{v}$. Prove that the K.E. of the rod is $\frac{1}{6}m(\mathbf{u}^2 + \mathbf{u}.\mathbf{v} + \mathbf{v}^2)$.

Three uniform rods OA, AB, BC each of mass m, are smoothly jointed together at A, B and hang in equilibrium under gravity from a fixed smooth joint at O. An impulsive couple G is applied to the rod AB in a vertical plane through O. Prove that the initial kinetic energy of the system is $57G^2/26ml^2$, where l is the length of AB. (L.U.)

The first part of this problem has been solved in Chapter 7, Section 7.8.

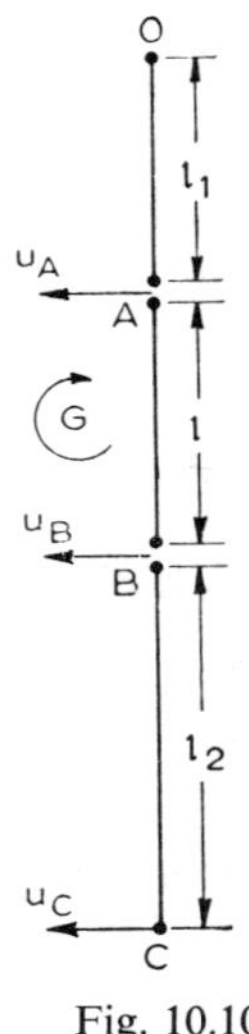

Fig. 10.10

Fig. 10.10 shows the system of rods, u_A, u_B, u_C being the velocities of A, B, C and δx_A, δx_B, δx_C, their corresponding virtual displacements. Then the K.E. is given by

$$T = \tfrac{1}{6}mu_A{}^2 + \tfrac{1}{6}m(u_A{}^2 + u_A u_B + u_B{}^2) + \tfrac{1}{6}m(u_B{}^2 + u_B u_C + u_C{}^2)$$
$$= \tfrac{1}{6}m(2u_A{}^2 + 2u_B{}^2 + u_C{}^2 + u_A u_B + u_B u_C) \; .$$

The generalized components of momentum are

$$\left.\begin{aligned} p_A &= \partial T/\partial u_A = \tfrac{1}{6}m(4u_A + u_B) \; , \\ p_B &= \partial T/\partial u_B = \tfrac{1}{6}m(u_A + 4u_B + u_C) \; , \\ p_C &= \partial T/\partial u_C = \tfrac{1}{6}m(u_B + 2u_C) \; . \end{aligned}\right\}$$

Since the virtual angular displacement of the rod AB in the direction of the impulsive couple G is $(\delta x_B - \delta x_A)/l$, the impulsive virtual work function is

$$\delta U = G(\delta x_B - \delta x_A)/l\ .$$

Thus the generalized impulsive forces are

$$J_A = -G/l\ ,\quad J_B = G/l\ ,\quad J_C = 0\ .$$

Lagrange's equations for impulsive motion are

$$\left.\begin{aligned}\tfrac{1}{6}m(4u_A + u_B) &= -G/l\ ,\\ \tfrac{1}{6}m(u_A + 4u_B + u_C) &= G/l\ ,\\ \tfrac{1}{6}m(u_B + 2u_C) &= 0\ .\end{aligned}\right\}$$

$$\therefore\ u_A = -27G/13ml\ ,\quad u_B = 30G/13ml\ ,\quad u_C = -15G/13ml\ .$$

Since T is a homogeneous quadratic function of the velocities,

$$\begin{aligned}2T &= p_A u_A + p_B u_B + p_C u_C\\ &= (-G/l)(-27G/13ml) + (G/l)(30G/13ml) + 0\ ;\end{aligned}$$

$$\therefore\ \underline{T = 57G^2/26ml^2}\ .$$

10.11 EQUILIBRIUM CONFIGURATIONS FOR CONSERVATIVE HOLONOMIC DYNAMICAL SYSTEMS

For a conservative holonomic dynamical system specified by generalized co-ordinates $q_i(j = 1, \ldots, n)$ with time explicitly absent, the K.E. has been shown in the last section to be a homogeneous quadratic function of velocities of the form

$$T = \tfrac{1}{2}\sum_{r=1}^{n}\sum_{s=1}^{n} a_{rs}\,\dot{q}_r\dot{q}_s\ ,$$

where $a_{rs} = a_{sr}$ and these coefficients are each functions of $q_1 \ldots, q_n$ for all r, s. Further the P.E. is of the form

$$V = V(q_1, q_2, \ldots, q_n)\ .$$

Thus Lagrange's equations for such a system are

$$d/dt(\partial T/\partial \dot{q}_j) - \partial T/\partial q_j = -\partial V/\partial q_j\ (j = 1, \ldots, n)\ . \qquad (1)$$

Now $\partial T/\partial q_j = \tfrac{1}{2}\sum_{r=1}^{n}\sum_{s=1}^{n}\{(\partial a_{rs}/\partial q_j)\dot{q}_r\dot{q}_s\}$. Also,

$$\begin{aligned}\partial T/\partial \dot{q}_j &= \tfrac{1}{2}\sum_{r=1}^{n}\sum_{s=1}^{n} a_{rs}(\dot{q}_r\partial\dot{q}_s/\partial\dot{q}_j + \dot{q}_s\partial\dot{q}_r/\partial\dot{q}_j)\\ &= \tfrac{1}{2}\sum_{r=1}^{n} a_{rj}\dot{q}_r + \tfrac{1}{2}\sum_{s=1}^{n} a_{js}\dot{q}_s = \sum_{r=1}^{n} a_{rj}\dot{q}_r\ ;\end{aligned}$$

$$\therefore\ (d/dt)(\partial T/\partial \dot{q}_j) = \sum_{r=1}^{n} \{a_{rj}\ddot{q}_r + \dot{q}_r(\dot{q}_1 \partial a_{rj}/\partial q_1 + \ldots + \dot{q}_n \partial a_{rj}/\partial q_n)\}\ .$$

Now suppose the system is in equilibrium for a configuration specified by the co-ordinates $q_j = \alpha_j (j = 1, \ldots, n)$. For these values $\dot{q}_j = 0$, $\ddot{q}_j = 0$ $(j = 1, \ldots, n)$.

Hence it follows that $\partial T/\partial q_j = 0$, $(d/dt)(\partial T/\partial \dot{q}_j) = 0$, $(j = 1, \ldots, n)$ for these values of the co-ordinates. Thus Lagrange's equations (1) reduce to

$$\partial V/\partial q_j = 0 \qquad (j = 1, \ldots, n) \tag{2}$$

for $q_j = \alpha_j (j = 1, \ldots, n)$. It follows that *in order to find the equilibrium configurations for a system specified by n generalized co-ordinates* $q_j (j = 1, \ldots, n)$ *we express the potential energy V as a function of these co-ordinates and then solve the n equations* (2) *for the co-ordinates specifying the equilibrium configuration or configurations.*

Having found the configurations of equilibrium according to the above rule it is important to investigate whether these positions are *stable* or *unstable*. To this end, bearing in mind that for a conservative system $T + V = \text{const.}$, we know that if small perturbations are made from an equilibrium position, T must decrease or increase according to whether the position is stable or unstable, i.e. V must increase or decrease, respectively. *Thus at a stable position of equilibrium V is a minimum and at an unstable position V may be a maximum.*

Example 1

The ends of a uniform rod AB, of length $2a \cos 15°$ and weight W, are constrained to slide on a smooth circular wire of radius a, fixed with its plane vertical. The end A is connected by an elastic string, of natural length a and modulus $\frac{1}{2}W$, to the highest point of the wire. If θ is the angle which the perpendicular bisector of the rod makes with the downward vertical, show that the potential energy V is given by

$$V = -\tfrac{1}{2}Wa\{\cos(\theta - 75°) + 2\cos\tfrac{1}{2}(\theta + 75°)\} + \text{const.}$$

Verify that $\theta = 25°$ defines a position of equilibrium and investigate its stability. (L.U.)

Take C to be the highest point (Fig. 10.11).
Then $\angle ABO = \angle BAO = 15°$, $\angle AOB = 150°$,
$\angle OCA = \angle OAC = \frac{1}{2}\angle DOA = \frac{1}{2}(\theta + 75°)$.

Taking the horizontal through O as zero level of P.E., the depth of the centroid G of AB below this level is $OG \cos\theta$ or $a \sin 15° \cos\theta$. Hence the gravitational P.E. is $-Wa \sin 15° \cos\theta$.

Now $AC = 2a \cos \angle OCA = 2a \cos\frac{1}{2}(\theta + 75°)$. The stretch of the string is $2a\cos\frac{1}{2}(\theta + 75°) - a$. Hence the elastic P.E. of it is

$$\tfrac{1}{2}W \times a^2[2\cos\tfrac{1}{2}(\theta + 75°) - 1]^2/(2a) = \tfrac{1}{4}Wa[2\cos\tfrac{1}{2}(\theta + 75°) - 1]^2\,.$$

$$\therefore\ V = \tfrac{1}{4}Wa[2\cos\tfrac{1}{2}(\theta + 75°) - 1]^2 - Wa\sin 15°\cos\theta$$

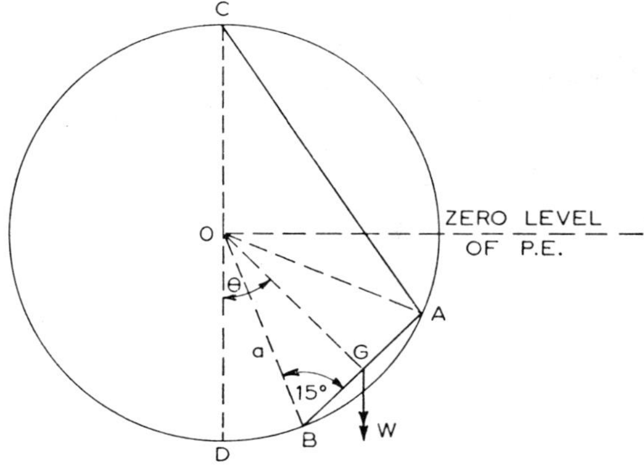

Fig. 10.11

$$= Wa[\cos^2\tfrac{1}{2}(\theta+75°) - \cos\tfrac{1}{2}(\theta+75°)] - Wa \sin 15° \cos\theta + \text{const.}$$
$$= \tfrac{1}{2}Wa[1+\cos(\theta+75°) - 2\cos\tfrac{1}{2}(\theta+75°) - \sin(15°+\theta) - \sin(15°-\theta)] + \text{const.}$$
$$= \tfrac{1}{2}Wa[\cos(\theta+75°) - 2\cos\tfrac{1}{2}(\theta+75°) - \cos(75°-\theta) - \cos(75°+\theta)] + \text{const.}$$
$$\therefore\ V(\theta) = -\tfrac{1}{2}Wa\{\cos(\theta-75°) + 2\cos\tfrac{1}{2}(\theta+75°)\} + \text{const.}, \qquad (1)$$
$$\therefore\ V'(\theta) = \tfrac{1}{2}Wa\{\sin(\theta-75°) + \sin\tfrac{1}{2}(\theta+75°)\}, \qquad (2)$$
$$V''(\theta) = \tfrac{1}{2}Wa\{\cos(\theta-75°) + \tfrac{1}{2}\cos\tfrac{1}{2}(\theta+75°)\}. \qquad (3)$$
$$\therefore\ V'(25°) = \tfrac{1}{2}Wa\{\sin(-50°) + \sin 50°\} = 0.$$
$\therefore\ \theta = 25°$ is an equilibrium configuration.
$$V''(25°) = \tfrac{3}{4}Wa\cos 50° > 0.$$

$\therefore$ V is a minimum and the configuration $\theta = 25°$ is one of stable equilibrium.

Example 2—Mathematical Model for Problem of Industrial Location.

Where should the warehouse servicing all the shops in a supermarket chain of stores be placed to secure performance at minimum cost? Under certain assumptions to be formulated this problem may be solved by means of a suitable analogy in mechanics involving the minimization of the potential energy of a system of particles.

Supposing the problem to be two-dimensional, let us introduce a system of rectangular Cartesian co-ordinates in which the warehouse has the undetermined co-ordinates (x, y) and each shop has the fixed co-ordinates (x_i, y_i), where $i = 1, 2, \ldots, n$, so that there are n shops. Then the distance d_i from the warehouse to the ith shop is $d_i = \{(x-x_i)^2 + (y-y_i)^2\}^{\frac{1}{2}}$. We assume, in accordance with managerial experience, that the cost of servicing the ith shop from the warehouse in proportional to: (i) the distance d_i; (ii) the demand w_i per unit time from this shop. Then the cost for the ith shop is $\alpha w_i d_i$ where α is a constant of proportionality. Hence the total cost C for the n shops is

$$C = \alpha \sum_{i=1}^{n} w_i d_i \ .$$

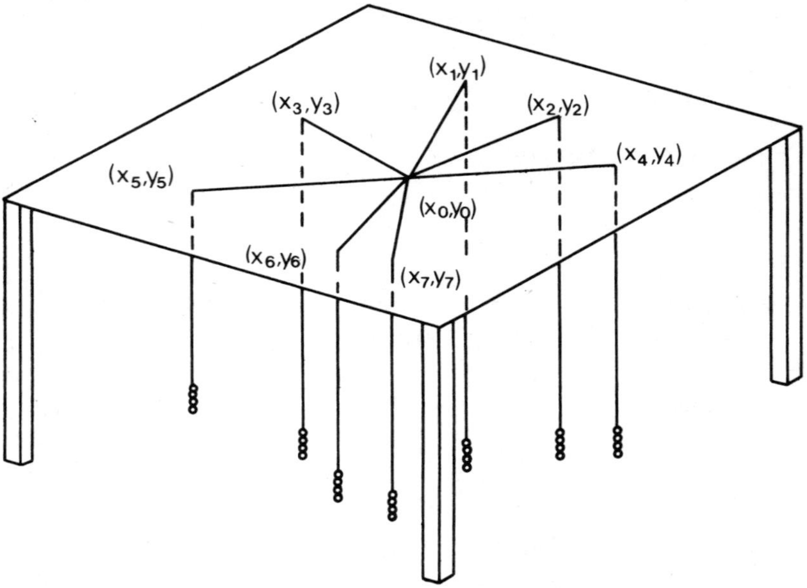

Fig. 10.12—Mechanical analogue.

Fig. 10.12 shows a suitable mechanical analogue which gives the desired solution. A map of the entire area under consideration is pasted on to a horizontal hardboard and small holes drilled at shop locations. Inextensible strings are passed through each hole, connected at (x, y) on the map and weights kw_i $(i = 1, 2, \ldots, n)$ hung from their free ends, where k is a positive constant. On release, the system will move into its equilibrium position in which its potential energy V will be a minimum. Referred to the hardboard as zero level of potential energy, if l_i is the length of the string supporting kw_i, then since w_i is distant $(l_i - d_i)$ below the zero level,

$$V = -\sum_{i=1}^{n} kw_i(l_i - d)$$

$$= -k \sum_{i=1}^{n} w_i l_i + kC/\alpha \ .$$

Since the first term on the right is constant and k, α are positive constants, the minimum potential energy V corresponds to the minimum cost C. Thus the management is able to see the solution immediately.

10.12 THEORY OF SMALL OSCILLATIONS OF CONSERVATIVE HOLONOMIC DYNAMICAL SYSTEMS

In the notation of the previous section, $q_j = \alpha_j (j=1, \ldots, n)$ are the co-ordinates which define an equilibrium configuration. Write $V_o = V(\alpha_1, \ldots, \alpha_n)$. Let the system undergo a small displacement from equilibrium so that at time t, $q_j = \alpha_j + \xi_j$ $(j=1, \ldots, n)$, the ξ_j and their time derivatives being *small*. Then since $\partial V/\partial q_j = 0$ for $q_j = \alpha_j (j=1, \ldots, n)$, by Taylor's theorem

$$V = V_o + \tfrac{1}{2}[(\xi_1 \partial/\partial q_1 + \ldots + \xi_n \partial/\partial q_n)^2 V]_o \ ,$$

to second order, the zero signifying that we put each $q_j = \alpha_j (j=1, \ldots, n)$ after performing the differentiations. Thus V assumes the form of a quadratic function of the ξ's (to second order) of the form

$$V = V_o + \tfrac{1}{2} \sum_{r=1}^{n} \sum_{s=1}^{n} c_{rs} \, \xi_r \xi_s \ ,$$

where $c_{rs} = c_{sr} = (\partial^2 V/\partial q_r \partial q_s)_o = \text{const.}$

Also $$T = \tfrac{1}{2} \sum_{r=1}^{n} \sum_{s=1}^{n} a_{rs} \dot{q}_r \dot{q}_s = \tfrac{1}{2} \sum_{r=1}^{n} \sum_{s=1}^{n} a_{rs} \, \dot{\xi}_r \dot{\xi}_s \ .$$

Let a^*_{rs} denote the value of each a_{rs} in the equilibrium configuration $q_j = \alpha_j (j=1, \ldots, n)$. Then, correct to second order,

$$T = \tfrac{1}{2} \sum_{r=1}^{n} \sum_{s=1}^{n} a^*_{rs} \, \dot{\xi}_r \dot{\xi}_s, \text{ where } a^*_{rs} = a^*_{sr}.$$

Thus to second order,

$$L = \tfrac{1}{2} \sum_{r=1}^{n} \sum_{s=1}^{n} (a^*_{rs} \dot{\xi}_r \dot{\xi}_s - c_{rs} \xi_r \xi_s) + \text{const.},$$

where each a^*_{rs} and each c_{rs} is constant.
Lagrange's equations are now

$$(d/dt)(\partial L/\partial \dot{\xi}_j) - \partial L/\partial \xi_j = 0 \qquad (j=1, \ldots, n) \ .$$

Now $$\partial L/\partial \xi_j = -\tfrac{1}{2} \sum_{r=1}^{n} \sum_{s=1}^{n} c_{rs} \, (\xi_r \partial \xi_s/\partial \xi_j + \xi_s \partial \xi_r/\partial \xi_j)$$

$$= -\tfrac{1}{2} \sum_{r=1}^{n} c_{rj} \xi_r - \tfrac{1}{2} \sum_{s=1}^{n} c_{js} \xi_s = - \sum_{s=1}^{n} c_{js} \xi_s \ .$$

Similarly $(d/dt)(\partial L/\partial \dot{\xi}_j) = \sum_{s=1}^{n} a^*_{js}\, \ddot{\xi}_s$.

Thus Lagrange's equations become

$$\sum_{s=1}^{n} (a^*_{js}\ddot{\xi}_s + c_{js}\, \xi_s) = 0 \qquad (j=1, \ldots, n) \ . \tag{1}$$

To solve equations (1), try to find constants $M_s (s=1, \ldots, n)$ such that

$$\xi_s = M_s \sin(\omega t + \epsilon) \qquad (s=1, \ldots, n) \tag{2}$$

is a solution. We have $\ddot{\xi}_s = -\omega^2 M_s \sin(\omega t + \epsilon)$, and so substituting in (1) we obtain the n equations

$$\sum_{s=1}^{n} (a^*_{js}\omega^2 - c_{js})M_s = 0 \qquad (j=1, \ldots, n) \ . \tag{3}$$

Eliminating the M_s from (3) leads to the determinantal equation

$$\det|a^*_{js}\omega^2 - c_{js}| = 0 \qquad (j, s=1, \ldots, n) \ . \tag{4}$$

Equation (4) is an nth degree equation for ω^2.
Let its n roots be

$$\omega^2 = \omega_s^{\,2} \qquad (s=1, \ldots, n) \ . \tag{5}$$

If these n roots are *positive*, we obtain n periodic expressions for the ξ_s showing that each corresponding configuration is *stable*. A negative root would imply an aperiodic solution showing the corresponding configuration to be unstable.

For stable equilibrium, the n roots of (4) will all be positive and will lead to the n periods of oscillation $2\pi/\omega_s$ $(s=1, \ldots, n)$. These are called the *normal periods of oscillation.*

From equations (3), we may derive the ratios of the amplitudes $M_1 : M_2 : \ldots : M_n$. For if, when $\omega = \omega_s$, the co-factors of any row or column in the determinant on the L.H.S. of (4) are $C_1(\omega_s), C_2(\omega_s), \ldots, C_n(\omega_s)$, then $M_1/C_1(\omega_s) = M_2/C_2(\omega_s) = \ldots = M_n/C_n(\omega_s) = a_s$ (say). (6)

Corresponding to $\omega = \omega_s$, we have the solutions

$$\xi_1 = C_1(\omega_s)a_s \sin(\omega_s t + \epsilon_s), \quad \xi_2 = C_2(\omega_s)a_s \sin(\omega_s t + \epsilon_s), \ldots$$
$$\xi_n = C_n(\omega_s)a_s \sin(\omega_s t + \epsilon_s) \ .$$

Since the differential equations (1) are linear, their general solutions are found by *superposition* giving

$$\left.\begin{aligned}\xi_1 &= \sum_{s=1}^{n} a_s C_1(\omega_s)\sin(\omega_s t+\epsilon_s)\ ,\\ \xi_2 &= \sum_{s=1}^{n} a_s C_2(\omega_s)\sin(\omega_s t+\epsilon_s)\ ,\\ &\text{— — — — — —}\\ \xi_n &= \sum_{s=1}^{n} a_s C_n(\omega_s)\sin(\omega_s t+\epsilon_s)\ .\end{aligned}\right\}\qquad (7)$$

These solutions contain appropriately the $2n$ integration constants a_1, $a_2, \ldots, a_n$; $\epsilon_1, \epsilon_2, \ldots, \epsilon_n$.

The following alternative approach is rather more illuminating. We observe that the forms for T, $(V-V_o)$, viz.

$$T=\tfrac{1}{2}\sum_{r=1}^{n}\sum_{s=1}^{n} a^*_{rs}\,\dot{\xi}_r\dot{\xi}_s\ ; \qquad V-V_0=\tfrac{1}{2}\sum_{r=1}^{n}\sum_{s=1}^{n} c_{rs}\xi_r\xi_s\ ,$$

are both homogeneous quadratic forms, the first in the velocities $\dot{\xi}_r$, the second in the co-ordinates ξ_r. The reader familiar with the processes of linear algebra* will appreciate that in general one can transform the co-ordinates $\xi_r(r=1, \ldots, n)$ to a new system of co-ordinates $\eta_r(r=1, \ldots, n)$ by making substitutions of the form

$$\xi_r=b_{r1}\eta_1+b_{r2}\eta_2+\ldots+b_{rn}\eta_n \qquad (r=1, \ldots, n) \qquad (8)$$

so that, with suitable choice of b's, T and $(V-V_o)$ assume the forms

$$\left.\begin{aligned}T&=\tfrac{1}{2}(\dot{\eta}_1^2+\dot{\eta}_2^2+\ldots+\dot{\eta}_n^2)\ ,\\ V-V_o&=\tfrac{1}{2}(\gamma_1\eta_1^2+\gamma_2\eta_2^2+\ldots+\gamma_n\eta_n^2)\ .\end{aligned}\right\}\qquad (9)$$

Then $L=\frac{1}{2}\sum_{r=1}^{n}(\dot{\eta}_r^2-\gamma_r\eta_r^2)-V_o$ and Lagrange's equations for the new generalized co-ordinates $\eta_r(r=1, \ldots, n)$ become

$$\ddot{\eta}_r+\gamma_r\eta_r=0 \qquad (r=1, \ldots, n)\ . \qquad (10)$$

Writing $\gamma_r=\omega_r^2(r=1, \ldots, n)$, we obtain the solutions of (10):

$$\eta_r=A_r\sin(\omega_r t+\epsilon_r) \qquad (r=1, \ldots, n)\ . \qquad (11)$$

The co-ordinates $\eta_r(r=1, \ldots, n)$ so determined are called the *normal co-ordinates* of the system. Substitution into (8) gives the solution for the original co-ordinates $\xi_r(r=1, \ldots, n)$. Clearly this solution is a linear superposition of the normal co-ordinates. If the $(n-1)$ normal co-ordinates $\eta_1, \eta_2, \ldots, \eta_{r-1}, \eta_{r+1}, \ldots, \eta_n$ are all permanently zero for a system vibrating about a configuration of stable equilibrium, and if $\eta_r\neq 0$, then the system is said to execute a *normal mode* of vibration. The motion of each particle

* See W. L. Ferrar, *Algebra*, p. 151, Oxford University Press.

is periodic, passing twice through the equilibrium configuration in every complete vibration and the velocity of each particle vanishes at the same instant twice in every complete oscillation. In design engineering, it is often important to know the normal periods of a structure so that any applied periodic forces do not have these periods, for otherwise there is danger of large amplitudes building up due to resonance effects.

Example

A uniform rod AB, of mass $2m$ and length $2a$, swings freely about a horizontal axis through the end A. One end of a light elastic string is attached to the end B of the rod and carries a particle of mass m at its other end. When the system is in stable equilibrium, the string is of length $4a/3$, its extension being ϵ. If the system performs small oscillations in the vertical plane through AB, the rod and the string making angles θ, ϕ respectively with the downward vertical, and the length of the string being $x+4a/3$, show that

$$x, \qquad \phi+2\theta\ , \qquad 2\phi-3\theta$$

are normal co-ordinates for the system, and the lengths of the corresponding simple pendulums are ϵ, $8a/3$, $a/3$. (L.U.)

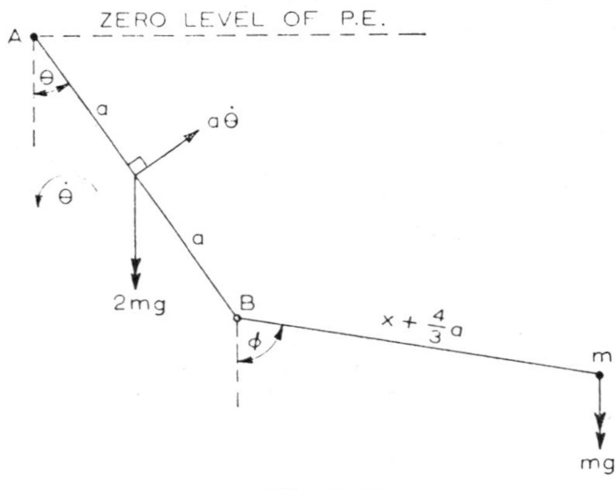

Fig. 10.13

Let λ be the modulus of elasticity of the string. In the equilibrium position, mg is the tension in the string and so

$$mg=\lambda\epsilon/(\tfrac{4}{3}a-\epsilon) \quad \text{or } \lambda=mg(4a-3\epsilon)/3\epsilon\ .$$

The K.E. of the system is

$$T=\tfrac{1}{2}(2m)(a\dot{\theta})^2+\tfrac{1}{6}(2m)a^2\dot{\theta}^2$$
$$+\tfrac{1}{2}m[\![[\{d/dt\}\{2a\sin\theta+(x+\tfrac{4}{3}a)\sin\phi\}]^2$$
$$+[\{d/dt\}\{2a\cos\theta+(x+\tfrac{4}{3}a)\cos\phi\}]^2]\!]$$

$$=\tfrac{4}{3}ma^2\dot\theta^2+\tfrac{1}{2}m\{(2a\dot\theta\cos\theta+\dot x\sin\phi+(x+\tfrac{4}{3}a)\dot\phi\cos\phi)^2$$
$$+(-2a\dot\theta\sin\theta+\dot x\cos\phi-(x+\tfrac{4}{3}a)\dot\phi\sin\phi)^2\}$$
$$=\tfrac{10}{3}ma^2\dot\theta^2+\tfrac{1}{2}m\dot x^2+\tfrac{1}{2}m(x+\tfrac{4}{3}a)^2\dot\phi^2+2ma\dot\theta\dot x\sin(\phi-\theta)$$
$$+2ma\dot\theta\dot\phi(x+\tfrac{4}{3}a)\cos(\phi-\theta)\ .$$

Correct to the second order, this gives

$$\underline{T=\tfrac{10}{3}ma^2\dot\theta^2+\tfrac{1}{2}m\dot x^2+\tfrac{8}{9}ma^2\dot\phi^2+\tfrac{8}{3}ma^2\dot\theta\dot\phi\ .}$$

The gravitational P.E., referred to the horizontal through A as zero level, is

$$-2mga\cos\theta-mg[2a\cos\theta+(x+\tfrac{4}{3}a)\cos\phi]$$
$$\simeq-mgx+2mga\theta^2+\tfrac{2}{3}mga\phi^2+\text{const., to 2nd order.}$$

Since the stretch of the string is $x+\epsilon$, the elastic P.E. is

$$\tfrac{1}{2}\lambda(x+\epsilon)^2/(\tfrac{4}{3}a-\epsilon)=\tfrac{1}{2}mg(x+\epsilon)^2/\epsilon=\tfrac{1}{2}mg(2x+x^2/\epsilon)+\text{const.}$$

Thus, to the second order, the total P.E. is

$$\underline{V=(mgx^2/2\epsilon)+2mga\theta^2+\tfrac{2}{3}mga\phi^2+\text{const.}}$$

The Lagrangian is $L=T-V$ or $L=\tfrac{1}{2}m\dot x^2+\tfrac{10}{3}ma^2\dot\theta^2+\tfrac{8}{9}ma^2\dot\phi^2+\tfrac{8}{3}ma^2\dot\theta\dot\phi$

$$-(mgx^2/2\epsilon)-2mga\theta^2-\tfrac{2}{3}mga\phi^2+\text{const.}$$

Lagrange's equations are $(d/dt)(\partial L/\partial\dot x)-\partial L/\partial x=0$, etc.,

i.e.

$$\ddot x+gx/\epsilon=0\ , \qquad (1)$$
$$\tfrac{20}{3}a\ddot\theta+\tfrac{8}{3}a\ddot\phi+4g\theta=0\ , \qquad (2)$$
$$\tfrac{8}{3}a\ddot\theta+\tfrac{16}{9}a\ddot\phi+\tfrac{4}{3}g\phi=0\ , \qquad (3)$$

(1) shows x is a normal co-ordinate, the length of the corresponding simple equivalent pendulum being ϵ.

In (2), (3), put

$$\theta=\theta_o\sin(\omega t+\epsilon)\ , \qquad \phi=\phi_o\sin(\omega t+\epsilon)\ .$$

Then, on simplifying,

$$(5a\omega^2-3g)\theta_o+2a\omega^2\phi_o=0\ , \qquad (4)$$
$$6a\omega^2\theta_o+(4a\omega^2-3g)\phi_o=0\ . \qquad (5)$$

From (4), (5) the eliminant of θ_o, ϕ_o is

$$(5a\omega^2-3g)(4a\omega^2-3g)-12a^2\omega^4=0\ .$$

This has roots $\omega^2=\omega_1{}^2$, $\omega^2=\omega_2{}^2$, where

$$\underline{\omega_1{}^2=3g/a\ , \qquad \omega_2{}^2=3g/8a\ .}$$

From (4),

$$\theta_o/(2a\omega^2/g)=-\phi_o/(5a\omega^2/g-3)\ .$$

When $\omega=\omega_1$, these are satisfied by

$$\theta_o=C_1\ , \qquad \phi_o=-2C_1\ .$$

When $\omega=\omega_2$, solutions are

$$\theta_o=2C_2\ , \qquad \phi_o=3C_2\ .$$

Hence the general solution is

$$\theta=C_1\sin(\omega_1 t+\epsilon_1)+2C_2\sin(\omega_2 t+\epsilon_2)\ , \tag{6}$$
$$\phi=-2C_1\sin(\omega_1 t+\epsilon_1)+3C_2\sin(\omega_2 t+\epsilon_2)\ . \tag{7}$$

From (6), (7),

$$\left.\begin{aligned}2\theta+\phi&=7C_2\sin(\omega_2 t+\epsilon_2)\ ,\\ 2\phi-3\theta&=-7C_1\sin(\omega_1 t+\epsilon_1)\ .\end{aligned}\right\}$$

Thus $(2\theta+\phi)$, $(2\phi-3\theta)$ are expressible as single sinusoidal terms and are normal co-ordinates. Their periods are

$$2\pi/\omega_2=2\pi(8a/3g)^{\frac{1}{2}}\ ;$$
$$\underline{2\pi/\omega_1=2\pi(a/3g)^{\frac{1}{2}}\ .}$$

Alternative procedure

The normal co-ordinates are found directly from equations (2) and (3) as follows. Written more simply these equations are

$$5a\ddot{\theta}+3g\theta+2a\ddot{\phi}=0\ , \tag{2'}$$
$$6a\ddot{\theta}+4a\ddot{\phi}+3g\phi=0\ . \tag{3'}$$

Form $\lambda\times(2')+\mu\times(3')$ to give

$$a(5\lambda+6\mu)\ddot{\theta}+3g\lambda\theta+2a(\lambda+2\mu)\ddot{\phi}+3g\mu\phi=0\ . \tag{8}$$

Choose the ratio $\lambda:\ \mu$ so that

$$\text{coeff. of }\ddot{\theta}/\text{coeff. of }\theta=\text{coeff. of }\ddot{\phi}/\text{coeff. of }\phi\ .$$
$$\therefore\ a(5\lambda+6\mu)/(3g\lambda)=2a(\lambda+2\mu)/(3g\mu)\ ,$$

i.e. $\lambda/\mu=-3/2$ or $\lambda/\mu=2$.

When $\lambda/\mu=-3/2$, (8) gives

$$-3a\ddot{\theta}-9g\theta+2a\ddot{\phi}+6g\phi=0$$

or $(D^2+3g/a)(2\phi-3\theta)=0 \qquad (D\equiv d/dt)$.

$$\therefore\ \underline{2\phi-3\theta=A_1\sin[(3g/a)^{\frac{1}{2}}t+\epsilon_1]\ .}$$

When $\lambda/\mu=2$, (8) gives

$$16a\ddot{\theta}+6g\theta+8a\ddot{\phi}+3g\phi=0$$
$$\therefore\ (D^2+3g/8a)(2\theta+\phi)=0$$
$$\therefore\ \underline{2\theta+\phi=A_2\sin[(3g/8a)^{\frac{1}{2}}t+\epsilon_2]\ .}$$

10.13 EULERIAN ANGLES

Suppose a rigid body rotates in any manner about a fixed point O. Then the motion of the body as a whole is known if that of a spherical octant $OABC$ is known (Fig. 10.14). To investigate this motion, perform in order the following system of operations:

(i) Rotate the octant $OABC$ about OC through an angle ϕ to assume the new configuration $OA_1B_1C_1$ where $C_1 \equiv C$;

(ii) Rotate the octant $OA_1B_1C_1$ about OB_1 through an angle θ to assume the new configuration $OA_2B_2C_2$, where $B_2 \equiv B_1$;

(iii) Rotate the octant $OA_2B_2C_2$ about OC_2 through an angle ψ to assume the configuration $OA_3B_3C_3$, where $C_3 \equiv C_2$.

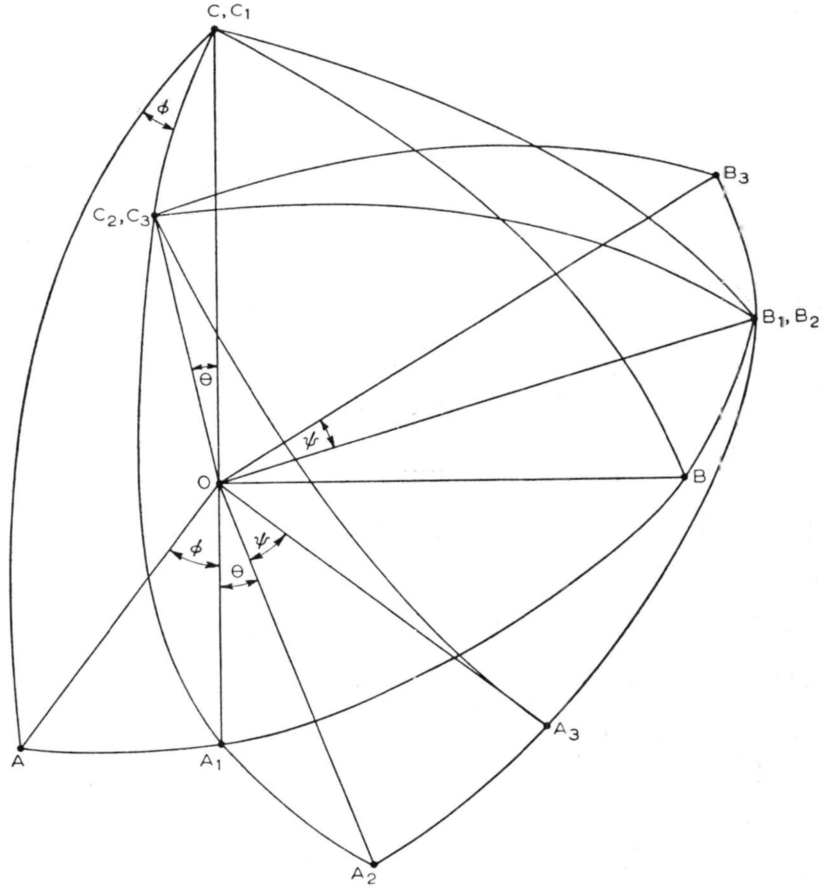

Fig. 10.14

A little thought will convince the reader that starting from any given configuration $OABC$ one can obtain any other configuration $OA_3B_3C_3$ by suitably choosing θ, ϕ, ψ. Thus the motion of the octant $OA_3B_3C_3$ of the rigid body, and hence that of the body as a whole, is known relative to some initial configuration $OABC$ if we know the angles θ, ϕ, ψ. They are called *Eulerian angles*.

Suppose the above system of rotations is made in time t, and suppose the original axes OA, OB, OC coincide with axes fixed in space. Then we have the following angular velocities:

(i) $\dot{\phi}$ about $OC \equiv \dot{\phi}$ about OC_1

$$\equiv \dot{\phi} \cos\theta \text{ about } OC_2 - \dot{\phi} \sin\theta \text{ about } OA_2$$
$$\equiv \dot{\phi} \cos\theta \text{ about } OC_3 - \dot{\phi} \sin\theta \cos\psi \text{ about } OA_3 + \dot{\phi} \sin\theta \sin\psi \text{ about } OB_3;$$

(ii) $\dot{\theta}$ about $OB_1 \equiv \dot{\theta}$ about OB_2

$$\equiv \dot{\theta} \cos\psi \text{ about } OB_3 + \dot{\theta} \sin\psi \text{ about } OA_3;$$

(iii) $\dot{\psi}$ about $OC_2 \equiv \dot{\psi}$ about OC_3.

Let $[\omega_1, \omega_2, \omega_3]$ be the resultant spin components about OA_3, OB_3, OC_3. Then

$$\left.\begin{aligned} \omega_1 &= \dot{\theta} \sin\psi - \dot{\phi} \sin\theta \cos\psi \ , \\ \omega_2 &= \dot{\theta} \cos\psi + \dot{\phi} \sin\theta \sin\psi \ , \\ \omega_3 &= \dot{\psi} + \dot{\phi} \cos\theta \ . \end{aligned}\right\} \tag{1}$$

If we further suppose the moving axes OA_3, OB_3, OC_3 are *principal axes* for O about which A, B, C are the principal moments of inertia, then the K.E. is

$$T = \tfrac{1}{2}\{A\omega_1{}^2 + B\omega_2{}^2 + C\omega_3{}^2\} \ . \tag{2}$$

Regarding θ, ϕ, ψ as the generalized co-ordinates and Q_θ, Q_ϕ, Q_ψ, as the generalized forces associated with them, the virtual work function is

$$\delta W = Q_\theta \delta\theta + Q_\phi \delta\phi + Q_\psi \delta\psi \ . \tag{3}$$

Thus Lagrange's equation for co-ordinate ψ is

$$(d/dt)(\partial T/\partial \dot{\psi}) - \partial T/\partial \psi = Q_\psi \ . \tag{4}$$

Now $\quad \partial T/\partial \dot{\psi} = \partial T/\partial \omega_3 = C\omega_3 \ ;$

$$\therefore \ (d/dt)(\partial T/\partial \dot{\psi}) = C\dot{\omega}_3 \ .$$

Also, $\quad \partial T/\partial \psi = (\partial T/\partial \omega_1)(\partial \omega_1/\partial \psi) + (\partial T/\partial \omega_2)(\partial \omega_2/\partial \psi)$

$$= A\omega_1(\dot{\theta} \cos\psi + \dot{\phi} \sin\theta \sin\psi) + B\omega_2(-\dot{\theta} \sin\psi + \dot{\phi} \sin\theta \cos\psi)$$
$$= A\omega_1\omega_2 - B\omega_2\omega_1 \ .$$

Thus (4) becomes

$$C\dot{\omega}_3 - (A - B)\omega_2\omega_1 = Q_\psi \ .$$

Let $[L_1, L_2, L_3]$ be the moments of the external forces about the moving axes OA_3, OB_3, OC_3. Then $Q_\psi = L_3$ and so

$$C\dot{\omega}_3 - (A-B)\omega_2\omega_1 = L_3 \ . \tag{5}$$

This is one of Euler's dynamical equations; the others are easily written down from symmetry.

10.14 MOTION OF A SYMMETRICAL TOP

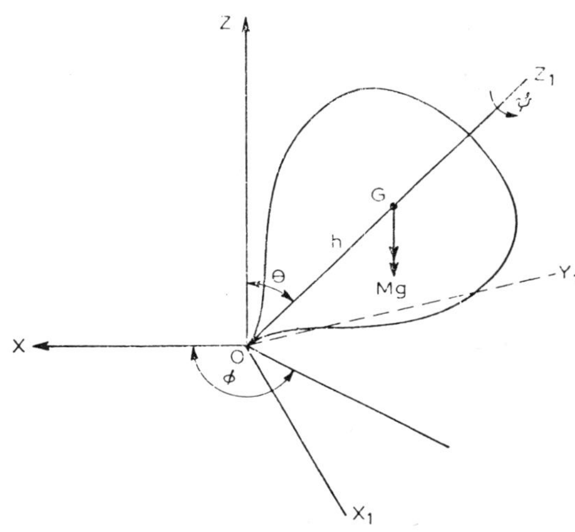

Fig. 10.15

A symmetrical top is a rigid body which is a solid of revolution about its axis of symmetry. Fig. 10.15 shows such a top rotating about a fixed point O on its axis of symmetry OZ_1. At time t, suppose OZ_1 makes angle θ with the upward vertical OZ, and let ϕ be the angle which the plane ZOZ_1 makes with a fixed horizontal direction OX. Further, let ψ be the rotation about OZ_1. Then θ, ϕ, ψ are the Eulerian angles for the system. Let OX_1 be taken in the plane ZOZ_1 at right angles to OZ_1 and OY_1 perpendicular to the plane ZOZ_1 so that the moving axes OX_1, OY_1, OZ_1 form a right-handed system. The spin system comprises $\dot{\phi}$ about OZ, $\dot{\theta}$ about OY_1, $\dot{\psi}$ about OZ_1. Now

$$\dot{\phi} \text{ about } OZ \equiv \dot{\phi} \cos\theta \text{ about } OZ_1 - \dot{\phi} \sin\theta \text{ about } OX_1 \ .$$

Hence the resultant angular velocity components about OX_1, OY_1, OZ_1 are

$$[-\dot{\phi}\sin\theta, \dot{\theta}, \dot{\psi}+\dot{\phi}\cos\theta] \ .$$

Now OX_1, OY_1, OZ_1 are principal axes for the top. Let A, A, C be the corresponding moments of inertia about them. Then the K.E. is

$$T=\tfrac{1}{2}A(-\dot{\phi}\sin\theta)^2+\tfrac{1}{2}A\dot{\theta}^2+\tfrac{1}{2}C(\dot{\psi}+\dot{\phi}\cos\theta)^2$$
$$=\tfrac{1}{2}A(\dot{\phi}^2\sin^2\theta+\dot{\theta}^2)+\tfrac{1}{2}C(\dot{\psi}+\dot{\phi}\cos\theta)^2\ .$$

The P.E. of the top referred to the horizontal through O as zero level is

$$V=Mgh\cos\theta\ ,$$

where M is the mass of the top and h the distance of the centroid G from O. Thus

$$L=\tfrac{1}{2}A(\dot{\phi}^2\sin^2\theta+\dot{\theta}^2)+\tfrac{1}{2}C(\dot{\psi}+\dot{\phi}\cos\theta)^2-Mgh\cos\theta\ .$$

Taking θ, ϕ, ψ as the generalized co-ordinates, Lagrange's equations are $(d/dt)(\partial L/\partial\dot{\theta})-\partial L/\partial\theta=0$, etc.,

i.e. $$A\ddot{\theta}-A\dot{\phi}^2\sin\theta\cos\theta+C\dot{\phi}\sin\theta(\dot{\psi}+\dot{\phi}\cos\theta)-Mgh\sin\theta=0,\quad (1)$$
$$(d/dt)[A\dot{\phi}\sin^2\theta+C\cos\theta(\dot{\psi}+\dot{\phi}\cos\theta)]=0\ ,\quad (2)$$
$$(d/dt)[C(\dot{\psi}+\dot{\phi}\cos\theta)]=0\ .\quad (3)$$

In addition we have, from the law of conservation of energy, $T+V=$ const., or

$$\tfrac{1}{2}A(\dot{\phi}^2\sin^2\theta+\dot{\theta}^2)+\tfrac{1}{2}C(\dot{\psi}+\dot{\phi}\cos\theta)^2+Mgh\cos\theta=\text{const.}\quad (4)$$

From (3), we have at once

$$\dot{\psi}+\dot{\phi}\cos\theta=n=\text{const.}\quad (3')$$

(3′) means that the total angular velocity n about the axis of symmetry stays constant. Physically this follows from the constancy of angular momentum about this axis.

Integrating (2) now gives

$$A\dot{\phi}\sin^2\theta+Cn\cos\theta=D=\text{const.}\quad (2')$$

Equations (1), (4) may now be written

$$A\ddot{\theta}-A\dot{\phi}^2\sin\theta\cos\theta+Cn\dot{\phi}\sin\theta-Mgh\sin\theta=0\ ,\quad (1')$$
$$A\dot{\theta}^2+A\dot{\phi}^2\sin^2\theta+2Mgh\cos\theta=E=\text{const.}\quad (4')$$

The motion due to change in θ is called *nutation*; that due to change in ϕ is called *precession*. The general motion of the top about its fixed apex O is a combination of these.

N.B. (2′) expresses constant angular momentum about OZ.

(i) *Case of steady motion*

When the top executes steady motion, $\theta=\alpha=$ const., and from (2′) $\dot{\phi}=\Omega=$ const. Provided $\theta\neq 0$, (1′) now gives

$$A\Omega^2\cos\alpha-Cn\Omega+Mgh=0\ .\quad (5)$$

(5) shows that there is a pair of real and distinct precessional angular velocities Ω_1, Ω_2 provided

$$C^2n^2>4AMgh\cos\alpha\ .$$

Thus for a sufficiently great spin n about the axis of symmetry, the top can execute a steady motion $\theta=\alpha$ about the vertical under two possible precessional angular velocities Ω_1, Ω_2.

(ii) *Stability investigation*

(1′) may be written

$$A\ddot{\theta}+f(\theta)=0 \ ,$$

where $f(\theta)=-A\dot{\phi}^2\sin\theta\cos\theta+Cn\dot{\phi}\sin\theta-Mgh\sin\theta$. Since $f(\alpha)=0$ (from (5)), if we put $\theta=\alpha+\epsilon$ where ϵ is small, then $f(\theta)\simeq\epsilon f'(\alpha)$ to the first order, so that (1′) becomes

$$A\ddot{\epsilon}+\epsilon f'(\alpha)=0 \ . \tag{1''}$$

If we can show $f'(\alpha)>0$, this equation will reduce to the form $\ddot{\epsilon}+\omega^2\epsilon=0$ showing stability about $\theta=\alpha$ for small perturbations.

From (2′), $\dot{\phi}=(D-Cn\cos\theta)/A\sin^2\theta$. Hence
$Af(\theta)\sin^3\theta=-\cos\theta(D-Cn\cos\theta)^2+Cn\sin^2\theta(D-Cn\cos\theta)-A\,Mgh\sin^4\theta$.
Differentiating both sides of this, putting $\theta=\alpha$ and using $f(\alpha)=0$,

$$\begin{aligned}Af'(\alpha)\sin^3\alpha&=\sin\alpha(D-Cn\cos\alpha)^2-2Cn\cos\alpha\sin\alpha(D-Cn\cos\alpha)\\&\quad+2Cn\sin\alpha\cos\alpha(D-Cn\cos\alpha)+C^2n^2\sin^3\alpha-4AMgh\sin^3\alpha\cos\alpha\\&=(D-Cn\cos\alpha)^2\sin\alpha+C^2n^2\sin^3\alpha-4AMgh\sin^3\alpha\cos\alpha \ .\end{aligned}$$

From (2′), (5),

$$\begin{aligned}D-Cn\cos\alpha&=A\Omega\sin^2\alpha \ ,\\ Cn&=A\Omega\cos\alpha+Mgh/\Omega \ .\end{aligned}$$

Hence $Af'(\alpha)=A^2\Omega^2\sin^2\alpha+(A\Omega\cos\alpha+Mgh/\Omega)^2-4AMgh\cos\alpha$

$$=A^2\Omega^2-2AMgh\cos\alpha+(Mgh/\Omega)^2 \ .$$

Thus (1″) gives

$$\ddot{\epsilon}+[\Omega^2-2(Mgh/A)\cos\alpha+(Mgh/A\Omega)^2]\epsilon=0 \ . \tag{6}$$

In (6), the coefficient of ϵ

$$\begin{aligned}&=[\Omega-(Mgh/A\Omega)]^2+2(Mgh/A)[1-\cos\alpha]\\&>0, \text{ whenever } \alpha\neq 0 \ .\end{aligned}$$

Thus, for $\alpha\neq0$, the motion is of S.H.M. type showing that the position $\theta=\alpha$ is one of *stability*.

Example

Obtain the equations of motion of a symmetrical top moving under gravity about a fixed point of its axis.

A top consists of a thin uniform spherical shell of radius a and centre C and is free to move about a fixed point O of its surface. The radius OC makes an angle θ with the upward vertical OZ and ϕ is the angle between the plane ZOC and a fixed vertical plane. Initially, $\theta=\frac{1}{2}\pi$, $\dot{\theta}=0$, $\dot{\phi}=\frac{2}{5}n$, where $n=\sqrt{(15g/2a)}$ is the spin of the top about its axis. Show that subsequently $\cos\theta=\tanh^2\{(3g/10a)^{\frac{1}{2}}t\}$ and that C traces out a spiral path on a fixed sphere of centre O and radius a. Sketch the path of C.

(L.U.)

In the notation of the text,

$$C=\tfrac{2}{3}Ma^2\ ,\quad A=\tfrac{5}{3}Ma^2\ ,\quad h=a\ ,\quad n=\sqrt{(15g/2a)}\ .$$

The energy equation (4′) gives

$$\dot{\theta}^2+\dot{\phi}^2\sin^2\theta+(6g/5a)\cos\theta=\text{const.}=(2n/5)^2=6g/5a\ .$$

$$\therefore\ \underline{\dot{\theta}^2+\dot{\phi}^2\sin^2\theta=(6g/5a)(1-\cos\theta)\ .} \qquad \text{(i)}$$

Equation (2′) gives

$$\underline{\dot{\phi}\sin^2\theta+\tfrac{2}{5}n\cos\theta=\text{const.}=\tfrac{2}{5}n\ .} \qquad \text{(ii)}$$

From (ii), $\dot{\phi}=\frac{2}{5}n/(1+\cos\theta)$ and (i) becomes

$$\dot{\theta}^2=6g\cos\theta(1-\cos\theta)/5a(1+\cos\theta)\ .$$

Substituting initial values into (1′), the reader will see that initially $\ddot{\theta}<0$ and so $\dot{\theta}<0$ for $t>0$ since $\dot{\theta}=0$ for $t=0$. Thus we have

$$\dot{\theta}=-\{6g\cos\theta(1-\cos\theta)/5a(1+\cos\theta)\}^{\frac{1}{2}}\ ,$$

$$\therefore\ t\sqrt{\left(\frac{6g}{5a}\right)}=-\int_{\pi/2}^{\theta}\left\{\frac{1+\cos\theta}{\cos\theta\,(1-\cos\theta)}\right\}^{\frac{1}{2}}d\theta$$

$$=-\int_{\pi/2}^{\theta}\frac{\sin\theta\,d\theta}{(1-\cos\theta)\sqrt{\cos\theta}}\ .$$

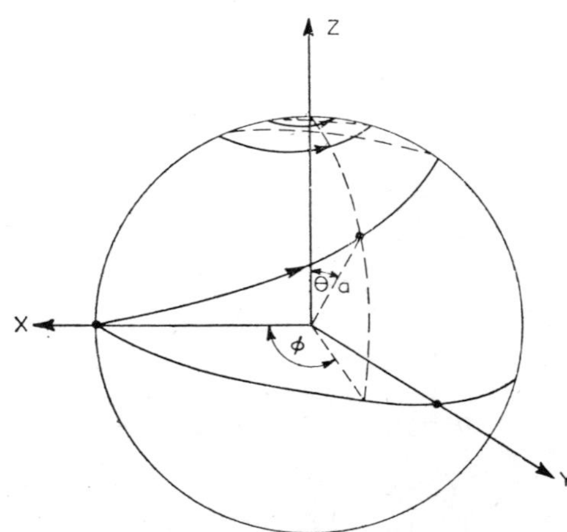

Fig. 10.16

Put $\cos\theta = y^2$. Then

$$t\sqrt{\left(\frac{6g}{5a}\right)} = +\int_0^{\sqrt{\cos\theta}} \frac{2y\,dy}{y(1-y^2)} = 2\,\mathrm{th}^{-1}(\sqrt{\cos\theta})\ .$$

$$\therefore\ \underline{\cos\theta = \mathrm{th}^2[t(3g/10a)^{\frac{1}{2}}]}\ .$$

As t increases from 0 to ∞, the above shows $\cos\theta$ increases from 0 to 1 or θ decreases from $\frac{1}{2}\pi$ to 0. Let $\phi = 0$ when $t = 0$, i.e. when $\theta = \frac{1}{2}\pi$. Fig. 10.16 shows the path of C.

10.15 MOTION OF A SPHERE ON A SPHERE

The type of analysis used in the treatment of the motion of a top is here extended to a slightly different type of motion.

Example

A uniform solid sphere of radius a rolls on the perfectly rough outer surface of a sphere of radius b. Set up the equations of motion and show that the moving sphere can spin stably in contact with the highest point of the fixed sphere if its spin exceeds $[35g(a+b)/a^2]^{\frac{1}{2}}$. (L.U.)

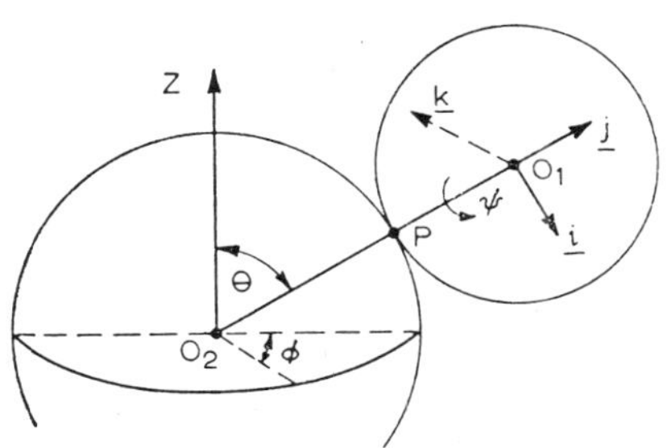

Fig. 10.17

O_1 is the centre of the rolling sphere (of mass M), O_2 that of the fixed sphere (Fig. 10.17). O_2Z is the upward vertical through O_2, $\angle ZO_2O_1 = \theta$ and the vertical plane ZO_2O_1 makes angle ϕ with a fixed vertical plane through O_2Z. Let ψ be the angle of rotation about O_2O_1. Let $\mathbf{j}$ denote the unit vector in $\overline{O_2O_1}$, $\mathbf{i}$ the unit vector through O_1, in the plane ZO_2O_1 perpendicular to $\mathbf{j}$ and in the sense of θ increasing, and let $\mathbf{k}$ be defined by $\mathbf{k} = \mathbf{i}\wedge\mathbf{j}$. Denote the unit vector in $\overline{O_2Z}$ by $\hat{\mathbf{z}}$ so that $\hat{\mathbf{z}} = \cos\theta\mathbf{j} - \sin\theta\mathbf{i}$. The total spin of the upper sphere is $\boldsymbol{\omega}$ where

$$\begin{aligned}\boldsymbol{\omega} &= \dot{\phi}\hat{\mathbf{z}} - \dot{\theta}\mathbf{k} + \dot{\psi}\mathbf{j}\\ &= -\dot{\phi}\sin\theta\mathbf{i} + (\dot{\psi} + \dot{\phi}\cos\theta)\mathbf{j} - \dot{\theta}\mathbf{k}\ .\end{aligned}$$

Since P is instantaneously at rest and $\overline{PO_1} \equiv a\mathbf{j}$, the velocity of the centroid O_1 of the upper sphere is

$$\mathbf{v} = \boldsymbol{\omega}_\wedge a\mathbf{j} = a[\dot{\theta}\mathbf{i} - \dot{\phi} \sin \theta \, \mathbf{k}] \ .$$

Now $\mathbf{i}$, $\mathbf{j}$, $\mathbf{k}$ are principal axes for the sphere at O_1 and the M.I. about each is $\frac{2}{5}Ma^2$. Hence the K.E. is

$$\begin{aligned} T &= \tfrac{1}{2}M\mathbf{v}^2 + \tfrac{1}{5}Ma^2[\dot{\phi}^2\sin^2\theta + (\dot{\psi} + \dot{\phi} \cos \theta)^2 + \dot{\theta}^2] \\ &= \tfrac{7}{10}Ma^2(\dot{\theta}^2 + \dot{\phi}^2\sin^2\theta) + \tfrac{1}{5}Ma^2(\dot{\psi} + \dot{\phi} \cos \theta)^2 \ . \end{aligned}$$

Referred to the horizontal through O_2 as zero level, the P.E. is $V = Mg(a+b)\cos \theta$. Hence the Lagrangian $L = T - V$ is

$$L = \tfrac{7}{10}Ma^2(\dot{\theta}^2 + \dot{\phi}^2\sin^2\theta) + \tfrac{1}{5}Ma^2(\dot{\psi} + \dot{\phi} \cos \theta)^2 - Mg(a+b) \cos \theta.$$

Forming Lagrange's equations for the generalized co-ordinates θ, ϕ, ψ we obtain on simplification

$$7a^2\ddot{\theta} - 7a^2\dot{\phi}^2\sin \theta \cos \theta + 2a^2\dot{\phi} \sin \theta(\dot{\psi} + \dot{\phi} \cos \theta) - 5g(a+b)\sin \theta = 0 \ , \quad (1)$$

$$7\dot{\phi} \sin^2\theta + 2 \cos \theta(\dot{\psi} + \dot{\phi} \cos \theta) = \text{const.}, \quad (2)$$

$$\dot{\psi} + \dot{\phi} \cos \theta = n = \text{const.} \quad (3)$$

From (2) and (3),

$$7\dot{\phi} \sin^2\theta + 2n \cos \theta = \text{const.} = 2n \ ,$$

or

$$\dot{\phi} = \tfrac{2}{7}n/(1 + \cos \theta) \ .$$

Substituting for $\dot{\phi}$ from this in (1) gives, for small θ,

$$7a^2\ddot{\theta} - 7a^2\theta(\tfrac{1}{7}n)^2 + 2a^2\theta(\tfrac{1}{7}n^2) - 5g(a+b)\theta = 0 \ ,$$

retaining only first order terms, or

$$49\ddot{\theta} + [n^2 - 35g(a+b)/a^2]\theta = 0 \ .$$

Stability about $\theta = 0$ requires $n > [35g(a+b)/a^2]^{\frac{1}{2}}$.

10.16 THE GYROSCOPIC COMPASS

A uniform heavy flywheel rotating about an axle which is an axis of symmetry comprises a *gyroscope*. When such a gyroscope is placed so that its axis is free to move in a horizontal plane, we shall show that if the gyroscope is at rest relative to the earth, then its axis points in the direction of the north – south geographical meridian through the considered station on the earth's surface. Used in this way, the instrument comprises a *gyroscopic compass*.

Fig. 10.18 shows the earth rotating about its axis with angular velocity Ω. P is a point on its surface at which angle of latitude is λ, $\overline{PV}$ signifies the upward vertical direction through P and $\overline{PN}$ the meridian direction towards the north pole. A gyroscopic compass whose axis is free to rotate horizontally is placed at P. Fig. 10.19 shows the axis of the compass at an

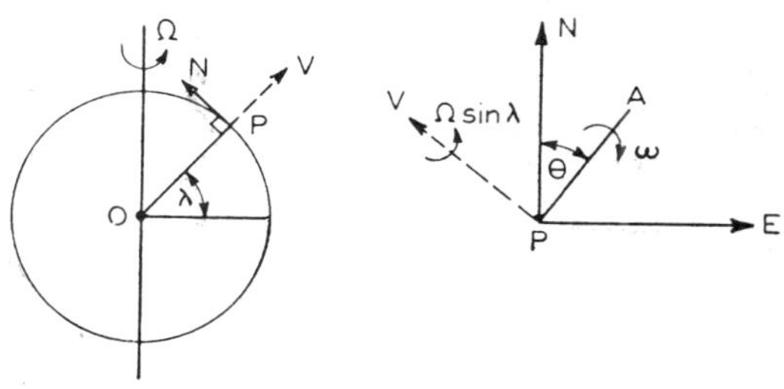

Fig. 10.18 Fig. 10.19

angle θ to PN, ω being the angular velocity of the wheel about its axis PA and PE the eastern direction through P.

The total angular velocity vector of the gyroscope has components $(\Omega \sin \lambda - \dot{\theta})$ about PV, $\Omega \cos \lambda$ about PN and ω about PA. An equivalent representation is $[\Omega \sin \lambda - \dot{\theta},\ \Omega \cos \lambda \sin \theta,\ \omega + \Omega \cos \lambda \cos \theta]$ about the mutually perpendicular (principal) axes PV, an axis through P mutually perpendicular to PV and PA, and PA respectively. Suppose A, A, C are the principal moments of inertia about these axes. Then the K.E. is

$$T = \tfrac{1}{2}\{A(\Omega \sin \lambda - \theta)^2 + A\Omega^2 \cos^2\lambda \sin^2\theta + C(\omega + \Omega \cos \lambda \cos \theta)^2\}\ .$$

This is also the expression for the Lagrangian L if we take the zero level of P.E. through P. Suppose the wheel rotates through an angle ϕ in time t so that $\phi = \omega$. The reader will easily verify that Lagrange's equation for co-ordinate ϕ is

$$(d/dt)(\omega + \Omega \cos \lambda \cos \theta) = 0$$

or

$$\omega + \Omega \cos \lambda \cos \theta = n \tag{1}$$

and that for co-ordinate θ is

$$A\ddot{\theta} - A\Omega^2 \cos^2\lambda \sin \theta \cos \theta + C\Omega \cos \lambda \sin \theta(\omega + \Omega \cos \lambda \cos \theta) = 0\ . \tag{2}$$

Substituting (1) into (2), assuming θ to be small and neglecting Ω^2, we obtain

$$A\ddot{\theta} + (Cn\Omega \cos \lambda)\theta = 0\ . \tag{3}$$

The form of (3) shows that $\theta = 0$ is a position of equilibrium of the gyroscope and that the period of small oscillations about it is $2\pi\sqrt{(A/Cn\Omega \cos \lambda)}$.

EXERCISE 10

1. A particle on a smooth horizontal table is attached to a string passing through a small hole in the table and carries an equal particle hanging vertically. If at time t the particle is distant r from the hole and the string makes an angle θ with some fixed line in the table, obtain Lagrange's equations of motion of the system in terms of r, θ and their time-derivatives.

The particle on the table is initially projected at right angles to the string with velocity $\sqrt{(2gh)}$ when distant a from the hole. Prove that

$$(dr/dt)^2 = gh(1-a^2/r^2) + g(a-r) \ .$$

2. A light string passes over a fixed smooth pulley. It carries a mass of $6M$ at one end, the other end being attached to a smooth pulley of mass $3M$ over which passes a light string whose ends carry masses $2M$ and M. Assuming the system starts from rest and neglecting moments of inertia of the pulleys, obtain expressions for the velocities and accelerations of the movable pulley and masses.

3. A uniform rod AB of mass m and length $2a$, attached to a fixed point O by a light elastic string OA of natural length l and modulus of elasticity λ, moves in a vertical plane through O with the string taut. At time t, OA, AB make angles θ, ϕ with the downward vertical and the length of OA is $l+x$. Prove that the kinetic energy of the rod is

$$\tfrac{1}{2}m[\dot{x}^2 + (l+x)^2\dot{\theta}^2 + \tfrac{4}{3}a^2\dot{\phi}^2] + ma\dot{\phi}[(l+x)\dot{\theta}\cos(\phi-\theta) - \dot{x}\sin(\phi-\theta)] \ .$$

Derive Lagrange's equations of motion of the system, if in addition to gravity, an external force with horizontal and vertical components X, Y acts on the rod at B. (L.U.)

4. The ends A, B of a light rod of length $2a$ are constrained to move along two fixed smooth wires OC, OZ, respectively; OC is horizontal and OZ vertically downwards. A particle of mass m is attached to each end of the rod, and a bead of mass M can slide freely along the rod. The whole system is made to rotate about OZ with constant angular velocity ω, and at time t the bead is distant $a+x$ from B, while the rod makes an angle θ with the upward vertical. Prove that the kinetic energy of the system is

$$2ma^2[\dot{\theta}^2 + \omega^2\sin^2\theta] + \tfrac{1}{2}M[\dot{x}^2 + (a^2 + 2ax\cos 2\theta + x^2)\dot{\theta}^2 + 2a\dot{x}\dot{\theta}\sin 2\theta + (a+x)^2\omega^2\sin^2\theta] \ .$$

Show, by using Lagrange's equations, that a steady motion in which $x=0$, $\theta=\alpha(=\text{const.})$, is possible only if $a\omega^2 = g\cot\alpha\,\text{cosec}\,\alpha$ and

$$(M+2m)\tan^2\alpha = M+4m \ .$$

(L.U.)

5. A uniform rod AB of mass m and length $2a$ is at rest at time $t=0$, suspended from a fixed smooth horizontal rail by means of a small ring at A of mass m. A force F is applied at A along the rail. If x is the distance moved by A from its initial position O and $\angle OAB = \tfrac{1}{2}\pi - \theta$, obtain the equations of motion of the system.

If A is constrained to move with constant acceleration g, show that the rod oscillates between the vertical ($\theta=0$) and the horizontal ($\theta=\tfrac{1}{2}\pi$), and that

$$F = \tfrac{1}{8}mg[19 - 12\sin\theta + 9(\sin 2\theta - \cos 2\theta)] \ .$$

(L.U.)

6. A uniform circular wheel of radius a has moment of inertia A about its fixed horizontal axle round which the wheel can turn freely. Initially the wheel is at rest and a beetle of mass m is at rest on the rim at its lowest point. The beetle then begins to crawl along the rim with constant tangential acceleration f relative to the rim. Show that, if $f=2ma^2g/\pi A$, during the subsequent motion the level of the insect will oscillate between that of its starting point and that of the axle. (L.U.)

7. A pair of uniform disc wheels, each of mass m and radius a, connected by a light axle normal to each wheel through its centre, rolls with the axle horizontal down a rough plane of angle α. A light rod, of length $l(<a)$, smoothly hinged at one end to the midpoint of the axle so that it can swing freely in a plane perpendicular to the axle, carries a mass $2m$ at the other end. Initially the system is held at rest with the rod normal to the inclined plane and below the axle. If, at time t after the system is released from rest, ϕ is the angle turned through by the wheels and θ the angle turned through by the rod (in the opposite sense), show that

$$5a\phi+2l\sin\theta=2gt^2\sin\alpha\ ,$$

and that the rod swings over an arc $2\tan^{-1}(\frac{1}{5}\tan\alpha)$. (L.U.)

8. AB, BC, CD are three equal uniform rods each of mass m, freely jointed at B, C. They lie at rest in a straight line on a smooth horizontal table. If a blow J is administered at B in a horizontal direction perpendicular to the rods, find the initial velocities of A, B, C, D and show that the angular velocities of the rods are in the ratios 7:–6:2.

9. AB, BC, CD, DE are four equal uniform rods freely jointed at B, C, D and at rest in a straight line on a smooth horizontal table. If an impulsive couple G is applied to the rod BC in the plane of the table, find the initial velocities of the ends of the rods, taking m to be the mass and $2a$ the length of a rod.

10. A square plate has mass M and side $2a$. It is freely suspended from a corner by a string of length $2\sqrt{2}a$ and receives a blow J in the plane of the plate at its lowest point. Show that the plate will begin to rotate with angular velocity $3J/Ma\sqrt{2}$.

11. A framework consists of four equal uniform rods, smoothly jointed at their ends so as to form a square. It is turning in its own plane about its centre with angular velocity ω when the midpoint of one of the rods is suddenly fixed. Find the instantaneous changes in the angular velocities of the rods.

Show also that the new angular velocity is $\frac{4}{7}\omega$ in the corresponding problem for a similar, but rigid, framework. (L.U.)

12. A uniform rod of mass m is moving in a plane, its longitudinal velocity being w, and the transverse velocities of its end points being u and v. Show that the kinetic energy of the rod is

$$T=\tfrac{1}{6}m(u^2+v^2+uv+3w^2)\ .$$

Three uniform rods AB, BC, CD of equal masses m, smoothly jointed together at B and C, lie on a smooth table, with AB and CD perpendicular to BC and on opposite sides of it. An impulse I is applied to D in the plane of the rods. If T is the kinetic energy generated, prove that

$$8/21 \leqslant mT/I^2 \leqslant 19/12 \ .$$

(L.U.)

13. The end points of a uniform rod of mass m are moving perpendicular to the line of the rod and in the same direction with velocities u, v; prove that the kinetic energy of the rod is $\frac{1}{6}m(u^2+uv+v^2)$.

Three rods A_1A_2, A_2A_3, A_3A_4, each of mass m, smoothly hinged together at A_2, A_3, hang in equilibrium under gravity from a fixed smooth hinge at A_1. A horizontal impulse is applied to a point of A_1A_2. If there is no impulsive reaction on the hinge A_1 show that $A_1P/PA_2 = 26/7$. (L.U.)

14. Three uniform rods AB, BC, CD each of mass m and equal length, are smoothly jointed at B and C and are placed in a straight line. The rods are set in motion from rest by an impulse I applied at A at right angles to AB. Show that the kinetic energy generated is $26I^2/15m$ if the rods are free, and is less than this by an amount $I^2/390m$ if the end D is fixed. State the general theorem illustrated by this result. (L.U.)

15. A rhombus $ABCD$ consisting of four freely jointed rods each of mass m and length $2a$ is moving in any manner on a smooth horizontal table. Denoting by u, v the component velocities of the centre of mass of the system in the directions AC, BD and by ω_1, ω_2 the angular velocities of CD, AD, show that the kinetic energy of the rhombus is $2m(u^2+v^2)+$ $+\frac{4}{3}ma^2(\omega_1{}^2+\omega_2{}^2)$. The rhombus receives a blow I in the direction CA at the point C. Calculate the corresponding generalized components of I if the angle $\angle BAC$ is α.

The rhombus, in the form of a square, is moving without rotation with velocity U in the direction AC when C strikes a vertical, inelastic wall at right-angles. Prove that after impact the angular velocities of the rods are all $3U/5a\sqrt{2}$. Find the impulse I at C. (L.U.)

16. Determine the periods of the normal modes of small oscillation in a horizontal plane of a system consisting of three particles of equal mass m, attached at equal intervals l, to a light inextensible string of length $4l$ fixed at its end and stretched to a tension P. (I.C.S.)

17. A uniform heavy rod AB of mass m and length $12a$ is suspended from a fixed point O by a light string OC of length $5a$ attached to a point C of the rod so that $AC = 4a$. The system makes small oscillations about its position of equilibrium in a vertical plane through O, and OC, AB make angles q_1, q_2 respectively, with the vertical at time t. Prove that the kinetic energy of the system is

$$\tfrac{1}{2}ma^2(25\dot{q}_1{}^2+20\dot{q}_1\dot{q}_2+16\dot{q}_2{}^2) \ ,$$

and that the normal periods of oscillation are $2\pi/p_1$ and $2\pi/p_2$, where $3ap_1{}^2 = g$ and $10ap_2{}^2 = g$. (L.U.)

18. Two double pendulums ABC, $A'B'C'$, smoothly jointed at B and B' are freely suspended from the fixed points A A' at the same level and smoothly

connected by a rod BB'. The lengths AB, BC, BB', $B'C'$, $B'A'$, AA' are each equal to $2a$. The rods AB, BB', $B'A'$ are each uniform and of mass m, and the rods BC, $B'C'$ are each uniform and of mass $4m$. The rods AB, BC, $B'C'$ make angles θ, ϕ, ψ with the downward vertical. Prove that $\phi-\psi$ is a normal co-ordinate for small oscillations of the system (in a vertical plane through AA') about the equilibrium configuration, the length of the equivalent simple pendulum in the corresponding normal mode of oscillation being $4a/3$.

Show that the remaining normal co-ordinates are $5\theta-2\phi-2\psi$ and $4\theta+\phi+\psi$ and that the corresponding equivalent simple pendulums are of lengths $a/3$, $44a/15$. (L.U.)

19. A bead of mass m is threaded on a smooth uniform circular wire of mass M and radius a; one point O of the circumference of the wire is fixed and the wire can rotate freely about O in a fixed vertical plane. If the system is disturbed slightly from its configuration of stable equilibrium show that the periods of small vibrations are $2\pi/p_1$, $2\pi/p_2$, where

$$2ap_1{}^2=g\ , \qquad Map_2{}^2=(M+m)g\ .$$

(L.U.)

20. Explain in detail the Lagrangian method for finding the periods of the normal modes of small oscillation about a position of stable equilibrium for a holonomic system with n degrees of freedom.

A bead of mass $3m$ is free to slide on a smooth thin uniform wire of mass $2m$ bent into the shape of a circle of radius a. One point of the wire is fixed and the system is performing small oscillations in a vertical plane about its position of stable equilibrium. Show that the periods of the normal modes are

$$2\pi(2a/g)^{\frac{1}{2}}\ ,\quad 2\pi(2a/5g)^{\frac{1}{2}}\ .$$

(L.U.)

21. A uniform rod BC of mass m and length $2a$ is attached to a fixed point A by a light elastic string AB of natural length b. The modulus of elasticity of the string is such that its stretched length is a when the rod hangs from A in equilibrium under gravity. The system makes small oscillations about the configuration of equilibrium in a vertical plane through A. Find the kinetic and potential energies of the system if at time t the length of AB is $a+x$ and AB, BC make angles θ, ϕ, respectively, with the vertical. Show that one normal period of vibration is $2\pi[(a-b)/g]^{\frac{1}{2}}$ and find the other two periods. (L.U.)

22. Two identical rods BC, CD each of mass m and length $2a$ are freely jointed at C and rest on a smooth horizontal table. Two light elastic strings AB, DE are attached to fixed points A, E respectively so that $ABCDE$ is a straight line with $AB=DE=2a$. When the system is at rest the tension in AB, DE is F. The system performs small, free oscillations so that C moves in a line at right angles to AE. Calculate the kinetic energy of the system correct to the second order of velocities and show that if x, y, z denote the

displacements of B, C, D respectively, perpendicular to AE, the potential energy of the system is

$$V=(F/2a)(x^2+y^2+z^2-xy-yz) .$$

Prove that the frequency of the normal mode of oscillation that is anti-symmetrical about C is $(3F/ma)^{\frac{1}{2}}/2\pi$. Prove also that the symmetrical modes have frequencies $p_1/2\pi$, $p_2/2\pi$, where

$$map_1{}^2=F(4+\sqrt{13}) , \qquad map_2{}^2=F(4-\sqrt{13}) .$$

(L.U.)

23. A top, symmetrical about its axis, is mounted so that it can turn freely about a point O on its axis. The mass of the top is M and its centre of mass G is at a distance h from O. The principal moments of inertia at O are A, A and C. The top is given a spin n about its axis when the axis is at rest and horizontal. Show that, if $4AMgh=C^2n^2\epsilon$ where second order terms in ϵ are negligible, then at any subsequent time OG makes an angle θ with the downward vertical where

$$4A^2\sin^2\theta\ \dot{\theta}^2=C^2n^2\cos\theta(\epsilon-2\cos\theta)(2+\epsilon\cos\theta) .$$

Find the greatest value of the precessional velocity of G in the subsequent motion. (L.U.)

24. A thin uniform spherical shell of radius a and centre C moves freely about a fixed point O of its surface under gravity. Prove that the resolute n of the angular velocity of the shell about the diameter OC remains constant and, if OC makes an angle θ with the upward vertical, find an equation for $\dot{\theta}^2$, interpreting any further constants used.

If $n^2=15g/2a$, and $z=\cos\theta$, and the motion starts with OC moving horizontally with an angular velocity $\sqrt{(6g/5a)}$, prove that

$$\dot{z}^2=(6g/5a)\ z(1-z)^2 ,$$

and hence that

$$z=\tanh^2\{(3g/10a)^{\frac{1}{2}}t\} .$$

(L.U.)

25. A top consists of a uniform circular disc of mass $4m$ and radius $2l$ and a uniform thin rod AB of length $2l$ and mass m. The centre of the disc is rigidly attached to the end B of the rod which is perpendicular to the plane of the disc and the other end A is fixed. Originally AB is vertical with B above A and the top has a spin $3(g/l)^{\frac{1}{2}}$. Show that if the top is slightly displaced AB descends until it makes an angle $\pi/3$ with the vertical.

(L.U.)

Chapter 11

Variational Methods in Mechanics

11.1 THE CALCULUS OF VARIATIONS

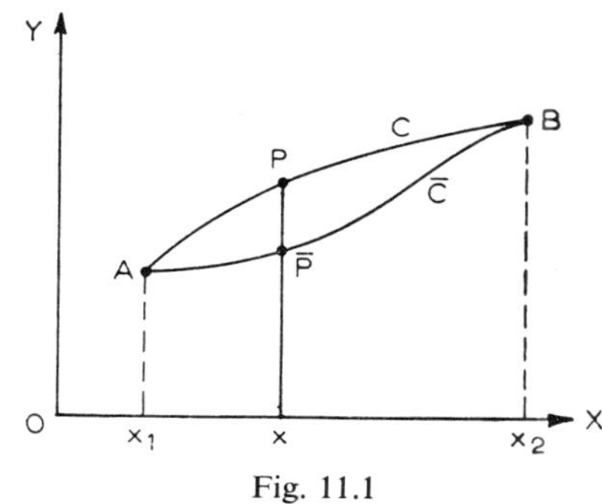

Fig. 11.1

Suppose A, B are the fixed points (x_1, y_1), (x_2, y_2) in a Cartesian co-ordinate plane (Fig. 11.1). Also suppose $f(x, y, y')$ is a known functional form of the variables x, y, y' $(=dy/dx)$. Then if C is a curve joining A to B and having equation $y=y(x)$, the integral

$$I \equiv \int_{x_1}^{x_2} f(x, y, y')dx \tag{1}$$

has a definite value whenever the function $y(x)$ is prescribed. The value of I will change as we vary the form of the curve C through A, B. Consequently, we may expect that in general there will be some curve $\bar{C}$ through these fixed points such that the value of I taken along it is *stationary* (in general a maximum or minimum) compared with the value along neighbouring paths C. The calculus of variations is concerned amongst other things with finding the form of $y(x)$ for which this stationary property holds.

Let the path $\bar{C}$ which renders I stationary have equation $y=\bar{y}(x)$ and let the equation of a neighbouring curve C be $y=\bar{y}(x)+\epsilon\eta(x)$, where ϵ is small and $\eta(x)$ is an arbitrary continuous differentiable function of x satisfying

$\eta(x_1)=0$, $\eta(x_2)=0$, to ensure that the curve passes through A and B. Then in the diagram $\overline{P}P=\epsilon\eta(x)$. The value of I taken along C is thus a function of ϵ of the form

$$I(\epsilon)=\int_{x_1}^{x_2} f(x,\bar{y}+\epsilon\eta,\bar{y}'+\epsilon\eta')dx \ . \tag{2}$$

Since the curve $\overline{C}$, for which $\epsilon=0$, is to render I stationary, we require the condition $I'(0)=0$. Formally differentiating (2) with respect to ϵ gives

$$I'(\epsilon)=\int_{x_1}^{x_2} [\eta f_y(x,\bar{y}+\epsilon\eta,\bar{y}'+\epsilon\eta')+\eta' f_{y'}(x,\bar{y}+\epsilon\eta,\bar{y}'+\epsilon\eta')]dx \ ,$$

where $f_y(x,\bar{y}+\epsilon\eta,\bar{y}'+\epsilon\eta')=(\partial/\partial y)f(x,y,y')$ when $y=\bar{y}+\epsilon\eta$, $y'=\bar{y}'+\epsilon\eta'$, with a similar meaning for $f_{y'}(x,\bar{y}+\epsilon\eta,\bar{y}'+\epsilon\eta')$. The stationary requirement $I'(0)=0$ now gives

$$\int_{x_1}^{x_2} [\eta f_y(x,\bar{y},\bar{y}')+\eta' f_{y'}(x,\bar{y},\bar{y}')]dx=0 \ . \tag{3}$$

Now

$$\int_{x_1}^{x_2} \eta' f_{y'}(x,\bar{y},\bar{y}')dx=[\eta f_{y'}(x,\bar{y},\bar{y}')]_{x_1}^{x_2}-\int_{x_1}^{x_2} \eta \frac{d}{dx} f_{y'}(x,y,y')dx$$

$$=-\int_{x_1}^{x_2} \eta \frac{d}{dx} f_{y'}(x,y,y')dx \ ,$$

since $\eta(x_1)=0=\eta(x_2)$. Thus (3) becomes

$$\int_{x_1}^{x_2} \eta(x)[f_y(x,\bar{y},\bar{y}')-\frac{d}{dx} f_{y'}(x,\bar{y},\bar{y}')]dx=0 \ . \tag{4}$$

Since $\eta(x)$ is arbitrary, subject to its being differentiable and vanishing at A, B, (4) implies that

$$f_y(x,\bar{y},\bar{y}')-d/dx\, f_{y'}(x,\bar{y},\bar{y}')=0 \ , \tag{5}$$

This follows from the result that if

$$\int_{x_1}^{x_2} \eta(x)\phi(x)dx=0 \ .$$

for all continuous and differentiable functions $\eta(x)$ satisfying $\eta(x_1)=0=\eta(x_2)$, then $\phi(x)=0$ for all x satisfying $x_1<x<x_2$.*

* For a proof, see *An Introduction to the Calculus of Variations*, by C. Fox (Oxford University Press, 1950).

We have thus established that the equation $y=\bar{y}(x)$ of the curve $\overline{C}$ along which I is stationary is given by the solution of the differential equation

$$\frac{d}{dx}\frac{\partial f(x, y, y')}{\partial y'} - \frac{\partial f(x, y, y')}{\partial y} = 0 \ . \tag{6}$$

This equation (6) is a second order differential equation for y and is known as the *Euler-Lagrange equation.* It may readily be expressed in the equivalent form

$$y'' f_{y'y'} + y' f_{y'y} + f_{y'x} - f_y = 0 \ , \tag{7}$$

the suffixes denoting partial differentiations. The corresponding curve $\overline{C}$ determined by (7) is called an *extremal.*

The case in which the independent variable x is explicitly absent from the form of f, so that $f=f(y, y')$, is of especial interest in practice. In these circumstances, (6) may be written

$$y' f_y - y' d/dx\, f_{y'} = 0 \ ,$$

or

$$(y' f_y + y'' f_{y'}) - \{y'' f_{y'} + y'(d/dx) f_{y'}\} = 0 \ .$$

This integrates at once to give

$$f - y' f_{y'} = \text{const.} \tag{8}$$

(8) is a first integral of the Euler-Lagrange equation for this special case: it may permit of further integration according to the nature of f.

11.2 THE BRACHISTOCHRONE PROBLEM

Suppose A and B are any two points in space but not in the same vertical line. Let A and B be connected by a plane curve C and suppose A is at a higher level than B (Fig. 11.2). The problem of finding the form of C which ensures that the time of fall from A to B under gravity of a smooth particle along it is a minimum, starting from rest at A, is called the *brachistochrone problem.* The problem was first solved by different methods by the brothers John and James Bernoulli.

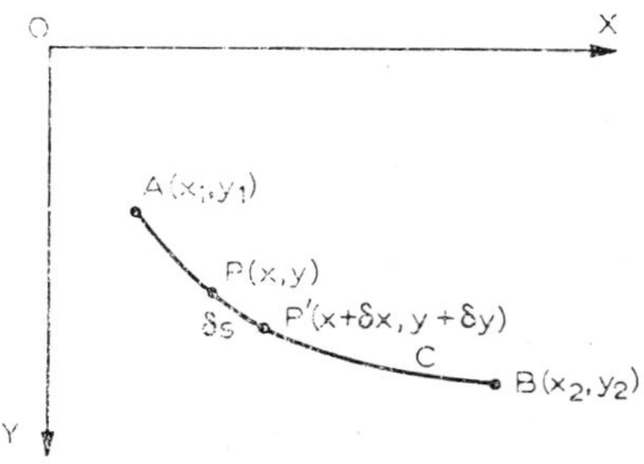

Fig. 11.2

Take co-ordinate axes OX horizontally and OY vertically downwards and in the plane of A and B so that these points have co-ordinates (x_1, y_1), (x_2, y_2) respectively. Let v be the velocity of the particle when it has reached $P(x, y)$ on C between A and B starting from rest at A. Then, taking m to be the mass of the particle, since the vertical distance fallen from A to B is $y-y_1$, we have

$$\tfrac{1}{2}mv^2 = mg(y-y_1) \ ,$$

in virtue of the principle of conservation of energy. Thus $v=\sqrt{\{2g(y-y_1)\}}$. Let P' be the neighbouring point $(x+\delta x, y+\delta y)$ on C and such that the arc length PP' is δs. Then the time of transit of the particle from P to P' is δt where

$$\delta t = \delta s/v = (y-y_1)^{-\frac{1}{2}}(1+y'^2)^{\frac{1}{2}}\delta x/\sqrt{(2g)} \ ,$$

since $\delta s = (1+y'^2)^{\frac{1}{2}}\delta x$, to the first order. Thus the total time of transit from A to B is T where

$$T = \frac{1}{\sqrt{2g}} \int_{x_1}^{x_2} (y-y_1)^{-\frac{1}{2}}(1+y'^2)^{\frac{1}{2}} dx \ .$$

If we require to choose C to ensure that T be a minimum, subject to its passing through the fixed points A and B, then the attendant variational problem is

$$\int_{x_1}^{x_2} f(y, y')dx = \text{stat.},$$

where $f(y, y') = (y-y_1)^{-\frac{1}{2}}(1+y'^2)^{\frac{1}{2}}$. Since x is explicitly absent from the form of f, a first integral is

$$f - y'f_{y'} = \text{const.}$$

This leads to

$$(1+y'^2)^{\frac{1}{2}}(y-y_1)^{\frac{1}{2}} = A(=\text{const.}).$$

Writing $y' = \tan\psi$, we find

$$\begin{cases} y-y_1 = \frac{1}{2}\, A^2(1+\cos 2\psi) \ , \\ x-B = -\frac{1}{2}\, A^2(2\psi + \sin 2\psi) \ . \end{cases}$$

These are the parametric equations of a *cycloid.*

11.3 EXTENSIONS OF THE VARIATIONAL METHOD

Suppose the n co-ordinates $q_r(r=1, 2, \ldots, n)$ are each functions of an independent variable t and that we require the solution of the variational problem

$$I \equiv \int_{t_1}^{t_2} f(q_1, q_2, \ldots, q_n; \dot{q}_1, \dot{q}_2, \ldots, \dot{q}_n; t)dt = \text{stat.};$$

where f is of known functional form, t_1 and t_2 are fixed, and each q_r is to be determined.

Following the procedure for the case of a single independent variable, we make variations of the form

$$q_r = \bar{q}_r + \epsilon_r \eta_r \qquad (r = 1, \ldots, n) ;$$

where the η_r are arbitrary functions of t only and vanish at $t = t_1$, $t = t_2$. Then I becomes a function of the n variables ϵ_r $(r = 1, \ldots, n)$ and we require the solutions of the n equations

$$(\partial I / \partial \epsilon_r) = 0 \qquad (r = 1, \ldots, n) ,$$

with $\qquad \epsilon_1 = \epsilon_2 = \ldots = \epsilon_n = 0$.

These lead to the n *Euler – Lagrange equations* of the variational problem

$$\frac{\partial f}{\partial q_r} - \frac{d}{dt}\frac{\partial f}{\partial \dot{q}_r} = 0 \qquad (r = 1, \ldots, n)$$

for the n co-ordinates $q_r (r = 1, \ldots, n)$.

11.4 HAMILTON'S PRINCIPLE

Suppose T, V are the kinetic and potential energies of a conservative holonomic dynamical system defined by n generalized co-ordinates $q_r (r = 1, \ldots, n)$ at time t. Writing $L = T - V$, we know that Lagrange's equations for the motion of the system are

$$d/dt(\partial L / \partial \dot{q}_r) - \partial L / \partial q_r = 0 \qquad (r = 1, \ldots, n) .$$

The previous section shows that these are the n Euler-Lagrange equations arising from the variational problem,

$$\int_{t_1}^{t_2} L(q_1, q_2, \ldots, q_n; \dot{q}_1, \dot{q}_2, \ldots, \dot{q}_n; t)dt = \text{stat.};$$

where t_1, t_2 are fixed. We have thus established that *during the motion of a conservative holonomic dynamical system over a fixed time interval, the time-integral over that interval of the difference between the kinetic and potential energies is stationary*. This is *Hamilton's Principle*.

11.5 THE PRINCIPLE OF LEAST ACTION

The *action* A of a dynamical system over the interval $t_1 < t < t_2$ is defined to be

$$A = \int_{t_1}^{t_2} 2T \, dt ,$$

where T is the kinetic energy.

In deriving Hamilton's Principle, we kept t_1 and t_2 fixed. Now suppose t is no longer the independent variable but that the motion is dependent upon some other independent variable r which is to assume fixed values at the end points. Then

$$A = \int_{r_1}^{r_2} 2T\frac{dt}{dr}\,dr \qquad (r_1, r_2 \text{ fixed}) .$$

We further suppose that the law of conservation of energy

$$T + V = c \,(= \text{const.})$$

holds and that T is *independent of time*. Then

$$L = T - V = 2T - c$$

and L is also independent of time.

When r is allowed to vary to $r + \delta r$, A varies to $A + \delta A$ so that

$$\delta A = \delta \int_{r_1}^{r_2} 2T\frac{dt}{dr}\,dr$$

$$= \int_{r_1}^{r_2} \left[\delta(2T)\frac{dt}{dr} + 2T\delta\left(\frac{dt}{dr}\right)\right] dr .$$

But $\delta(dt/dr) = d(\delta t)/dr = (d(\delta t)/dt)(dt/dr)$, and $\delta(2T) = \delta L$. Hence

$$\delta A = \int_{r_1}^{r_2} \left[\delta L + 2T\frac{d}{dt}(\delta t)\right]\frac{dt}{dr}\,dr .$$

Further, $\delta L = \sum_{k=1}^{n} (\partial L/\partial q_k)\delta q_k + \sum_{k=1}^{n} (\partial L/\partial \dot{q}_k)\delta \dot{q}_k$.

Now denoting total differentiations with respect to r by primes,

$$\begin{aligned}\delta\dot{q}_k &= \delta(q_k{}'/t') = \delta q_k{}'/t' - (q_k{}'/t'^2)\delta t' \\ &= (1/t')d(\delta q_k)/dr - (q_k{}'/t'^2)d(\delta t)/dr \\ &= d(\delta q_k)/dt - (q_k{}'/t')d(\delta t)/dt \\ &= d(\delta q_k)/dt - \dot{q}_k d(\delta t)/dt .\end{aligned}$$

$$\therefore \sum_{k=1}^{n} (\partial L/\partial \dot{q}_k)\delta\dot{q}_k = \sum_{k=1}^{n} (\partial L/\partial \dot{q}_k)d(\delta q_k)/dt - \sum_{q=1}^{n} (\dot{q}_k \partial L/\partial \dot{q}_k)d(\delta t)/dt .$$

But $\partial L/\partial \dot{q}_k = \partial T/\partial \dot{q}_k$ and since T is a homogeneous quadratic function of the velocities,

$$\sum_{k=1}^{n} (\dot{q}_k \partial L/\partial \dot{q}_k)d(\delta t)/dt = \Big(\sum_{k=1}^{n} \dot{q}_k \partial T/\partial \dot{q}_k\Big)d(\delta t)/dt = 2Td(\delta t)/dt .$$

$$\therefore\ \delta L=\sum_{k=1}^{n}(\partial L/\partial q_k)\delta q_k+\sum_{k=1}^{n}(\partial L/\partial \dot{q}_k)\delta \dot{q}_k$$

$$=\sum_{k=1}^{n}\{(\partial L/\partial q_k)\delta q_k+(\partial L/\partial \dot{q}_k)d(\delta q_k)/dt\}-2Td(\delta t)/dt\ .$$

$$\therefore\ \delta A=\int_{r_1}^{r_2}\sum_{k=1}^{n}\{(\partial L/\partial q_k)\delta q_k+(\partial L/\partial \dot{q}_k)d(\delta q_k)/dt\}\frac{dt}{dr}\,dr$$

$$=\int_{r_1}^{r_2}\sum_{k=1}^{n}\left\{\delta q_k\frac{d}{dt}\left(\frac{\partial L}{\partial \dot{q}_k}\right)+\frac{\partial L}{\partial \dot{q}_k}\frac{d(\delta q_k)}{dt}\right\}\frac{dt}{dt}\,dr$$

$$=\int_{r_1}^{r_2}\sum_{k=1}^{n}\frac{d}{dt}\left\{\delta q_k\frac{\partial L}{\partial \dot{q}_k}\right\}\frac{dt}{dr}\,dr$$

$$=\left[\sum_{k=1}^{n}\delta q_k\frac{\partial L}{\partial \dot{q}_k}\right]_{r_1}^{r_2}\ .$$

If we allow $\delta q_k=0$ $(k=1,\ldots,n)$ at each end point $r=r_1$, $r=r_2$, then $\delta A=0$. This shows that A is stationary. Such a result is known as the *Principle of Least Action*. Here we have only established the stationary character of A, but it can be shown to be a minimum.

For a single particle of mass m moving along a curve with velocity $\mathbf{v}$,

$$A=\int_{t_1}^{t_2} mv^2dt=\int_{s_1}^{s_2} mv\,ds\ ,$$

since $ds=vdt$. Thus the principle of least action becomes

$$\int_{s_1}^{s_2} v\,ds=\text{stationary.}$$

This form was originally developed by Maupertuis.

11.6 DISTINCTIONS BETWEEN HAMILTON'S PRINCIPLE AND THE PRINCIPLE OF LEAST ACTION

In establishing Hamilton's Principle the time interval t_2-t_1 was prescribed. In establishing the principle of least action, however, the time interval from one configuration to another was in no way restricted. In the latter case, we chose fixed end points $r=r_1$, $r=r_2$ and prescribed the total energy $T+V$ of the system.

To illuminate the two principles, consider the motion under gravity of the particle as depicted in Fig. 11.2 of this chapter. Then at $P(x, y)$,

$$T=\tfrac{1}{2}m(\dot{x}^2+\dot{y}^2)\ , \qquad V=-mgy\ ,$$

if we take OX to be the zero level of P.E. Hence

$$L=\tfrac{1}{2}m(\dot{x}^2+\dot{y}^2)+mgy\ .$$

Hamilton's Principle requires $\int_{t_1}^{t_2} L dt$ to be stationary and leads to Lagrange's equations

$$\begin{cases}(d/dt)(\partial L/\partial\dot{x})-\partial L/\partial x=0\ ,\\ (d/dt)(\partial L/\partial\dot{y})-\partial L/\partial y=0\end{cases}$$

or

$$\begin{cases}\ddot{x}\qquad=0\ ,\\ \ddot{y}-g=0\ .\end{cases}$$

These are the equations of motion of a projectile moving *in vacuo* under gravity. No constraints are imposed on the path and at no stage has the principle of conservation of energy been used. Such equations always hold throughout any prescribed time interval.

When A and B are joined by a smooth curve, the conservation of energy law holds and Maupertuis' form of the principle of least action gives

$$\int_A^B v\, ds=\text{stat.},$$

or, since $\tfrac{1}{2}mv^2=mg(y-y_1)$ and $ds=(1+y'^2)^{\frac{1}{2}}dx$,

$$\int_{x_1}^{x_2}(y-y_1)^{\frac{1}{2}}(1+y'^2)^{\frac{1}{2}}dx=\text{stat.}$$

In this case x is explicitly absent from the integrand and the reader can readily confirm the solution for y is a parabolic trajectory through the end points.

EXERCISE 11

1. Use the variational method to show that the shortest curve joining two fixed points is a straight line.
2. Show that the area of the surface of revolution of a curve $y=y(x)$ is

$$2\pi\int_{x_1}^{x_2} y(1+y'^2)^{\frac{1}{2}}dx\ .$$

Hence show that for this to be a minimum, the curve must be a catenary.

3. State Hamilton's Principle.

By using Lagrange's equations, or otherwise, establish the Principle for a conservative, holonomic system.

A particle of unit mass moves along OX under a constant force f, starting from rest at the origin at time $t=0$. If T and V are the kinetic and potential energies of the particle, calculate $\int_0^{t_0} (T-V)dt$. Evaluate this integral for the varied motion (in the Hamiltonian sense) in which the position of the particle is given by

$$x=\tfrac{1}{2}ft^2+\epsilon ft(t-t_0) \;,$$

where ϵ is a constant, and show that the result is in agreement with Hamilton's Principle. What are the essential features of the varied motion that ensure this agreement? (L.U.)

4. A particle moves on a surface under no forces. Show that it describes a geodesic on the surface. (A curve through any two given points on a surface having a minimum length is called a *geodesic*.)

Appendix I

TABLE SHOWING CENTROID POSITIONS FOR STANDARD DISTRIBUTIONS

Body	*Centroid position*
Triangular lamina	At intersection of medians
Triangle of three uniform rods	At centre of in-circle
Solid pyramid or tetrahedron	At the point on the line joining the vertex to the centroid of the base dividing it in the ratio 3:1
Solid right circular cone	On the axis at $\frac{1}{4} \times$ altitude from base
Hollow circular cone	On the axis at $\frac{1}{3} \times$ altitude from base
Circular arc, radius a, subtending angle 2α at centre	On axis of symmetry at $(a \sin \alpha)/\alpha$ from centre
Circular sector, radius a, subtending angle 2α at centre	On axis of symmetry at $(2a \sin \alpha)/(3\alpha)$ from centre
Semicircular lamina, radius a	At $4a/3\pi$ from centre and on A.O.S.
Solid hemisphere, radius a	On A.O.S. at $\frac{3}{8}a$ from base
Hemispherical shell, radius a	On A.O.S. at $\frac{1}{2}a$ from base

Appendix II

TABLE SHOWING SQUARE OF THE RADIUS OF GYRATION (k^2) FOR STANDARD DISTRIBUTIONS

(N.B. Moment of inertia about axis = mass of distribution × k^2 about axis.)

Body	*Axis*	*k^2 about axis*
Uniform thin rod length $2a$	Perpendicular bisector	$\frac{1}{3}a^2$
Uniform rectangular lamina $2a \times 2b$	Line through centroid parallel to side $2b$	$\frac{1}{3}a^2$
Uniform rectangular parallelepiped $2a \times 2b \times 2c$	Line through centroid parallel to edge $2a$	$\frac{1}{3}(b^2+c^2)$
Uniform ring of radius a	Diameter	$\frac{1}{2}a^2$
Uniform circular disc of radius a	(i) Normal through centre (ii) Diameter	$\frac{1}{2}a^2$ $\frac{1}{4}a^2$
Uniform elliptic lamina of semi-axes a, b	(i) Axis of length $2a$ (ii) Axis of length $2b$	$\frac{1}{4}b^2$ $\frac{1}{4}a^2$
Uniform solid sphere of radius a	Diameter	$\frac{2}{5}a^2$
Uniform solid hemisphere of radius a	Diameter	$\frac{2}{5}a^2$
Uniform solid ellipsoid of semi-axes $2a$, $2b$, $2c$	Axes	$\frac{1}{5}(b^2+c^2)$, $\frac{1}{5}(c^2+a^2)$ $\frac{1}{5}(a^2+b^2)$
Uniform spherical shell, radius a	Diameter	$\frac{2}{3}a^2$
Uniform solid right circular cylinder of radius a	Axis of cylinder	$\frac{1}{2}a^2$

Solutions

Exercise 1

5. $(\lambda x_2 - \mu x_1)/(\lambda - \mu)$, $(\lambda y_2 - \mu y_1)/(\lambda - \mu)$, $(\lambda z_2 - \mu z_1)/(\lambda - \mu)$.
12. Vector moment $= [2, 0, 2]$; physical moment $= 2$.
17. $\phi = \frac{1}{2}(x^2 + y^2 + z^2) + \text{const.} = \frac{1}{2}r^2 + \text{const.}$
    (Use $d\phi = (\partial\phi/\partial x)\,dx + (\partial\phi/\partial y)\,dy + (\partial\phi/\partial z)\,dz$.)

Exercise 2

3. min. vel. $= bV/\sqrt{(b^2 + d^2)}$ at $\frac{1}{2}\pi + \tan^{-1}(b/d)$ to direction of approach.
4. min. time $= 2AB(V^2 - v^2 \sin^2\phi)^{\frac{1}{2}}/(V^2 - v^2)$; bearing at α to AB, where $\sin\alpha = (v/V)\sin\phi$.
8. Body centrode is a parabola, focus A, S.L.R. $= a$, axis perpendicular to AB.
13. Orbit is $A/r = V_2[1 - (V_1/V_2)\sin\theta]$.
    (i) $V_1 > V_2$ — hyperbola;
    (ii) $V_1 = V_2$ — parabola;
    (iii) $V_1 < V_2$ — ellipse.

Exercise 3

3. Time $= (M/kF)^{\frac{1}{2}} \tan^{-1}[\frac{1}{2}(P/kM)^{\frac{1}{2}}(kM/F)^{\frac{1}{2}}]$
5. $\frac{1}{2}a\sqrt{(a/k)}$.
6. (i) Particle tends ultimately to rest at a distance $u^{2-n}/k(2-n)$ after an infinite time;
   (ii) Particle never comes to rest in a finite time or in a finite displacement.
7. $$(\text{Speed at return})^2 = \frac{2ga}{2k-1}\left\{1 - \left(1 + \frac{k}{a}\right)^{-(2k-1)}\right\}$$
   $$= \frac{2ga}{2k-1}\left\{1 - \left[1 + \frac{u^2(2k+1)}{2ag}\right]^{\frac{1-2k}{1+2k}}\right\}$$
10. Impulse $= 2e_1e_2/bv$: K.E. gained $= 2e_1{}^2e_2{}^2/mb^2v^2$.
    max. displacement $= 2e_1e_2/kmbve$.
11. Condition is $n > k$: $p = (n^2 - 2k^2)^{\frac{1}{2}}$.
15. Distance fallen $= (1/k)\log\cosh[t\sqrt{(gk)}]$
16. Max. speed $= gT(\log 2 - \frac{1}{2})$.

Exercise 4

2. Point of concurrence rises with acceleration $= g$.
10. Eccentricity $= U/V$.
14. $r = (2t^2 + 2\sqrt{6}t + 4)^{\frac{1}{2}}$
15. Time to reach $O = (2/\omega \sin \alpha)\text{ch}^{-1}(4/3)$.
18. Period of small oscns. $= 2\pi\sqrt{(37a/32g)}$.
19. Period of small oscns. $= 2\pi\sqrt{(17a/24g)}$.

Exercise 5

3. Apsidal angle $= \pi\sqrt{\{c/(a+c)\}}$.
8. Eccentricity $= (1 - e^2 + e^4)^{\frac{1}{2}}$.
9. Time $= \frac{5}{4}\sqrt{(5/\mu)}\pi a^{3/2}$.
14. Q ultimately moves along axis of hyperbola.
16. Hodograph is $(\dot{x}/an)^2 + (\dot{y}/u)^2 = 1$.
19. Lengths of axes for relative motion: a; $v/\sqrt{3}n$.
21. $\dot{a} = na(2a/r - 1)$; $\dot{e} = n(1/e - e)(1 - a/r)$.

Exercise 6

1. Distance $= (v^2/g)\sin 2\alpha$.
2. Tension $= 3\sqrt{3}\ mg/8$.
5. Tension $= 27mg/23$; reaction $= \sqrt{3}\ mg/23$.
6. Rel. vel. $= 2\sqrt{\{gl(M+m)/(4M+m)\}}$.
7. At time t, the centroid moves a distance $\frac{1}{2}ut$ and the rod rotates through angle $ut/2a$.
8. Impulsive tension $= 2\sqrt{3}mv/7$; impulsive reaction $= \sqrt{3}\ mv/7$.
11. Angle $= \frac{1}{3}\pi$; time $= (\frac{2}{3}\pi + \sqrt{3})a/V$ from release.
12. $V \cos 2\alpha(1 + 2\sin^2\alpha)$.
14. Tension $= I\sin^2\alpha/ml$.
18. $\frac{3}{2}I$, $\frac{3}{16}I$, where $I = m\sqrt{(2gh)}\cos\alpha$.
22. Greatest depth $= \frac{1}{5}(1 + \sqrt{11})a$.

Exercise 7

2. The P.As. are the diagonals.

Exercise 8

6. $x = 0$ or $3a/2$.
8. $R = 80mg/49$, $F = 11mg/49$.
16. Vel. of board $= 7Mv/(7M+2m)$; vel. of sphere $= 2Mv/(7M+2m)$.
17. Ang. vel. $= \sqrt{(3g/5a)}$.
19. $R = \frac{1}{3}mg[2k\theta + 7\cos\theta - 4]$.
27. Ang. vels. are $2V/a\sqrt{3}$; $V/a\sqrt{3}$. K.E. $= 7mV^2/9$.
28. $\frac{3}{8}m\sqrt{(2gh)}\sin 2\alpha$.
29. $J = 28ma\omega/3$.
30. $9J^2/ml$.

Exercise 9

1. Semi-angle $=45°$; ang. vel. $=\Omega$.
2. $\frac{2}{3}\Omega(\cos\theta-\sin\theta)$; $\frac{2}{3}\Omega(\cos\theta+\sin\theta)$; $\frac{1}{3}\Omega$, where $\theta=\frac{1}{5}\Omega t$.
5. Angle with downward vertical $=\cos^{-1}(g/2a\omega^2)$;
   period $=4\pi a\omega(4a^2\omega^2-g^2)^{-\frac{1}{2}}$.
7. Ang. vels.: n sech$(\sqrt{5}\,nt)$; $n\sqrt{5}$ th$(\sqrt{5}\,nt)$; $3n$ sech$(\sqrt{5}\,nt)$.
15. Period $=\pi\sqrt{14/\omega}$.
16. Velocity $=(5a\Omega+24u)/29$.
18. K.E. $=7J^2/2M$; vel. of $B=J/M$; vel. of $C=7J/M$.
19. Initial accn. of $P=Hl\Omega^2/(A+B)$.
23. Eastward deviation $\simeq\frac{1}{3}\omega gt^3\cos\lambda-\omega vt^2\sin(\alpha-\lambda)$.

Exercise 10

2. Downward accelerations of $6M$, $3M$, M, $2M$ are respectively $g/35$, $-g/35$, $-13g/35$, $11g/35$.
3. $m\{\ddot{x}-a\ddot{\phi}\sin(\phi-\theta)-a\dot{\phi}^2\cos(\phi-\theta)-(l+x)\dot{\theta}^2\}$
   $=-\lambda x/l+mg\cos\theta+X\sin\theta+Y\cos\theta$;
   $m\{(l+x)^2\ddot{\theta}+2(l+x)\dot{x}\dot{\theta}+a\ddot{\phi}(l+x)\cos(\phi-\theta)-a\dot{\phi}^2(l+x)\sin(\phi-\theta)\}$
   $=-mg(l+x)\sin\theta+X(l+x)\cos\theta-Y(l+x)\sin\theta$;
   $m\{\frac{4}{3}a\ddot{\phi}+2\dot{x}\dot{\theta}\cos(\phi-\theta)+(l+x)\ddot{\phi}\cos(\phi-\theta)+(l+x)\dot{\theta}^2\sin(\phi-\theta)$
   $-\ddot{x}\sin(\phi-\theta)\}$
   $=-mg\sin\phi+2X\cos\phi-2Y\sin\phi$.
5. $m(2\ddot{x}-a\ddot{\theta}\cos\theta+a\dot{\theta}^2\sin\theta)=F$,
   $4a\ddot{\theta}-3\ddot{x}\cos\theta=-3g\sin\theta$.
8. Vels. are $-14J/15m$, $28J/15m$, $-8J/15m$, $4J/15m$.
9. Vels. are $33G/56ma$, $-33G/28ma$, $9G/ma$, $-9G/28ma$, $9G/56ma$.
11. Ang. vels. are ω, $\frac{2}{5}\omega$.
15. $I=\frac{8}{5}mU$.
16. Periods are $2\pi/p$, where $p^2=2P/ml$ or $p^2=(2\pm\sqrt{2})P/ml$.
21. Periods are $2\pi/p$, where $p^2=\frac{1}{2}(7\pm\sqrt{37})g/a$.
23. Greatest angular velocity of precession $\simeq n+\epsilon Mgh/C$.

Exercise 11

3. Integral $=\frac{1}{3}f^2t_o{}^3(1+\epsilon^2)$. This is a minimum for $\epsilon=0$, appropriately. The varied motion is such that x has the same values as for the unvaried motion at the end points $t=0$, $t=t_o$.
4. Under no forces, v is constant and so Maupertuis' form of the principle of least action becomes

$$v\int_A^B ds=\text{min.},$$

or

$$\int_A^B ds=\text{min.},$$

i.e. the required path is the geodesic through A and B on the surface.

Index

A

Acceleration, 27
  components along a curve, 26
  polar co-ordinates, 29
  uniform, 59
Action, 259
Activity, 54
Amplitude, 62
Angular momentum, 98
  of rigid body, 152
Angular velocity, 34
Aperiodic motion, 64
Aphelion, 110
Apse, 110
Apsidal angle, 110

B

Body centrode, 44
Brachistochrone problem, 257

C

Calculus of variations, 255
Central axis, 37
Central force, 99
Centrode, 44
Centroid, 21
Changing mass problems, 70
Coefficient of restitution, 144
Conservation
  of angular momentum, 99
  of linear momentum, 135
  of energy, 57
Conservative force, 55
Constrained particle motion, 93
Coplanar vectors, 24
Couple, 21
Curl of a vector, 23
Cycloid, 103
Cylindrical polars, 42

D

Damped harmonic motion, 64
Differentiation of vectors, 22
Disturbed orbits, 120
Divergence of a vector, 23

E

Earth, rotating, 202
Elastic
  impact, 144
  string, 65
Elliptic harmonic motion, 122
Equimomental systems, 159
Escape velocity, 133
Eulerian angles, 241
Euler-Lagrange equation, 257
Euler's dynamical equations, 191
Euler poles, 48
Extremal, 257

F

Field
  conservative, 55
  scalar, 23
  vector, 23
Forced vibrations, 64
Foucault pendulum, 205

G

General motion of rigid body, 35, 164
Generalized
  co-ordinates, 212
  force, 215
  impulse, 224
  momentum, 224
  velocity, 214
Gradient of a scalar, 23
Gravitational constant, 118
Gyroscopic compass, 248

H

Hamilton's principle, 259
Herpolhode, 193
Hodograph, 31
Holonomic system, 213
Hooke's law, 65

I

Impulse of force, 58
Impulsive force, 58
Industrial warehouse, 233
Instantaneous
  axis, 34, 44
  centre, 44
Isochronous pendulum, 104

K

Kepler's laws, 117
Kinematics of rigid body motion, 33
Kinetic energy, 54
  single particle, 54
  system of particles, 137
  rigid body, 157

L

Lagrange's equations, 215
  for impulsive motion, 226
Least action principle, 259

M

Mass, 53
Modulus of elasticity, 65
Momental
  ellipse, 161
  ellipsoid, 157
Moment of inertia, 151
Moving axes, 40
Moving origin, 138

N

Newton's laws
  of motion, 53
  of restitution, 144
Normal co-ordinates, 237
  modes, 237
  periods, 237
Non-holonomic systems, 213
Nuclear repulsion, 117
Nutation, 244

O

Orbital motion, 110

P

Parabola of safety, 89
Parallel axes theorem, 152
Parallelogram of vectors, 12
Pedal co-ordinates, 113
Pendulum
  compound, 175
  simple, 93
Perihelion, 110
Perpendicular axes theorem, 152
Phase, 62
Plate tectonics, 47
Polhode, 193
Polygon of vectors, 13
Potential energy, 55
Power, 54
Precession, 244
Principal
  axes, 153
  moments of inertia, 154
Product of inertia, 151
Projectile
  in resisting medium, 91
  in vacuo, 84
Pursuit curves, 30

R

Radial and transverse
  acceleration, 29
  velocities, 29
Rectilinear motion, 58
Reciprocal polars, 111
Reduction of system of forces, 25, 51
Relative velocity and acceleration, 30
Resisted motion, 59
Resonance, 65
Rocket motion, 70, 73
Rigid body, 33
Rotating
  axes, 40
  earth, 202

S

Satellite orbits, 123
Scalars, 11
Scalar
  field, 23
  product, 14
  triple product, 18
Screw motion of rigid body, 37
Shot putt analysis, 85
Simple harmonic motion, 62
Small oscillations, 95
Space centrode, 44
Spherical polars, 43
Stability of orbit, 112
Stable equilibrium, 232
Staged rockets, 73
Systems of particles, 134

T

Three-dimensional problems, 191
Top
  stability, 245
  steady motion, 244
Transient, 64
Triangle of vectors, 12
Two-dimensional problems, 84, Ch. 8

U

Uniform acceleration, 59

V

Variational methods, 255
Vector, 11
  calculus, 23
  field, 23
  moment, 19
  product, 16
  triple product, 19
Virtual work, 214

W

Work, 54
Work done in stretching an elastic string, 65
Wrench, 51